THE CONSUMPTION READER

Consumption affects every aspect of the contemporary world, from the most intimate moments of everyday life to the great geopolitical struggles that have been set in train by the forces of globalization. Consumer culture has recast the world in its own image, and we are only just beginning to make sense of the enormous social, political, economic, moral, and environmental implications. This Reader offers an essential selection of the best work on the consumer society.

By drawing from the arts, humanities, and social sciences, *The Consumption Reader* presents the history and geography of consumer societies, the social and political aspects of consumer culture, and the discourses and practices of consumption. It reaches back to the consumer revolutions of the eighteenth century and spans the globe. *The Consumption Reader* focuses in particular on the histories, geographies, subjects, and objects of consumer societies. It also provides an accessible introduction to the key theoretical debates that surround consumption and consumerism: from Marx, Mauss, and Simmel to Barthes, Bourdieu and Baudrillard.

Students will appreciate *The Consumption Reader* for its scope, clarity, and ease of use. The material is arranged so that it will develop the student's knowledge through a logical progression, but it may also be read selectively so that the student can rapidly get to grips with key issues, ideas, and authors. It brings together a diverse range of topics and theoretical perspectives in a way that is engaging, surprising, and thought provoking. Each section is introduced by the editors, who also provide a substantial General Introduction.

David B. Clarke is Senior Lecturer in Human Geography at the University of Leeds. **Marcus A. Doel** is Professor of Human Geography at University of Wales Swansea. Having lectured in Human Geography at the Universities of Leeds and Southampton, **Kate M.L. Housiaux** is now a marketing professional at the John Hansard Gallery, University of Southampton.

The Consumption Reader

Edited by

David B. Clarke,

Marcus A. Doel and

Kate M.L. Housiaux

Routledge
Taylor & Francis Group
LONDON AND NEW YORK

First published 2003 by Routledge
11 New Fetter Lane, London EC4P 4EE

Simultaneously published in the USA and Canada
by Routledge
29 West 35th Street, New York, NY 10001

Routledge is an imprint of the Taylor & Francis Group

Typeset in Amasis and Akzidenz Grotesk by Keystroke, Jacaranda Lodge, Wolverhampton
Printed and bound in Great Britain by Bell & Bain Ltd, Glasgow

British Library Cataloguing in Publication Data
A catalogue record for this book is available from the British Library

Library of Congress Cataloging in Publication Data
The consumption reader/[edited by] David B. Clarke, Marcus A. Doel &
Kate M.L. Housiaux.
p. cm.
Includes bibliographical references and index.
1. Consumption (Economics)–Social aspects. 2. Consumer behavior.
I. Clarke, David B., 1964– II. Doel, Marcus A. 1966– III. Housiaux, Kate M.L., 1971–
HC79.C6 C677 2003
339.4′7–dc21 2002152421

ISBN 0–415–21376–2 (hbk)
ISBN 0–415–21377–0 (pbk)

Contents

GENERAL INTRODUCTION

Introducing 'consumption' is a daunting task, not simply because there has been an explosion of writing and research on the topic over recent decades, but also because most people already feel they understand the term well enough. Most of us are 'consumers' on a daily basis, and we all know what it means to 'consume' something or other. Yet consumption is a particularly tricky concept to define. Sometimes people reductively equate consumption with the act of purchasing. Sometimes consuming is regarded as synonymous with using. We would like to suggest, however, that consumption always involves much more than simply purchasing, obtaining, and using goods and services. However neat a definition of 'consumption' one might offer, consumption itself (so to speak) seems to have different ideas. Consumption invariably spirals out in all directions, affecting how society is structured, how we see ourselves as individuals, how the course of history develops, what the world we live in is like, and so on. So, let us begin with a formal assessment of what consumption actually is. For doing so will allow us to orientate you to the *purpose* of this Reader, to give you a general sense of *why* people with various interests in society have begun increasingly to explore the territory labelled 'consumption', and why *you* should think about doing so too.

Let us begin by considering the origins of the word. While the English language uses the Latinate term 'consumption' far more than its sister word 'consummation', both of these turn out to be crucially important. Consumption derives from the Latin '*consumere; con sumere*', which means to use up entirely, to destroy (as in the old-fashioned name for the lung disease tuberculosis, and the expression 'consumed by fire'). By contrast, 'consummation' derives from the Latin '*consumare; con summa*', which means to sum up, to bring to completion (as in the old-fashioned expression for sex: 'the *consummation* of a relationship'). In some languages, such as French, these two words have been rolled into one, so that the French word '*consommer*' carries the *dual* implication of both fulfilment and annulment. This semantic ambivalence might be extrapolated to suggest that the English-language word 'consumption' entails both an act of *destruction* (which explains why it often seems to make more sense to speak of 'consuming a hamburger' than 'consuming a cathedral') and an act of *creation* (bringing to a climax, reaching a peak, achieving a promised fulfilment). Indeed, recent work on consumption and the consumer society is based on precisely this kind of duplicity. Somewhat paradoxically, then, 'consumption' means both 'destroying' (using up) and 'creating' (making full use of). Like the capitalist system of which it has become a fundamental part, consumption should be thought of as a form of 'creative destruction'. So, while economists have often taught us that consumption is simply the opposite of production and that demand should be coupled with supply, 'consumption' actually retains a far richer set of meanings. For now, suffice to say that it carries a considerable degree of *ambivalence*, since two *opposed* values are rolled into one. Indeed, as many of the extracts in this Reader demonstrate, one of the most interesting – and infuriating – things about the consumer society is its ambivalence in almost every respect: socially, culturally, aesthetically, politically, economically and morally. Hence all of the heated and often polarized debates across a host of disciplines.

If all this sounds a little abstract, let us make things more concrete by pointing out that eating a cream cake involves both destroying it and obtaining pleasure from it at one and the same time. The 'two sides' of the ambivalent nature of consumption appear to come together remarkably easily in much of what we do as consumers. This is why so many people want to consume so much. Yet this underlying ambivalence has caused all kinds of theoretical problems and conceptual difficulties in what has become known as 'consumption studies'. Consider, for instance, the sense in which consumption has often been tied to needs. One hears it said that we have to consume simply in order to survive. And who would deny the fact? Yet this is not a particularly helpful observation, insofar as it sets up too strong an opposition between *real* needs and *artificial* needs, and between *basic* goods and *luxury* goods. It is perfectly reasonable to make such distinctions, and they often prove useful (even though knowing where to draw the dividing line is notoriously difficult). Rather than an *opposition*, however, we would argue that it is far more useful to think about the *duality* of all goods. This is how we would like to get you to think about consumption: not in terms of all-or-nothing (either/or), but in terms of complexity (both/and). There are no easy solutions to the problems raised by the advent of the consumer society.

For example, as far as keeping a person alive is concerned, food and drink would seem to respond to *needs*. But certain foods and drinks – such as champagne and caviar – are generally regarded as having far more to do with *wants* or even *desires* than with *needs*. But what if a latter-day Robinson Crusoe were shipwrecked on a desert island with nothing to consume but the champagne and caviar rescued from the galley of a stricken motor launch? Luxury goods can certainly be called upon to address *needs* if circumstances so dictate. Similarly, while we may think of certain goods as basic rather than luxurious – as always responding to needs rather than wants or desires – they usually turn out to operate in a way that is surprisingly similar to luxury goods. Bread and water may sound like basic goods – but is the bread in question white sliced, granary, wholemeal or ciabatta? And is the water tap water, spring water, spring water with a hint of lime, muddy water from a partially dried-up river bed or what? Even goods that are, on the face of it, basic – that is, necessary to survival – cannot be easily prised away from their potential 'higher' functions. To 'inhale yourself fitter', you can even buy servings of pure and flavoured oxygen in a new breed of 'oxygen bars' that are springing up among the restaurants and cafés in a number of cities! Goods that are needed for *physical* survival also tend to be needed for *social* survival: for expressing similarity with and demonstrating difference from others; for seeking social solidarity and asserting social distinction. And this goes for the most humble of goods as well as the most extravagant. Those who have tried to formulate a once-and-for-all distinction between needs and luxuries invariably encounter insurmountable difficulties. Not only have what count as needs and luxuries changed over time, they also vary from place to place and from culture to culture. Fur, salt and pepper are obvious examples of how the distinction between luxury and necessity is far from straightforward (Schivelbusch, 1993).

It appears that the term 'consumption' not only denotes the meeting of basic needs but also relates to things that are plainly *not* vital for physical survival – but which make life that much more *enjoyable, comfortable, sociable* and *pleasurable*. Note that the word 'consume' does not, in and of itself, distinguish between necessities and luxuries. Of course, it is still possible to mark such a distinction by saying that some people more or less *subsist*, while others enjoy a far better standard of living. Subsistence seems to imply a particularly *restricted* kind of consumption, while 'consumption' per se often carries connotations of something *beyond* mere subsistence. One might object that there is nothing to imply this in the strict definition of the word – and that a term such as *consumerism* is far more apt when referring to situations involving consumption *above and beyond* basic needs (to 'excessive' levels of consumption or whatever). Even so, consumption probably remains semantically closer to consumerism than it does to (mere) subsistence; tied to abundance and affluence rather than to dearth and scarcity.

MAPPING CONSUMPTION: A DISCIPLINARY SKETCH

As the above discussion has hopefully managed to suggest, consumption is a *wide-ranging, contentious* and *contested* concept. In thinking about consumption, there is always much that is up for grabs. People use the term in different ways in different contexts. They play on one side of the term's ambivalence more than on the other. They contrast it with different terms in order to draw out different distinctions. This is particularly evident in relation to the varying concerns of different academic disciplines. While this Reader aims to be truly interdisciplinary, unconstrained by established conventions and set expectations, it is nonetheless useful to cast an eye over a range of disciplines that have sought to make consumption their own. Given the limited space available, we have chosen to focus on five disciplinary engagements. The first two of these are economics and marketing, which have effectively been concerned with consumption as an 'economic' category. The next three areas are anthropology, sociology and cultural studies, and they have investigated the much broader 'social' and 'cultural' implications of consumption. Finally, history and geography have demonstrated the importance of the temporal and spatial context of consumption. As we shall see, other concerns cut across these disciplinary engagements, particularly with regard to politics, ethics and aesthetics.

Economics and marketing

Economics may seem the most obvious contender for the accolade of the discipline with the most authority and legitimacy to speak about consumption – although this depends, of course, on what one thinks 'consumption' is. Certainly, economics has focused in considerable detail on consumer demand and on how this connects with production (the so-called 'supply side' of the economy). However, its development into a supposedly rigorous science has meant that economics has, for a considerable time, found itself in a somewhat isolated position. In the main, only certain maverick economists have managed to succeed in speaking to a wider audience than professional economists alone. While economics has often shown a great deal of ingenuity in dealing with consumption, it has tended to maintain a rather restrictive set of assumptions and approaches. To ascertain as much, let us briefly trace the way in which economics detached itself from its earlier manifestation as 'political economy'; turned in on itself, to develop an ever more refined and abstract style of analysis; and ultimately generated internal disaffection (giving rise, in the process, to the more practically orientated treatment of consumption within the discipline of marketing).

Adam Smith's *Inquiry into the Nature and Causes of the Wealth of Nations*, first published in 1776, almost single-handedly managed to carve out a niche for the new science of 'political economy'. Smith was one of the first to recognize that the world had changed fundamentally with the advent of modernity – that henceforth the *economic* sphere (the 'invisible hand' of the market) rather than the *political* sphere (the heavy-handed authority of the state) would be the determinant factor in shaping the overall life of society. Towards the end of his momentous study, Smith (1937, 625) announced that 'Consumption is the sole end and purpose of all production and the interest of the producer ought to be attended to, only so far as it may be necessary for promoting that of the consumer'. Although many were later to castigate Smith's work as being little more than an *ideological* justification of the market economy – with scant regard for social justice, and every concern for giving free rein to rich and powerful business interests – it is important to note that Smith believed that 'free enterprise' was destined to increase the level of wealth as a whole and therefore, potentially at least, to improve the lot of everyone. Along with other political economists, such as David Ricardo and John Stuart Mill, Smith more or less envisaged the burgeoning market system as a *one-off* shift, as society geared itself up to a new situation. There would be ample time for redistributive concerns in the longer run, once the new situation had become firmly established. For the time being, the most pressing concern was to allow the 'natural laws' of the economy to work themselves out, avoiding undue government interference. The role reserved for political economy was merely to determine what the natural laws of the economy were, in order to ascertain how they could best be nurtured and allowed to take their course. While some political economists had more of a social conscience than others, it was left to Karl Marx, in his monumental *Critique of Political Economy*, to point out that these supposedly 'natural' laws were in fact the historically and geographically specific consequences of human actions, and that what humans had made might be made *otherwise*, to the greater benefit of all humankind (Marx, 1967). For Marx, the purportedly 'natural' laws of the market rested on the *exploitation* of one class by another: productive labour by unproductive capitalists, financiers and landed gentry. But all this revolutionary talk was by the by as far as political economy was concerned – for political economy was about to undergo a revolution of its own. The so-called 'marginalist revolution' of the late nineteenth century would see the would-be science of economics dispense with the appellation 'political' and recast itself in the mould of the natural sciences: objective, disinterested and free of irrational human influence. As far as the new science of 'economics' was concerned, Marx was merely a minor post-Ricardian thinker.

The technical detail of the marginalist revolution, which gave rise to the tradition of *neoclassical economics*, and the accompanying disquiet of self-consciously *political* economists, need not concern us unduly here (although see Mátyás, 1985). We note briefly that the earliest attempt to account for consumption along these new lines was made by Gossen, in 1854, and was later independently achieved by Jevons, in 1871, and Walras, in 1874. Of most significance to our present purpose is the way in which this new approach reduced consumption to the remarkably narrow set of concerns that has characterized orthodox economics ever since. Rejecting the radical tradition that Marx represented, the neoclassical theory of consumption founded itself firmly on Bentham's doctrine of utilitarianism. Bentham held that it is human nature to seek pleasure and avoid pain. Consumption can be explained easily in such terms. An orange, for example, has a basic 'utility'. It supplies nutrition, satisfies hunger, and is generally pleasurable to eat. As rational human beings, consumers will attempt to maximize their overall levels of utility. This simple fact, according to the neoclassical economic tradition, underlies consumption in all of its infinite complexity. Working from this basis, Bentham himself argued that a just society is one that achieves 'the greatest happiness of the greatest number' (a phrase originally coined by Hutcheson, but which became associated especially with Bentham). This kind of principle hovers over the supposedly 'value-free' science of economics, however, and has the effect of ruling out of court the sorts of objections that Marx was able to raise about the essential unfairness of the market. From its seemingly innocuous basis, the 'objectivity' of economic science manages to exclude a whole host of issues relating to social inequality and social justice. One of the central paradoxes of Bentham's philosophy is the notion that 'utility' is capable of being understood 'objectively' only by accepting that it is an indisputably *subjective* matter – a matter of personal preference. In other words, whatever someone feels is good *is* good – by definition. It is no use my pointing out that your favourite pop-star is a talentless nobody. If you like her she is good in your eyes – and that's all that matters, at least as far as utility theory is concerned. One can certainly see the rationale behind this principle. It restates the old adage that 'one man's meat is another man's poison', and the even older dictum that 'there's no accounting for taste' (*de gustibus non est disputandum*). However, utility theory was developed precisely as a means of 'accounting for taste', and predicting consumer behaviour. It dismisses the possibility of reaching judgements on aesthetic, political, moral or any other criteria by elevating one kind of judgement on a premise it declares to be 'beyond challenge', namely, *the subjective judgement of the individual*. For many social scientists, the intimate relationship between economics and utilitarianism ensures that economics merely legitimizes a given economic system, rather than being either the neutral, objective science it regards itself as being or the critique of political economy that it perhaps should be.

The earliest economic theories assumed that 'utility' could be measured in cardinal terms, so that, for example, eating an orange might give someone five units of satisfaction (or 'utils'), drinking from a frosted glass in a revolving cocktail lounge high above Downtown might provide ten utils, while being chauffeur-driven in a Rolls-Royce might offer twenty utils. While it may be tempting to think that utils are an innate property of things, they actually exist only in terms of the relationship between a subject and an object (this is also true of use-values). Thus, while one might consume exactly the same thing over and over again, the more one does so the less satisfying it invariably becomes. On this basis, one could obtain a precise measure of the *marginal* utility of each *additional* thing. For example, I derive five utils from the first juicy orange I devour; three utils from the second; and just one util from the third. By now I am sick of eating oranges, and judge subsequent fruits in increasingly negative terms. As well as fuelling a great many moral panics, this seemingly universal principle of *diminishing* marginal utility allowed economists to predict how many commodities would be demanded if they were made available at a particular price (on the assumption that consumers are rational, knowledgeable and competent beings who work within limited budgets to maximize their *total* satisfaction by choosing *between* the array of marginal utilities set out before them). Knowing both production costs *and* consumer utilities, as well as their diminishing marginal values, economists could put the two sides of the economy together to predict the *equilibrium* price and quantity of each particular thing, as in Marshall's Partial Equilibrium Analysis (Endres, 1991), or of *all* goods in an entire economy, as in Walras' General Equilibrium Analysis.

This basic approach to consumer demand remained central to economics until the Keynesian revolution and the development of *macroeconomic* theory. However, because utils remained unmeasurable in *practical* terms, economists turned increasingly difficult analytical somersaults in an attempt to get around the problem. One of the most effective manoeuvres was to adopt an *ordinal* rather than a cardinal approach to the measurement of utility. For if the economist could not be sure precisely how much satisfaction one derives from things, she can be fairly sure about one's relative preferences. For example, one might prefer oranges to bananas, and bananas to apples. So-called 'indifference analysis' – which originated with Edgeworth in 1881, but was developed principally by Pareto in 1906 – rested on this approach, which Hicks and Allen (1934) took to dizzying new heights of mathematical sophistication. A 'revealed preference' approach was also developed, which determined utility on the basis of past preferences: the ordinal utility of oranges can be determined by their relative position (or rank) in past patterns of consumer choice (Samuelson, 1948). Finally, the notion of 'subjective expected utility' – first conceived by Bernoulli in 1738, and formally explicated by von Neumann and Morgenstern in 1947 – has been used to develop a statistical approach to measuring utility and modelling consumer choice known as 'random utility theory' (cf. Herden *et al.*, 1999). However, despite all of this analytical ingenuity, the economic theory of consumer demand has remained firmly mired in the dubious logic of utilitarianism (Mishan, 1961). Many economists still go to extreme lengths to maintain its unassailable rationality, arguing, for instance, that dangerously addictive drugs are 'good' if individuals find them 'good', and that no one else is in any position to say otherwise (Becker and Murphy, 1988). This kind of limiting premise means that the neoclassical theory of consumption is generally unable to deal with anything outside of certain narrow confines. As the supposedly most 'scientific' of the social sciences, economics frequently has little to say about society at all. It has remained an inward-looking discipline, more concerned with the dogged defence of its own premises than anything else. This is, admittedly, a sweeping generalization. There have been some innovative and interesting developments stemming from the economic tradition, often displaying considerable feats of technical ingenuity. But, as generalizations go, it is not especially inaccurate.

In this light, it is not unsurprising to find that economics has consistently spawned discontent from within. This is most evident in relation to a second stream of orthodox economic thought: the macroeconomics initiated by Keynes, which we will review briefly below. It is also evident, however, in terms of a string of heterodox economists – from Veblen to Galbraith – who have challenged conventional economic wisdom in the most fundamental terms. Finally, one can see the initial development of consumer research in the field of marketing as an attempt to overcome the restrictive assumptions of the neoclassical account of the consumer. While Keynes effectively spawned a new orthodoxy, which has since brought about a *rapprochement* between macro- and microeconomics, heterodox economics and marketing point in the direction of a more holistic approach to consumption, which is what this Reader also seeks. Before developing this suggestion further, let us consider the impact of Keynes.

Keynes' (1936) virtuosic 'General Theory' dispensed with much of what had gone before in an attempt to develop a radically new approach to economics. In certain respects, it harked back to the *classical* tradition of political economy, dismissing the entire edifice of neoclassical economics as a misguided turn up a narrow cul-de-sac. Where the neoclassical tradition had overturned the classical approach by opting to begin from the individual (in terms of consumer preferences, for example), and then working to build up a picture of the entire economy, Keynes reverted to dealing in aggregate quantities from the outset: total national income, aggregate demand, and so on. This approach recalls one of the earliest representations of the modern economy which appears in the *Tableau Économique* of the eighteenth-century French Physiocrats. Keynes rejected explicitly the neoclassical model of demand because of its patently unrealistic assumptions regarding consumer preferences (Drakopoulos, 1992). Having done so, he was able to show that economic stagnation could arise as a result of a lack of 'effective demand' (demand backed up by money). Writing in the wake of the Great Depression, Keynes revealed that the 'market equilibrium' that was confidently predicted by neoclassical economic theory might not be achieved automatically by the free market. In such circumstances, it would make sense to set people to work – even on pointless tasks such as digging holes and filling them in again – and to pay them from government funds, since they would then spend at least a portion of the income received stimulating the economy in the process. If the government collects taxes in an economy that is working at less than full strength, spends the money thus obtained on public works, creating employment and increasing the overall level of effective demand in the process, such intervention might eventually get the economy back to working at full strength. This kind of state-sponsored 'demand management' was, for a time, put into practice, and appeared to work in the short term – until unforeseen problems began to arise, leading to its unceremonious fall from political favour in the 1970s and 1980s at the hands of monetarism and neoliberalism.

At the heart of Keynes' theory was the relationship between income and consumption. Keynes (1936, 96) held that the dependence of consumption on income was a stable linear function, reflecting a 'fundamental psychological law'. He suggested that as incomes rise people will always consume more, though at a *decreasing* rate (in formal terms, the marginal propensity to consume is always positive but less than one). Keynes also suggested that the *marginal* propensity to consume will be less than the *average* propensity to consume, which means essentially that if income levels were to fall relative to recent levels, people would not lower their established consumption standards proportionately (they would not be able to give up everything they had grown used to consuming). Similarly, a sudden rise in income would not see a *proportionate* rise in consumption. People can only eat so much food, wear so many shoes, and so on. On the basis of this eminently reasonable reasoning, Keynes developed a picture of the economy as an integrated whole, and thereby worked out a basis on which one could tinker with the system, using various policy measures to get the economy back to working at full strength (at the 'full-employment' level).

While the monetarist school of economics was later to criticize Keynes' reliance on fiscal (taxation) policy, exposing some basic difficulties with his theory of money, of more pertinence to our present concerns is the fact that Keynes' account of consumption seemed to be contradicted by the evidence from the start. As we have seen, Keynes suggested an essential non-proportionality of consumption with respect to income. Yet this was contradicted by longitudinal (time-series) data, which suggested a *proportional* relationship between consumption and income levels (Kuznets, 1946). Cross-sectional evidence (that is, evidence from households at a variety of living standards) on differences in income and consumption levels also suggested a proportional relationship between the two. In an attempt to resolve the apparent contradiction between non-proportionality in theory and proportionality in practice, most macroeconomic work on consumption has preoccupied itself with stretching 'income' to become something much more than 'current income' alone, such as one's average expected income over an entire lifetime. The most notable examples are Duesenberry's (1949) 'relative-income theory'; Friedman's (1957) 'permanent-income theory'; and the 'life-cycle hypothesis' (Ando and Modigliani, 1963; Modigliani and Brumberg, 1955). While Keynes' influence was tarnished by the long-term problems created by demand management, allowing the monetarist backlash of the Thatcherite and Reaganite 1970s and 1980s to reinstate neoclassical economic principles as the basis for macroeconomic theory, Keynes' antagonism towards neoclassical theory at least encouraged a number of developments along different lines.

One such development has been the approach which, taking its cue from Veblen (1994), rejects the idea that consumption is a purely utilitarian matter and that 'tastes' and 'preferences' are a wholly individual affair. Duesenberry's (1949) work rested on such a rejection (McCormick, 1983) and, as Pollack's (1970, 1976) technical elaborations reveal, the assumption of *interdependent* preferences (i.e. that people tend to prefer what others prefer) has particularly devastating consequences for the orthodox view founded upon independent preferences. Another line of approach, associated with Lancaster (1966a, 1966b), seeks to replace the simple notion that *consumers demand goods that give them satisfaction* with a more nuanced view: *people demand certain satisfactions, which certain characteristics of certain goods might supply*. For example, a television set might satisfy the desire for enjoyment *and* for information, while food and drink might satisfy the need for nutrition *and* for enjoyment. Consumers' demands for 'enjoyment' are met by television, food, drink, and so on. The complexity of this seemingly minor modification wreaks havoc with the conventional economic theory of consumption.

If these slight departures from the assumptions of neoclassical theory have such far-reaching implications, little wonder that economics has seen a whole series of discontented economists who have been damning of the unreality of its most basic simplifying assumptions. It is only relatively recently that economics as a discipline has begun to question its own propaganda that it represents a neutral science (McCloskey, 1985). Even so, economics has a long history of dissident thinkers. The first iconoclastic figure in this mould, as far as consumption goes, is **Veblen**, a sample of whose work on 'conspicuous consumption' is included in **Part 5** (cf. Hamilton, 1987). Veblen railed against the entire economic tradition established by Adam Smith, whom he accused of putting a consoling gloss on 'the messy truth of the money circuit of capital' (Dyer, 1997, 47). Ironically, however, Veblen's thought has often been caricatured in economics. Leibenstein's (1950) discussion of the 'Veblen effect' refers to an inversion of the usual situation dealt with by economics. Usually, a higher price leads to lower demand. With the 'Veblen effect', a higher cost paradoxically generates *greater* demand because consumers wish to display their wealth conspicuously. While this captures something of what Veblen was about, it is very much a reductive treatment of his work, as reading Veblen in the original (and the summary in **Part 5**) should reveal. Indeed, it might be said that Veblen's work, despite his disciplinary affiliation, rested on a recognition that consumption is as much social as it is economic. Consequently, his work has had a greater impact on other social science disciplines than it has had on economics itself. Another iconoclastic figure in much the same mould is Galbraith (1958, 1967). Galbraith sought to point out the fundamental absurdity of maintaining the assumption in economic *theory* that consumer demands were exogenous (or external) to the economic system, when the *reality* is that businesses spend vast amounts of money trying to influence consumer demand. Ironically, while Galbraith's arguments about the way in which advertising can condition consumer demand have become commonplace across the social sciences, they are probably least accepted in economics, which still clings to the 'accepted sequence': consumers have ingrained preferences that determine their demands, to which producers can only respond. The ghost of Adam Smith still well and truly haunts economics! Of course, the extent to which consumers can actually be manipulated by advertising remains a hot topic of debate, which several of the readings that follow pick up on. Many have argued that Galbraith's implicit model of the consumer as 'gullible dupe' is just as unrealistic as the neoclassical view of the consumer as 'rational utility maximizer'. Work in a similar vein to Galbraith's includes Scitovsky's (1976) indictment of the failure of economic growth to generate consumer satisfaction, Hirsch's (1976) treatise on the role of 'positional goods' (i.e. exclusive status symbols) in restricting satisfaction in an era of abundance, and Frank's (1999) more recent dazzling restatement of essentially similar themes.

What these latter contributions all reveal is a strong and persistent current emanating from within economics that has little truck with the restrictive assumptions of the orthodoxy. While the contributions we have mentioned seem to point towards **Veblen** as the source of this stream, we should not neglect the continuing significance of **Marx**, even if his marginalization within economics has tended to diminish his disciplinary significance (Fine, 1995; Fine and Leopold, 1993; **Mohun**, 1977: see **Part 3**; Preteceille and Terrail, 1985). The way in which an interdisciplinary approach to consumption has developed over the last few decades holds out the promise of making up for the neglect consumption has often suffered in the past, and Marx has a major role to play in this effort (although it is important to point out that Marx's work is susceptible to a number of points of critique from other traditions of work on consumption, as we will see in the next section). To conclude this section, we will make a few observations on the development of consumer research in a discipline that is more or less allied to economics: marketing (for a fuller review see Belk, 1995).

As Galbraith pointed out, at a certain point in the history of capitalism, goods became harder to sell than to manufacture: productive capacity exceeded consumptive capacity. In pondering the *practical* problems this situation created, business could find little of value in the economic theory of the consumer. As Scitovsky (1976) discusses in some detail, while economics and psychology share a common ancestry, economics became rapidly satisfied with a particularly simplistic psychological account of the consumer. The twain were destined to meet once more only when business itself began to call for a more pragmatic approach to understanding consumer behaviour. This, then, is our first point: much of the initial work in marketing and the subdiscipline of economic psychology aimed to open up the 'black box' of the utility-maximizing consumer in order to add a dose of behavioural and psychological realism to the evident simplifications and limitations of economic theory (Foxall, 1988; Howard and Sheth, 1969; Lunt and Livingstone, 1992). Our second point is that the subsequent development of consumer research in marketing has borrowed steadily from an increasing range of disciplines, covering an ever-expanding set of issues that are central to the social sciences as a whole. This development rests primarily on the growing recognition that consumption is as much a social as an economic phenomenon. While occupying a very particular position within the discipline of marketing, Holbrook's (1995) collection of essays gives a reasonable flavour of this trajectory, which has also become caught up in the general debate over postmodernism (Firat and Venkatesh, 1995). As paradoxical as it sounds, however, marketing has tended to develop its own, *disciplinary* version of interdisciplinarity – in stark contrast to the truly interdisciplinary approach to consumption that has since gathered pace across a network of different social science and humanity disciplines. Consequently, the discipline that we began by saying would appear, on the face of it, to possess the greatest authority to speak of consumption – economics, together with its offspring, in the shape of marketing – contributes least to this Reader. We shall see that this is nowhere near as odd as it might sound if we consider the consumption-related concerns of a second set of disciplines.

Sociology, anthropology and cultural studies

We began our survey of economics and consumption by considering Adam Smith and the tradition of political economy, noting that Smith was among the first to recognize the increasing importance of the market in governing social life. If we return briefly to this starting point, we can begin to appreciate how a different set of concerns to those of interest to the economist developed in other disciplines. Although Smith's recognition of the importance of the market signalled a significant change in thinking about the nature of society, the body of thought he developed harked back to the so-called 'Hobbesian problem'. The Hobbesian problem concerns how and why innumerable *individuals*, all possessed of free will, would choose to come together to function as a *society* – given that this would inhibit their original freedom to do whatever they wish with absolute impunity. Debate in political philosophy had waged long and hard over the necessity of a strict, authoritarian government as the only way to curb aggressive human impulses (e.g. Hobbes), as against the opposing view that individuals entered willingly into an unwritten 'social contract', forgoing a degree of individual freedom in exchange for a more secure, certain and stable social order (e.g. Locke and Rousseau). The economic liberalism proposed by the early political economists managed to side-step this debate very neatly by highlighting a situation where the self-interest of the individual and the collective interest of society coincide: economic exchange. The archetype of this situation is *barter*. For example: since I have potatoes but no eggs, you have eggs but no potatoes, and we would both like to eat egg and chips, we should swap some of each in order to ensure that we are *both* better off. Barter, it seems, presents us with a *win-win* scenario, a scenario that even applies in many seemingly conflict-ridden situations. As surprising as it may seem, acting in a purely self-interested way can turn out to be mutually beneficial: public virtue can be born of private vice, as Bernard de Mandeville famously put it in *The Fable of the Bees* (1714). Multiplied many times over, the principle of mutual advantage enshrined in barter approximates to the market system of *free enterprise* – the centre-piece of modern Western society as Smith saw it. Miraculously, under this system individual and society are no longer at odds. All that was needed to ensure an optimal distribution of things was to promote the free market and restrain government interference. However, one can easily read into the parable of barter the basis of an entirely different form of social arrangement. For what does it mean to say that *I have* some potatoes and *you have* some eggs? Is it not more to the point to say that *there are* eggs and potatoes, and *there are* people, *none* of whom have a God-given right to the ownership of eggs, potatoes or anything else, least of all land, labour or capital? It was on this basis that Proudhon famously declared that 'all property is theft'. Anarchism, socialism, Marxism and a variety of radical traditions of thought see the basic issue at the heart of society as one of arranging a fair distribution of eggs, potatoes, and so on, among the people without beginning from the dubious premise that any particular individual has a prior claim over anything at all. Little wonder, then, that political economy was seen by many as nothing more than an ideological justification for the iniquities of private property (such as exploitation, pauperization, dependency and alienation) and the class conflict that it sustains. Thus it is not surprising that a whole range of other perspectives should have found their way into consumption studies, and that the 'Hobbesian problem' should have become a point of departure (and a bone of contention) for a variety of other social science disciplines.

Let us begin to take stock of these by returning to Marx, since his legacy has had a major and long-standing significance for virtually all of the social sciences apart from economics. As we have noted previously, Marx was among the first to point out the limitations of the supposedly 'scientific' status of political economy. By treating private property and commodities as if they were a 'natural' state of affairs, political economy threw a veil over the underlying exploitative conditions of the modern economic system. People might well meet on an equal footing in the marketplace (the sphere of exchange), but hidden behind the apparent freedoms of the market are the profoundly unequal class relations that characterize the sphere of production: those who labour and those who profit from that labour. It is here, in the workplace, that people's true social situation is determined. For one class of people has nothing to exchange but its ability to work ('labour power'), while the dominant, capitalist class holds a monopoly over the 'means of production' (the tools and materials needed for production to take place). The social relations underlying market exchange, therefore, rest fundamentally on the *exploitation* of one class by another. In this light, it is hardly surprising that Marx should have paid more attention to production than to consumption. However, as the reading from the *Grundrisse* in **Part 5** reveals (Marx, 1973), **Marx** had a far more sophisticated conception of consumption than is often assumed. The importance Marx placed on the commodity, and the way in which he revealed the class inequalities that sustain the supposedly 'free' market, recurs in many of the readings included in this Reader.

Sociology has arguably been the discipline most enamoured of Marx's analysis. In relation to consumption, however, a somewhat caricatured view of Marx has often been deployed, suggesting that capitalism can simply pull the wool over people's eyes, creating 'false needs' that allow the expansion and perpetuation of the capitalist system. Although there is undoubtedly a sizeable grain of truth in this (Haug, 1986; Marcuse, 1964), work suggesting that consumers can be hoodwinked quite so easily has fallen from favour in recent years. Instead, a body of work that credits consumers with a little more 'nous' has been developed. Today, the influence of Marx is felt perhaps most strongly in sociological work that focuses on the effect of the changing nature of the economy on society as a whole. This is particularly true in relation to the putative transition from a *Fordist* economy (associated with mass production, mass consumption and Keynesian demand management) to a *post-Fordist* situation (characterized by flexible production, fragmented markets, and a neoliberal attitude towards regulation and policy) (Lee, 1992). While Marx figures strongly in many of the readings included here, sociologists have always been concerned to debate the emphasis Marx placed on the *material* basis of capitalism, as against Weber's insistence on the pre-eminent role of a particular *attitudinal* structure in explaining the origins of capitalism. This concern has manifested itself directly in relation to consumption.

Weber (1976) pointed to an 'elective affinity' between the spirit of capitalism and the Protestant religion. As far as Protestantism was concerned, *godliness* was next to *industriousness*. Weber maintained that the asceticism and denial of earthly pleasure promoted by Protestantism translated into a commitment to hard work, simple living and thriftiness which was ideally suited to the development of capitalism. The basic values that would promote capitalism just so happened to manifest themselves in Protestantism, particularly in the Calvinist doctrine of 'predetermination'. For the Calvinist – unlike the Catholic who could sin, confess and have the slate wiped clean – the soul was *indelibly* marked as either black or white. There was simply nothing one could do about it; not even performing good deeds and living a virtuous life could better one's chances of finally entering the Pearly Gates. However – and it is here that the value of Protestantism for capitalism comes into its own – one might be able to reveal (primarily to oneself but also to others) that one is numbered among the 'elect'. Living a good life was a sure sign that one's soul had *already* been saved, and the virtues of industry and thrift numbered among the best ways of demonstrating the purity of one's soul. Capitalism therefore might well operate in the way dissected by Marx, but the reasons behind its flourishing in the first place are to be found, according to Weber, in this kind of attitudinal complex. At first sight, all of this concern with hard work, asceticism, forgoing worldly pleasures, deferring gratification, and so on, might seem very distant from consumption – especially if consumption is regarded as involving more than mere subsistence. Indeed, although Protestantism and consumerism would seem to be extremely unlikely bedfellows, there are nevertheless some surprising connections. To begin with an apparently trivial example, one might point out that certain goods were ideally suited to the Protestant work ethic, such as coffee (Schivelbusch, 1993). In a more fundamental way, **Campbell** (1987) has suggested that it turned out to be a surprisingly short sequence of steps from Protestant asceticism to consumerist hedonism (see **Part 1**). Protestantism demanded *self-control*. Succumbing to one's emotions might easily lead away from contemplating the perfection of God towards sinful worldly pleasures. If controlling one's emotions, demonstrating one's empathy for others and cultivating one's sensibilities were initially directed towards other-worldly purposes, however, the way in which such practices nurtured the imagination also turned out to be central to the way of life of the modern, pleasure-seeking hedonist. A remarkably similar set of means ultimately came to serve a very different set of ends. While affording Weber's fable about the origin of capitalism less credibility than Campbell, Baudrillard (1998a), **Bauman** (1983: see **Part 1**) and Giddens (1994) all suggest that it was the essentially *compulsive* nature of Puritanism that facilitated the transition from ceaseless productive activity to interminable consumptive activity. The Protestant compulsion to work is much like the consumerist compulsion to seek pleasure. The proximity of Protestantism to consumerism is captured neatly in the American saying that 'the Quakers came to Pennsylvania to do good and ended up doing well'.

For the sake of completeness, let us note that Weber developed a number of other lines of argument that have been subsequently drawn upon to further our understanding of consumption. Not least is the emphasis Weber gave to *rationality* and *bureaucracy*. Ritzer (1992, 1997, 2002) has applied Weber's analysis of the relentless rationalization of modern society to the sphere of consumption, taking the fast-food industry as his example. However, Ritzer's insistence that such rationalization has faced little resistance and is leading to a wholesale homogenization of the globe has come in for sharp criticism (Smart, 1999).

Alongside Marx and Weber, Durkheim is taken to be the other 'founding father' of modern sociology. Durkheim emphasized that society was more than the sum of its parts. It is a decidedly *collective* affair, with discernible patterns and rules to the way it functions, which cannot be said to boil down to the sum total of the behaviour of society's individual members. Durkheim (1947, 338–339) stressed that the individual is 'far more a product of common life than its determinant'. In other words, the individual is a product of society, rather than society being a product of individuals. Although there are a variety of influences on the work of **Bourdieu** (1984), including Marx and Veblen, Bourdieu substantially inherited the mantle of Durkheim. Bourdieu's work is important in relation to consumption insofar as it demonstrates the extent to which the reproduction of a class society is crucially tied up with consumption and taste. Where economics typically regards the tastes and preferences of the consumer as a wholly individual affair, Bourdieu suggests that one's class position or social location influences one's tastes; and, reciprocally, one's tastes reinforce one's class position, insofar as taste is a marker of social class. The extract reproduced in **Part 5** details the kinds of mechanism at work in this situation, where the selections an individual makes from an existing structure of tastes serves simultaneously to classify that individual. Exercising his or her choice works to 'position' the individual in relation to others. In such a situation, as Bourdieu (1977, 72) says, 'structured structures [are] predisposed to function as structuring structures'. Many sociologists have found Bourdieu's work of considerable importance in understanding consumption, for at least two reasons. First, whereas a crude neo-Marxian stance implies that consumers are simply indoctrinated into wanting certain things they don't really need, Bourdieu manages to paint a far more credible picture of the consumer without abandoning the crucial point that consumption is implicated in the reproduction of the class structure of capitalist society. Second, while the stress Veblen placed on conspicuous consumption emphasized the way in which 'status symbols' serve to display social standing, Bourdieu effectively showed that *all* consumption, whether consciously conspicuous or not, reveals and reinforces one's class location.

In addition to his role in the emergence of sociology, Durkheim also influenced the development of anthropology in France. This is evident, for example, in the work of Durkheim's nephew **Mauss**, whose celebrated essay on *The Gift* (Mauss, 1990; see **Part 5**) has significantly influenced thinking on consumption (Berking, 1999; Carrier, 1991, 1994). Like Durkheim, Mauss developed an account of the sense in which society, as a *collective* form, is ineluctably *prior* to the individual – thereby overturning the presuppositions of classical political-economic thought that society develops out of some kind of rational arrangement between autonomous individuals. As anthropologists such as Dumont (1977, 1986) have subsequently elaborated, the conception of the 'individual' as a fully self-contained and independent 'unit' is not at all a reflection of a 'natural' state of affairs but a direct product of modern ways of thinking. Mauss' work on gift exchange sought to show that so-called 'archaic' (or pre-modern) societies reproduced themselves on the basis of an entirely different system to modern economic exchange. In the absence of private property rights, the modern trait of accumulating wealth was also absent. Instead, social groups tended to give away wealth – thus putting the recipients of such gifts under a bond of obligation. Moreover, if groups did not reciprocate then they were open to their chattels simply being taken by superior fighting force. In effect, giving puts one in a position of power, while receiving puts one in a state of debt. Mauss claimed that this logic underlay the organization of all societies until the development of the modern Western economic system. Mauss also noted that this situation sometimes gave rise to a ritualistic form of consumption (or destruction) of wealth: the potlatch ceremony. Here, rival tribes attempted to outdo one another by destroying immense quantities of their own possessions, as a means of *symbolically* demonstrating their wealth and power. **Bataille** (1988: see **Part 5**) developed and extended this aspect of Mauss' work, considering a variety of practices, such as sacrifice, in a diverse range of human societies. Like Mauss, he also sought to demonstrate the extent to which the logic of the gift has underpinned all societies prior to modern Western society. Bataille regarded modern society as simply *repressing* this logic, hiding it behind the supposedly 'rational' imperative to *accumulate* (rather than to *expend*) wealth. What today appears as irrational – as useless, excessive or destructive – is far more important to the way in which society coheres than we might care to imagine. Indeed, for Bataille, it is luxury and abundance, not need and scarcity, that are fundamental to the nature of human society. The modern economic system has not managed to dispense with this indispensable truth, even if it cares to imagine that it has. It simply disavows an inescapable principle, which is destined to manifest itself in all kinds of disturbing and unexpected ways. We would be far better off, Bataille maintains, in acknowledging the necessity of useless, excessive expenditure and destruction: in making ourselves aware of its inevitability.

Baudrillard's work may be related directly to Mauss', Bataille's, and Veblen's insistence that the principles modern society claims to have surpassed persist nonetheless. Baudrillard's (1981) distinction between the way in which objects can function as *symbols* – expressing social bonds in the manner suggested by Mauss – and the way in which commodities function as *signs* – which hide social bonds, reinforcing the sense in which the consumer is an isolated individual – forms the nub of his understanding of modern consumption (see **Part 5**). Baudrillard attempts to show that the consumer society disguises the underlying *social* logic governing human societies (in terms of class discrimination and the like) behind the *individualized* logic of economic rationality (the idea that goods meet the pre-existing needs of individuals). The way in which commodities function as 'signs' is an indictment of the modern economic system. Where objects once expressed and symbolized social relations directly, people in a consumer society relate first and foremost to objects as signs, in the hope that *possessing* them and *communicating* through them will allow individuals to forge a meaningful relationship with increasingly estranged others. This situation promotes an interminable desire to consume (to communicate and to consummate), since the sign can never achieve the kind of intimate social relations that were once symbolized in gift exchange.

In developing his argument about the inadequacy between sign-values and symbolic exchange, Baudrillard (1975, 1981) simultaneously developed a far-reaching and under-appreciated critique of Marx's conception of the commodity by deconstructing his fundamental distinction between use-value (usefulness) and exchange-value (money) (see **Part 5**; Baudrillard, 1998b). Marx saw 'use' as an intrinsic quality of things and 'exchange' as an artificial, distorting influence. Hence the Marxist dream of a classless society organized on the basis of usefulness rather than capital accumulation. However, Baudrillard, Derrida (1992, 1994) and Lyotard (1993, 1998) demonstrate how 'use' is a far from innocent category. Not only is it subject to the same calculations and ruminations as exchange-value and sign-value, but 'making use' has become a moral obligation for citizens of the consumer society. For Baudrillard in particular, usefulness is not a genuine reflection of human relations with objects, but a smokescreen thrown up by the modern economy. This smokescreen does double duty. Specifically, the cultivation of needs and uses (i.e. wants and desires) serves to perpetuate the absorption and squandering of vast quantities of manufactured things (for each of which a 'use' must obviously be found), while simultaneously masking the absence of the symbolic relations that objects once served to express (a task that is accomplished partly through the imposition of sign-values). By investing his revolutionary hopes in 'usefulness' and 'needs', Marx remained caught up in the ideological categories imposed by modern rationality in the service of capitalism (Deleuze and Guattari, 1984, 1988). This is why Baudrillard finds the kind of 'ambivalent' and 'formless' analysis presented by Mauss and Bataille far more radical than the ostensibly 'critical' analysis of Marx (Bois and Krauss, 1997).

Sadly, much of the force and ingenuity of Baudrillard's position has been overlooked because of confusion over the nature of signification and communication, neither of which even remotely approximates to the unthinking equation of consumption with efficacious speech acts (cf. **de Certeau**, 1984; see **Part 5**). Indeed, Baudrillard's (1996) insistence that consumption is structured like a language and that the consumer object is articulated as a sign-value fits perfectly with something like the media, legal and consumer reaction to the 'FCUK' campaign instituted by French Connection UK, which demonstrated just how difficult it is to 'say' anything using the all-too-blunt instruments of the consumer society (cf. Campbell, 1995, 1997; Warde, 1996). Be that as it may, **Douglas** (Douglas and Isherwood, 1996; **Douglas**, 1996; see **Part 3**) has developed this aspect of consumption in a highly convincing way, without suggesting simply that consumption 'communicates' in a literal sense. It occasionally does, but more often than not the cultural aspect of consumption is both a far more subtle *and* a far more ambivalent affair. It seems that the dissent expressed by Campbell and Warde largely relates to the way in which consumption has become embroiled with the debate over postmodernity, particularly in the work of writers such as Baudrillard (1996, 1998a). Their basic objection seems to be that such work is overly *culturalist* – that it goes too far in regarding consumption as a 'cultural' activity, consequently neglecting its 'material' aspects. This completely misses the point of Baudrillard's work, and misinterprets the tradition to which it belongs. For Baudrillard, Bourdieu, Douglas and a host of others, 'culture' cannot be understood as a relatively superficial aspect of social life (relating to insubstantial meanings and images, and detached from gritty material reality). Human societies are understood by such writers as ineluctably cultural. Meanings do not somehow 'float above' the social world; they are part and parcel of it. The social world is *constituted* by cultural categories, and can only be understood in such terms – both by the social actors involved and by social theorists. Trying to separate 'real' material issues from 'ethereal' cultural issues also risks disparaging the various advances that have been made in understanding consumption from the discipline of cultural studies (which might be regarded as the offspring of sociology in much the same way that marketing is the offspring of economics). As we shall see below, work on subcultures has often centred on the creative use of marginalized consumer goods and styles to enable some kind of collective resistance to 'mainstream' culture, in a sort of inverted version of Veblen's conspicuous consumption (taking pride in base culture rather than in high culture) and Marx's working-class subjects of history (drawing on the power of consumers rather than producers). Finally, Baudrillard continues to push things even further, arguing that radicality has passed out of the hands of producers *and consumers*. For him, radicalism belongs to the world of objects and events (a position that is similar in some respects to the view that while the working classes may have shirked their revolutionary responsibilities, Nature itself has increasingly risen to the challenge of resisting its enslavement to the will of capitalism and technoscience).

Since we began our discussion in this section with a consideration of Marx, Baudrillard's critique of Marx has brought us full circle. It is notable that the holy trinity of Marx, Weber and Durkheim – the three 'founding fathers' of modern sociology – has made its influence felt in the sociology of consumption. It is equally notable, however, that this conventional set of underpinnings begins to break down in the context of consumption studies. This is evident, for instance, in Baudrillard's suggestion that Mauss ultimately carries more radical implications than does Marx. It is also evident in the case of Veblen. For although Veblen was an economist by trade, his contribution to the sociology of consumption is arguably of equal importance and status to that of, say, Weber. Moreover, the work of **Simmel** (see Frisby and Featherstone, 1997; Wolff, 1950), along with others such as Benjamin (1999), might be considered as offering an equally important set of foundations for a sociology of consumption. Simmel, like Benjamin, focused on the nature of modern society and modern culture, noting the constant changes in tastes and fashions brought about by modernity (Simmel; see **Part 5**), the development of new commercial spaces in the city (see Buck-Morss, 1989), and the emergence of distinctively modern ways of living (such as that pursued by the *flâneur*, the free-wheeling 'city stroller': see Tester, 1994). It is not surprising, therefore, that a great deal of current research on consumption should have taken up the kind of concerns first expressed by Simmel. The legacy of Simmel and Benjamin has been rekindled in the wake of the 'cultural turn', which has made its influence felt across the social sciences over the past decade or so – to some extent bridging the gap between the social sciences and the humanities. A great deal of recent work on consumption, whether in sociology, anthropology and cultural studies or in English, history and other humanities subjects, has rested on the basic observation that consumption is as much a *cultural* as an economic phenomenon. A wide variety of recent work – such as McCracken (1990), MacClancey (1992), Lury (1996), Slater (1997), Storey (1999), and Hearn and Roseneil (1999) – signals the centrality of culture to consumption; as does work addressing specifically the gender dimension of consumption (Nava, 1992; Mort, 1996), work on postmodernism and consumption (Featherstone, 1991; Jameson, 1983; Bauman, 1987), and so on. Likewise, much of the recent interest in consumption from anthropology has developed in relation to a disciplinary concern with *material culture* (Miller, 1987). Simplifying to the extreme, one might suggest that two crucial issues emerge from this renewed engagement with consumption as a cultural phenomenon. One is the role of consumption in shaping personal identity and the character of the *consuming subject*. Another is the nature of the *object* in a consumer society. Although subject and object are obviously brought together in the act of consumption, let us tentatively separate them in order to consider each in a little more detail (since both receive fuller introductions in **Parts 3** and **4**, we can afford to be relatively brief in both cases).

As we have seen, one of the ways in which economics has reduced its effectiveness in dealing with consumption has been in the kind of abstractions it employs. For the economist, whether the object of consumption is a tomato, a car, a copy of the Bible or a pornographic magazine is, in principle, a matter of total indifference. Specific studies may consider the agricultural economy, the car industry, the economics of religion or pornography – but in theoretical terms, all goods are reduced to their lowest common denominator: they are regarded as being capable of satisfying needs. The concern with objects developed elsewhere in the social sciences affords far more significance to the distinctive nature of *particular* kinds of objects and their role in human society. Such an approach offers a far less *reductive* treatment. This is important, given the common tendency to simply ignore objects. For **Latour** (1992; see **Part 4**), objects count among the 'missing masses' of the social universe (cf. **Kiaer**, 1996; see **Part 4**). One of the most famous works on the nature of the object in a consumer society is **Barthes**' *Mythologies* (1973), a collection of short essays that reflect on all manner of objects – from steak and chips to Citroën cars via milk, plastic and toys (see **Part 4**). Barthes' aim is not simply to produce a phenomenological analysis of these particular objects. As in *The Fashion System* (Barthes, 1990), he attempts to read into objects the kind of social relations they speak of, and thereby delineate the contours of modern consumer culture. Baudrillard's (1996) *The System of Objects* follows closely in Barthes' footsteps, considering the significance of antiques, gadgets and collections alongside the kinds of everyday object that are usually taken for granted. While Barthes and Baudrillard are both associated with major advances in the 'science of signs' (aka semiotics and semiology) – a science inaugurated by de Saussure (1959) in the late nineteenth century, but which came into its own only in the latter half of the twentieth century under the guise of structuralism and poststructuralism – many other social scientists, adopting very different approaches and methods, have also sought to consider objects both in terms of their specificity *and* what the 'social life of things' might tell us about social relations.

To single out one example, Warde (1997) offers a sociological study of food consumption, based on detailed empirical analysis. There are similar studies of the consumption of clothing, sport, music, technology, and so on, many of which are usefully summarized in the growing number of textbooks on the sociology of consumption (Corrigan, 1997; Miles, 1998). Without denying that disciplinary differences remain, a variety of work on objects of consumption, from a variety of disciplinary perspectives, has alighted on a remarkably similar set of concerns – as contributions from anthropology (Appadurai, 1986), cultural studies (Hebdige, 1996) and social psychology (Csikszentmihályi and Rochberg-Halton, 1987; Dittmar, 1992) serve to confirm.

The social life of things, of course, is intimately related to the social life of people. Again, where the economist's consumer is an abstract agent – *homo economicus* or 'economic man' (see **Mohun**, **Part 3** for a critique) – other social science disciplines have sought to recover what economics has treated in a more or less reductive manner. This is especially evident in work emphasizing the *gendered* nature of consumption. Much of the impetus for this focus comes from the cross-disciplinary influence of feminism. Feminist thought has paid a great deal of attention to consumption – which is perhaps not surprising given the strong association of women with shopping and the domestic sphere in modern Western society (Jackson and Moores, 1995; **Miller**, 1983; see **Part 4**). Feminist thought has frequently placed many of the most basic issues connected with consumption under scrutiny. For example, if one believes that consumers are duped easily and inculcated with 'false needs', then given that women are associated most often with consumption (in theory, if not necessarily in practice), might this belief not, in fact, carry a significant gender bias? Would consumers have been so easily cast as 'dupes' if the majority of consumers were men? For instance, compare the typical characterization of the 'false needs' of consumers with the 'false consciousness' of workers, which did not prevent them from being a considerable power to be reckoned with. Few have said similar things about the consuming classes (although see Miller (1995) and the comments on consumer co-operation below).

As feminism has developed its engagement with consumption the hidden gender implications of a great deal of existing work have been given a thorough airing. Initially, feminist work in this area tended to focus on the extent to which the gendering of consumption pointed up yet another form of exploitation, which works to the benefit of both capitalism and patriarchy (Weinbaum and Bridges, 1979). From a less economistic perspective, and drawing on psychoanalytic insights into the nature of female desire (Bowlby, 1993), many feminists have suggested that a greater capacity for female pleasure explains the association of women with consumption. Whether this is unequivocally good, unequivocally bad, a matter of ambivalence or unacceptably essentialist in its characterization of women remains a moot point.

While most feminists remain critical of both capitalism and, of course, patriarchy, many feminist thinkers hold that consumption offers at least some opportunity for women's individual empowerment, creativity and pleasure – even if it does so in conditions that hardly grant women equality with men (**Bowlby**, 1985, 2001 see **Part 3**; Carter, 1984; McRobbie, 1997; Nava, 1992; Radner, 1995). If this broad shift from a concern with women's exploitation to an exploration of women's pleasures is apparent in the feminist literature as a whole, a number of internal debates, arrayed along these issues, have marked out opposing positions. One example is the debate between Wolff (1985) and Wilson (1992) over the role of the *flâneur* (the city-stroller). Citing the inequality of women in the nineteenth-century city, Wolff (1985) highlighted not only the presence of the (male) *flâneur* in voyeuristic and predatory terms, but also the impossibility of a female equivalent of the *flâneur*. Wilson (1992), however, has highlighted the way in which middle-class women consumers adopted a role akin to that of the *flâneuse*, especially in the new department stores of the period. At the same time, she has pointed to the 'impossibility' of the position the *flâneur* himself wished to occupy (which, she says, paralleled the uneasy existence of the female prostitute in the nineteenth-century city). Wilson, it seems, is more sensitized to the pleasures of consumption than is Wolff – though without necessarily accepting consumption as a 'good thing' in general terms. Needless to say, many feminist contributions have continued to debate the extent to which consumption in general (and shopping in particular) is a chore rather than a pleasure. It is all too easy to romanticize consumption, and whereas fashion shopping may give rise to one set of conclusions, shopping for items such as sanitary towels indicates an entirely different set of conclusions (George and Murcott, 1992). 'Is consumption pleasurable?' is the kind of question that needs to be specified more clearly before an unequivocal answer is given. Or again, it is the sort of question to which only an equivocal answer may be given. Although it is not couched in specifically gendered terms, **de Certeau**'s (1984) general perspective on consumption attempts to negotiate the same treacherous waters that much feminist analysis has charted, avoiding simplistic accounts of consumer manipulation while equally steering clear of an overly romanticized view of consumption as 'resistance'. In line with much feminist analysis, de Certeau recognizes the relative powerlessness of the consumer, but reveals how, in living their everyday lives, consumers *manage*, *make do*, and often (quite literally) *take liberties* (see **Part 5**).

Along with contributions from many other perspectives, feminist scholarship has also been central to the recent debate over consumption and *identity*. While the task of forging an identity for oneself has long been an important part of everyday life, the increasingly consumerist nature of society has altered this task fundamentally. Consumerism has seemingly managed to erode the extent to which identity is connected to class and other social categories; to elevate *personal* identity over collective identity; and even to undermine the logic of building a rigid identity – so that it increasingly makes sense to construct a *flexible* identity fit for an unstable world, rather than putting all one's eggs in one basket, so to speak (Bauman, 1996; though see **Crompton**, 1996; **Part 3**). The notion of 'lifestyle' captures the sense in which one can increasingly 'buy into' a particular identity and, indeed, buy into a *different* lifestyle, should the one adopted currently become unsatisfactory (Chaney, 1996; Shields, 1992). Much of the heated debate over postmodernity relates to the extent to which the consumer society offers the chance of constructing an identity at will or, more pointedly, the way in which the consumer society turns freedom into necessity, such that 'we have no choice but to choose' (Giddens, 1994, 75). While many have noted that identity is not entirely 'free-floating' – even if this postmodern image would suit the purposes of the market admirably – there is undoubtedly a strong sense in which identity is connected increasingly to consumption. Indeed, du Gay (1996) has shown how the various aspects of 'identity management' nurtured in the sphere of consumption have become sought-after skills in the job market.

It might appear that there are certain ways in which we can change our identities (with the willing assistance of the consumer market), and certain ways in which we cannot. While this may be true in some respects, the idea that identity has become more free-floating perhaps suggests something a little more subtle than an overly literal interpretation allows. For example, one may not be able to change one's class background *as such*, but money can certainly buy one a different lifestyle (even if it cannot guarantee happiness). While one's class origins might 'show through' in terms of one's tastes and dispositions (Bourdieu, 1984), it seems to be the case that how one *spends* one's money rather than how one *earns* it is becoming increasingly important. If, for the new middle classes, such as those employed in the burgeoning service industries, 'where you're at' matters more than 'where you're from', this has the capacity to act as a self-fulfilling tendency. If 'what it means to be from a working-class background' is widely regarded as being of less importance than 'what it means to be working in the finance industry' or 'what it means to be living a high-spending lifestyle', such a shift in attitudes alters social perceptions, behaviour and social structure in distinctive ways. Of course, this may not alter the fundamental class basis of society (as Crompton (Part 3) argues); nor is social mobility fluid enough to ensure that class background does not still significantly affect one's chances of being able to enjoy a high-spending lifestyle. Even so, for certain sections of the population, social class seems to be an increasingly poor predictor of consumption patterns, while society seems to be increasingly structured around consumption. It is not for nothing that a considerable amount of effort by the marketing industry is given over to attempting to predict consumption patterns in terms of 'lifestyle' types (**Goss**, 1995, **Part 4**; Lyon, 1994; Poster, 1990).

A similar situation is evident with respect to social categories other than class, such as ethnicity and gender. In a sense, 'what it means to be black', 'white', 'Asian', a 'woman', a 'man', 'gay', 'straight', 'bi', 'disabled', a 'teenager', and so on, has been profoundly affected by the impress of consumerism. For instance, in Western society men have traditionally been far less fashion-conscious than have women. Indeed, *not* being overly concerned with one's appearance and apparel has often been regarded as a typical masculine norm.

In the late twentieth century, however, it became far more acceptable, and even desirable, for men to pay similar levels of attention to their appearance as women have done traditionally (Edwards, 1997; Mort, 1996; Nixon, 1996). Consumption, in other words, has significantly influenced what it means to be 'a man'. Likewise with respect to ethnicity: what it means to be black, for instance, has been significantly affected by commercial culture – though whether this has had positive effects on black culture as a whole (Lamont and Molnár, 2001) or simply represents an instance of the commodification of 'race' (DuCille, 1994) remains hotly contested (Willis, 1990). Although representations of black people in advertising and on consumer-goods packaging have certainly shifted away from the overtly racist imagery that was once predominant (McClintock, 1994), many such representations continue to reinforce stereotyped images of 'blackness' (Jackson, 1994). Meanwhile, many socially constructed aspects of identity that are often (incorrectly) regarded as natural – such as gender, ethnicity, sexuality and the body (Falk, 1994; **Featherstone**, 1982; see **Part 3**) – are all caught up in, and being transformed by, consumer culture. The conditions promoted by consumerism often relate to issues that it is all too easy to regard as 'natural' and somehow prior to 'culture'. This is especially true in relation to the body. Yet many bodily conditions, such as obesity and anorexia, can be related directly to the impact of consumerism (Bordo, 1992). Such disorders are typically *medicalized* in today's society, and a wealth of research shows that such conditions relate to 'body image'. In conditions such as kleptomania, however, an interesting historical tension arose between medicalization and criminalization. As **Schwartz** (1989; see **Part 3**) shows, the temptations of the department store often led to stealing by the same middle-class women who comprised the store's clientele (**Abelson**, 1989; see **Part 4**). This resulted in such stealing being classified as a medical disorder, while similar behaviour by working-class women was more likely to be criminalized. The way in which identity has become increasingly embroiled in consumption has meant that 'identity crises' of all kinds are intertwined increasingly with consumption (Baker, 2000). Frequently, moreover, advertising and other marketing practices work in a manner that arouses fears and cultivates anxieties, only in order to offer the products that promise to curb such anxieties and quell such fears (Bauman, 1992, 2001; **Falk**, 1997; see **Part 4**).

If identity is increasingly a matter for the individual (insofar as identity construction is presented as a personal task and a personal responsibility), identities are, nonetheless, still sanctioned and validated *socially*. Not just any identity will do. While the social sanctioning of identity was once closely connected to class, a basic shift from this situation was first registered by those working within the field of cultural studies, particularly with respect to subcultures (**Hebdige**, 1979; see **Part 3**). Distinctive subcultures, particularly youth subcultures such as those associated with mods, rockers, punk, hip-hop, and so on, tended to assert group norms that were self-consciously different from those of 'mainstream' society. Belonging to a subculture involved little more than adopting the right style (in terms of clothing, hairstyle, musical tastes, and so on). This kind of situation, according to Maffesoli (1996), is increasingly prevalent and no longer a specifically *sub*cultural phenomenon: it is a situation that is common to consumer culture as a whole (**Bennett**, 1999; see **Part 3**). The individualized task of constructing an identity has given rise to a new, ephemeral, 'neo-tribal' form of social structure, which connects with the sense in which identities are also increasingly flexible rather than rigid entities. So, rather than dwelling on every aspect of one's lifestyle in detail, one can simply adopt a ready-made identity by joining a particular neo-tribe. Such a manoeuvre is a way of coping with an overwhelming set of choices. It cuts the options down to size and demands little long-term commitment. Indeed, such a situation entails that identities, fashions, lifestyles and neo-tribes themselves are all essentially ephemeral affairs – which is an ideal situation as far as the consumer market is concerned. The possibility of casting off an old identity and adopting a new one affords yet another opportunity for consumption. The neo-tribal situation described by Maffesoli involves a significant shift from the situation described by Veblen. Where Veblen saw all consumers attempting to 'keep up with the Joneses', mimicking or emulating the consumption habits of their social superiors, Maffesoli sees consumers attempting to 'keep different from the Joneses', so to speak: to assert and affirm their 'individuality' by selecting which set of social 'norms' best suits them (even if this 'individuality' is shared by all those who have made the same choice). The difference between Veblen's single 'social pecking-order' and Maffesoli's shifting, kaleidoscopic array of 'neo-tribes' is one of the basic differences between 'modern' and 'postmodern' society. We will return to this difference after considering the third component of our selection of disciplinary approaches to consumption.

History and geography

Given that history and geography are discussed in the introductions to **Parts 1** and **2**, it may seem superfluous to consider them here in advance. It is important to note, however, that many of the readings included in these parts of the Reader are not written by historians and geographers: topics and disciplines do not necessarily map on to one another very neatly. In this section, we aim to offer a succinct outline of the main consumption-related concerns of historians and geographers, although we cannot avoid making passing mention of how their concerns marry with other work on the histories and geographies of consumption. In one way or another, both history and geography are fundamentally concerned with *context*. What consumption means – its social significance, in other words – has varied greatly from one historical period to another, and still differs markedly from place to place. This is, perhaps, the single most important factor to bear in mind when considering the histories and geographies of consumption.

Consumption has become a significant focus of investigation in historical studies only relatively recently (Glennie, 1995). The key book to bring consumption centre-stage was McKendrick *et al.*'s (1982) *The Birth of a Consumer Society*, which identified a 'consumer revolution' in eighteenth-century England, prior to the 'industrial revolution' of the nineteenth century (see **McKendrick**; **Part 1**). Needless to say, pinning down a 'consumer revolution' to any particular period, let alone this one, has proved controversial. Many have insisted that a consumer revolution worthy of the name did not really occur until the mass-production techniques ushered in by industrialization made mass consumption possible (**Fine and Leopold**, 1990; see **Part 1**). Others have taken the opposite tack, accepting the ample evidence of increasing levels of pre-industrial consumption, but pushing back the origins of consumerism even further into the sixteenth and seventeenth centuries. Most work in the latter vein is centred firmly on the early-modern period, which undoubtedly witnessed a dramatic change in the significance being afforded to consumption (Appleby, 1978; **Appleby**, 1993; see **Part 1**; Bermingham and Brewer, 1995; Breen, 1988; Brewer and Porter, 1993; Shammas, 1990; Thirsk, 1978; Weatherhill, 1988). However, we should not lose sight of the fact that consumption has always been an important aspect of society. Sherratt (1998, 15), for instance, refers to changes in social organization in Europe in around 1300 BC as marking 'a consumer revolution', and one can identify any number of turning points in the history of consumption, most of which relate to other significant changes in social arrangements. In this light, it should be clear that many of the debates in the history of consumption amount to little more than semantic arguments. It is best to avoid pinpointing one single 'consumer revolution', and to accept that today's consumer society was brought about by a number of intersecting and overlapping changes.

While the historical record is extremely variable in terms of what is known about past patterns of consumption, the evidence provided by probate inventories (Moore, 1980; van der Woude and Schuurman, 1980) represents one of the prime historical sources. Probate inventories are contemporary valuations of the estate of a deceased person (or, more precisely, the movable assets of that person, excluding land and buildings). A great many such inventories survive for the period between *c.* 1540 and *c.* 1730 in England, and for far later periods in many other European countries and in North America. Simplifying to the extreme, what such sources tend to suggest is that people from many different walks of life have tended to consume an increasingly diverse array of goods the closer we get to the modern age. The early-modern period saw the previous doctrine of 'consumption by estates', whereby *sumptuary law* dictated precisely which items different social ranks were permitted to consume (**Hunt**, 1996a, 1996b; see **Part 1**), gradually give way to a far more fluid situation. While there is little evidence of 'individualistic' consumption in earlier periods (Dyer, 1989) the motives for consumption began to take on the character with which we are familiar today during the early-modern era.

The sumptuary regulations of the pre-modern period reveal a society forged in the image of the 'great chain of being' (Lovejoy, 1961), where social ranks were strictly fixed and tied together in a network of mutual obligation. The lowest social orders were involved in agricultural production, with the aristocracy commanding the right to appropriate a portion of the surplus in exchange for military protection. For a variety of reasons, this stable structure gradually began to give way, not least as trade increased and led to the development of a new bourgeois class with its own set of interests (Hill, 1955). This new class began to enjoy the fruits of its success in the sphere of consumption, promoting very different tastes to those of the established aristocracy (Schama, 1988), and generating new patterns and discourses of consumption, particularly in thriving urban areas (Glennie and Thrift, 1992; Muckerji, 1983). The fact that luxury items were no longer the preserve of an elite fuelled widespread fears over increasing levels of social disorder (Berry, 1994; Sekora, 1977). Although the bourgeoisie hardly saw the rise of mercantile capitalism and the establishment of new consumption patterns as a problem, the tide of disorder could not be entirely ignored. In the event, however, new ways of securing order were discovered, which led eventually to a new form of class society (Foucault, 1980). The pre-modern social order had seen the lower orders more or less left to their own devices, with the upper ranks being satisfied merely with extracting a proportion of the surplus product. In stark contrast, the new bourgeois order was based on the direct organization of the production process, and the imposition of *discipline* to ensure that the proletarianized workforce made maximum use of the means of production with which it was presented. The development of the factory system and the industrial revolution were the eventual upshot of this new system. However, the long-standing and important debate over the relative importance of demand- and supply-side factors in the emergence of industrial capitalism (**Fine and Leopold**, 1990; see **Part 1**; **McKendrick** *et al.*, 1982; Mokyr, 1977) should not be permitted to overshadow the importance of the changing class composition of consumption, particularly the gradual transformation of the working classes into a 'force of consumption' (Baudrillard, 1996). The expansion of capitalist production is functionally dependent on an expanding sphere of consumption. While the capitalist system might easily have sustained a pattern of bourgeois consumption and worker subsistence, especially if the Puritanical attitudes of the bourgeoisie gave way to modern hedonism (**Campbell**, 1987; see **Part 1**), circumstances conspired against this. As **Bauman** (1982, 1983; see **Part 1**) suggests, the initial reliance of industrial production on a 'labour aristocracy' of craft producers saw consumption become a means of *buying* compliance from workers. This established the basis for a fully fledged system of consumer capitalism, where the work effort no longer needed to be enforced by discipline, but was ensured by the seductive rewards on offer in the sphere of consumption.

While class conflict was exacerbated initially by the levels of consumption with which the bourgeoisie indulged themselves, the increasing embourgeoisement of a considerable portion of the working population in industrialized countries served ultimately to perpetuate consumerism as a socially sustainable form (Glickman, 1999; Tiersten, 1993). Consumerism's inherently seductive qualities developed rapidly during the nineteenth century, shaping society in its own image (Benson, 1994; Cross, 1993; Fox and Lears, 1983; Richards, 1991; Williams, 1982) and promoting itself as the solution to end all solutions. The 'annunciation of Acquisitive Man' would seemingly put paid to any more revolutionary form of social and political change (Thompson, 1963, 832). Significantly, however, consumption *itself* became the site of a potentially radical movement for a time in the nineteenth century. The consumer co-operation movement tried to forge a politics of consumption comparable to the working-class politics of labour (Furlough, 1991; Furlough and Strikwerda, 1999; Gurney, 1996; **Purvis**, 1998; see **Part 1**). Consumer co-operation was the mirror image of consumer society itself, attempting to create a socially responsible form of consumption fundamentally at odds with the unconstrained, compulsive and atomistic form of consumption unleashed by capitalism. Such 'responsible consumption' is still evident in green consumerism, ethical consumerism, and so on. Here as elsewhere, the current history of consumption contains many echoes of the past. In fact, our present-day consumer society actively recycles elements of the past: one can buy into various reinvented 'traditions' – from shamanic healing rituals to Celtic designs. It is a final irony of the consumer society that history itself has become just one more opportunity for consumption (Hewison, 1987; Lowenthal, 1996). Likewise with ethics, morality, religion, politics, and so on.

One of the significant changes brought about by the historical development of consumerism was the creation of specific places and spaces geared towards the promotion of consumption and the temptation of the consumer. While one might imagine that this is where geographers have focused their attention, it was, in fact, historians who initially devoted most attention to such spaces. The pre-eminent example is the nineteenth-century department store, wherein that archetype of the modern consumer, the *flâneuse*, was first subjected to the wonton display of commodities, the seductions of modern sales techniques, and the gaze of the store detective (**Abelson**, 1989; see **Part 4**). However, although the department store has often been characterized as vital to the development of a consumer culture, this is slightly misleading. For example, while this is true of France and North America, its role was less significant in England, where a diverse and vibrant consumer culture had developed at an earlier stage (Adburgham, 1981; Alexander, 1970; Davis, 1966; Fraser, 1981; Jefferys, 1954; Mui and Mui, 1989; Thrift and Glennie, 1993; Winstanley, 1983). Nonetheless, the department store certainly encapsulates many of the most important aspects of consumer culture, as an inordinate amount of work on the topic testifies (Benson, 1986; Bronner, 1983; Crossick and Jaumain, 1999; Laermans, 1993; Lancaster, 1995; Lawrence, 1992; Leach, 1984; Miller, 1981; Reekie, 1993).

It is only relatively recently that geographers have come to consider latter-day spaces of consumption in such terms. Geographers have certainly been involved in research into retailing over a long period of time (Berry, 1967; Berry and Parr, 1988; Dawson, 1980; Lowe and Wrigley, 2002), as well as exploring consumer purchasing behaviour (Shepherd and Thomas, 1980). As the discipline of geography has become more closely related to social theory, however, a better appreciation of the wider significance of consumption has emerged (Bell and Valentine, 1997; Clarke, 2003; Hartwick, 1998; Jackson and Thrift, 1995; Sack, 1993). Paralleling and complementing the historian's study of the department store, for instance, geographers have devoted a great deal of attention to shopping malls and large shopping centres, working alongside a host of others interested in the spaces of commerce (Crawford, 1992; Goss, 1993; Hopkins, 1991; Kowinski, 1985; Lehtonen and Mäenpää, 1997; Miller *et al.*, 1998; Morris, 1988; Shields, 1989, 1994; Williamson, 1992). As with the department store, however, there has been a tendency to overemphasize the most dramatic retail forms and to portray them as vital to consumer culture. In England, for instance, shopping malls on the American scale are largely conspicuous by their scarcity. In response to this one-sided focus, some have sought to bring to light alternative spaces of consumption, such as car-boot sales (Crewe and Gregson, 1997, 1998). One might also point out that the home is as much a space of consumption as the commercial spaces devoted to selling, yet there is relatively little attention to this in the geographical literature.

As well as creating specific commercial spaces and places, the consumer society has also had a more general influence on space as a whole. The entire urban environment has been affected by the impress of consumerism, and the city has become a prime locus of consumption-related activities (Clarke, 2003). As we have noted previously, the association between urbanism and consumerism is incredibly long-standing. One can find such an association in Ancient Greek texts. Even so, it has intensified considerably since the nineteenth century, not only in obvious ways, such as the development of the department store, but in much subtler ways as well – particularly in relation to light and vision (Charney and Schwartz, 1995; **Schivelbusch**, 1988; see **Part 2**). In more recent times, the processes of gentrification (Butler, 1997; Ley, 1996; **Redfern**, 1997; see **Part 2**; Smith, 1996) and the consumption-oriented regeneration of the city (Harvey, 1987; **Zukin**, 1988, 1990, 1995, 1998: see **Part 2**) have reinforced this association. Note, however, that the relation between consumption and the city varies over space, manifesting itself in different forms in different places (Clammer, 1997).

As we elaborate further in the Introduction to **Part 2**, it is important to emphasize the significance of *scale* in geographical analysis, in addition to stressing the importance of spatial variations. As **Taylor** (1996; see **Part 2**) demonstrates, the significance of consumption to the 'world-system' has changed over time, with a succession of countries taking control of the global pattern of consumption and the nature of the consumer culture in particular places. While this kind of influence was once strongly inflected by colonial relations (**Comaroff**, 1996; see **Part 2**), the influence of external forces on particular national cultures continues to take a variety of forms (**Wildt**, 1995; see **Part 2**). Today's post-colonial situation is especially complex in terms of the way in which processes operating at a variety of scales impact on particular places to produce an increasingly tangled (some would say 'messy' or 'hybrid') kind of consumer culture, where formerly place-related aspects of culture, such as regional and national cuisines, are likely to be adopted by other places and transformed in all manner of ways (**Cook and Crang**, 1996; see **Part 2**). Just as history has become an opportunity for consumption, thanks in no small part to the influence of the heritage industry, geography too is increasingly consumed (MacCannell, 1992; **Urry**, 1990, 1995; see **Part 2**). Geographers, therefore, have stressed the importance of spatial variations at a variety of scales in considering consumption, as well as pointing to the significance of particular commercial spaces. Most recently, geographers have been engaged in considering the effects of consumption on space as a whole, in much the same way that historians have begun to consider the long-term influences of consumption in shaping history itself. There can be little doubt, however, that the disciplinary divides we have been straining to restrict ourselves to in these brief sketches of different disciplinary traditions have been crumbling over recent years. Those with an interest in consumption are increasingly unlikely to be unaffected by the work of those operating outside of their own disciplines. Indeed, consumption lends itself to an interdisciplinary approach.

An interdisciplinary approach

Having conducted a wide-ranging survey of disciplinary concerns with consumption-related issues, let us stress that this Reader attempts to cut a swathe through the various disciplinary traditions and provide a valuable resource for the growing interdisciplinary interest in consumption. The range and variety of work uncovered in our disciplinary overviews – which do little more than scratch the surface of the range of material available – reveal that this is a formidable task. The best we can hope for is that the readings neatly packaged together here will provide a rich and varied set of insights into how a superficially simple term such as 'consumption' can be drawn upon to tell us a great deal about human societies: about how they have changed over time (**Part 1**), how they vary from place to place (**Part 2**), about what kind of people we consumers are (**Part 3**), how we relate to the material world (**Part 4**), and how we might best think about all these issues together (**Part 5**).

Yet this Reader is not concerned simply with consumption. First and foremost, it is concerned with the consumer society. Such a society has been defined in a variety of ways, and many different kinds of society have been regarded as 'consumer societies'. It is fair to note that most people reserve the term 'consumer society' for today's Western capitalist or market societies – such as the USA and many European countries. The phrase 'consumer society' typically implies a society in which people are defined as *consumers* as much as they are defined as producers (or mothers, fathers, workers, lovers, blacks, whites, citizens, subversives, aliens, lesbians, children, and so on). It implies, therefore, that 'membership' of society is defined first and foremost by the fact of being a *consumer*, and that membership of particular fractions of society (such as classes, lifestyles, subcultures, ethnicities, sexualities) is defined by the fact of being a particular *kind* of consumer. This is a distinctly odd way of defining social membership. It is not a legal or political definition, unlike such definitions as 'subjects' (of the monarch) or 'citizens' (of the state). It is an 'economic' definition in the sense that it relates to the market. But unlike earlier 'economic' definitions – such as 'worker' or 'capitalist' – it does not relate to one's *productive* role. It quite evidently relates to one's *consumptive* role – one's ability to buy, use, and above all *enjoy* the fruits of production. According to Bauman (1987, 1998), this is the crucial feature of postmodern society. In a *modern* society, people were primarily producers. In a *postmodern* society, they are first and foremost consumers (although in order to consume, most consumers still have to be producers in order to secure their purchasing power – often in the classic Marxist form of 'wage-slaves'). In addition, the way in which they consume (their level of consumption, the tastes they express and the lifestyles to which they aspire) seems to be increasingly significant to how society as a whole regards them. Meanwhile, those who cannot or will not consume – the poor, in other words – are more likely than ever to be thought of as 'second-class citizens' (Bauman, 1998). Those who are able to consume profligately and confidently seem far more like fully paid-up members of our present consumer society (however they earn their money or even whether they 'earned' it at all). In a consumer society, therefore, people are consumers more than anything else. It is their *duty* to consume; to want; to desire; to demand – all of which carries the most far-reaching consequences imaginable. Fortunately, we don't have to try to imagine these all by ourselves. The various extracts comprising this Reader bring together a lot of hard work in terms of trying to think them through.

MAPPING *THE CONSUMPTION READER*

We have divided *The Consumption Reader* into five broad areas which consider some of the most interesting consequences, implications and issues relating to consumption: **Part 1**, History; **Part 2**, Geography; **Part 3**, Subjects and Identity; **Part 4**, Objects and Technology; and **Part 5**, Theory. This structure is obviously not the only way of arranging the individual readings, and we make no apologies for all the things that are missing. This is necessarily a selective Reader, not an exhaustive encyclopedia. Our aim throughout is to give you an essential selection – or at least an enhanced flavour – of life in the consumer society.

Since the individual introduction to each particular part of the Reader provides a discussion of the key principles involved and the specific extracts themselves, we will explain here the rationale for our structure as briefly as possible. **Part 1** offers an introduction to the **History** of consumption. Not all the readings are by historians, but all of them seek to address the emergence of the consumer society, to discover what people consumed in the past, how they consumed, and what influence consumption had on past societies. Most of these readings provide for a variety of important reflections on today's consumer societies. Do today's consumer societies originate from a basic change in the *quantity* of goods people consumed or in the *type* of goods they consumed or from the *manner* in which they consumed them? Most of the readings suggest the origins of the consumer society lie in a fundamental *qualitative* change rather than in a simple *quantitative* one. But there are all kinds of debates about when, why, where and how this happened, what it meant at the time, what it means in retrospect, and to what extent it may have been *otherwise*. Let us say no more about the detail of such debates at this point; beyond pointing out that the history of consumption offers a perfect starting point because it serves to *defamiliarize* what we latter-day consumers – we denizens of the consumer society – usually take for granted.

If our initial foray into history is useful for stripping away our tendency to normalize the present state of affairs, our exploration of **Geography** in **Part 2** serves a similar purpose. Just as the consumer society is historically specific, it is also geographically specific. Things have not always been the way they are now – and 'the way they are now' depends on *where* one is talking about. Geography complicates almost everything because it shows that things are different in different places. Even so, many people have suggested that consumption generally acts to remake the world in its own image: McDonaldization, Disneyfication, Coca-colonization. Yet while there is an element of truth in talk of the geographically expansive nature of the consumer society, most of the readings included in this part of the Reader suggest that this is overly simplistic and caricatured. Again, let us leave the detail of the consequences of this for the Introduction to Part 2, and simply add that the geography of consumption demonstrates how many generalizations about consumption need to be pinned down and considered in detail within particular contexts. Geography teaches us that things happen differently in different places, and that this has knock-on effects in terms of other places. One final thing: not only does consumption take place over space; often, space and place are themselves consumed.

History and geography – time and space – together form a framework within which patterns, processes and practices of consumption unfurl. This is not a static or simple framework: the patterns, processes and practices of consumption reflexively influence both time (history) and space (geography). Nonetheless, the next two sections of the Reader focus in on these patterns, processes and practices. The readings in **Part 3 – Subjects and Identity** – focus on the nature of the 'consumer' and the identity of the 'consuming subject'. Those in **Part 4 – Objects and Technology** – concentrate on 'objects of consumption' and the various technologies associated with the way in which the consumer relates to the material world. As we have already noted, it is rather artificial to separate out these two dimensions of consumption, since consumption usually implies a relationship between 'subjects' and 'objects' – even if this relationship has more often than not been thought of in rather simplistic ways (since subjects have needs and objects have uses these objective uses may be employed to satisfy those subjective needs). Nonetheless, the readings in Part 3 maintain a distinct emphasis on the nature of the consumer, on how consumers act, and on how they use objects to say something about 'who they are' and construct their social identities. One of the most important aspects of this emphasis is the relationship between consumption and gender, and a number of readings in this section provide significant insights into this relationship. Again, though, this can never be entirely detached from the 'object-side' of the consumption process, and gender issues are hardly confined to this section of the Reader.

The emphasis on 'objects' in **Part 4** both complements and completes the emphasis on 'subjects' in Part 3. The readings contained in this section also point out the artificiality of the distinction. For example, alongside a number of explicit considerations of specific objects of consumption and technologies associated with particular consumption practices, a general recognition of the need to overcome the 'ontological apartheid' between 'subjects' and 'objects' is explicitly brought to the fore in many of the readings. Other readings point to the way in which the consumer society often tends to treat people as *things* while at the same time operating in ways that attribute objects a (social) life of their own. Let us not see the inevitable overlaps between Part 3 and Part 4 as problematic, therefore. They are better thought of as a set of resonances that reveal how consumption choreographs subjects and objects, people and things, into a sometimes harmonious – but often discordant – whole.

We have elected to devote **Part 5** to **Theory**. Given the theoretical sophistication of each of the separate parts of this volume, it is surely necessary to explain why we have chosen to do so, and why we have opted to put it at the end of the book. The first thing to say is that we do not wish to *isolate* theory – as if it were something to be handled with special care under conditions of restricted access, like a prized treasure or a dangerous substance. Theory should be neither revered nor reviled. It has as much or as little interest as anything else. Consequently, we have deliberately declined the temptation to put theory first. Rather than follow the customary practice and offer a speculative template through which the other parts could have been read off, we prefer to let all the extracts think for themselves! However, by positioning a collection of extracts on 'theory' at the end of the Reader, there is the equivalent danger of assuming that it stands as the *culmination* of the volume – as if the final truth of the world of consumption rests there. Theory is neither the origin nor the end of consumption studies. Instead, you should read this part as an exercise in *suspense*. To put it bluntly, the world of consumption holds our theoretical interest insofar as it forces us to rethink (and indeed to 'unthink') so many of the things that we continue to take for granted. The readings in this section are distinguished only to the extent that they are exemplary instances of attempts to engage in this kind of conceptual task.

Throughout the Reader we have selected extracts not only for their breadth of empirical interest, but also for the richness and profundity of their ideas. Whether our attention is turned towards history or geography, subjects or objects, or theory itself (as if this were ever a pure and simple matter), attempts to take seriously the myriad world of consumption have given rise to a massive amount of theoretical innovation, conceptual sophistication, and careful and detailed empirical investigation. Much of this work has become pivotal to cutting-edge research across huge swathes of the arts, humanities and social sciences. Consumption studies have provided a variety of new perspectives on the nature of society, not only by bringing to light facts, circumstances and evidence that no one really knew quite what to make of before, but also by prompting a re-evaluation of the ideas of writers whose names are probably less familiar than they deserve to be, such as Bataille and Mauss. The widespread interest in consumption today – conspicuous only by its absence a mere decade or two ago – has led to a vast amount of new research across a wide variety of disciplines, prompting a major re-evaluation of the contributions of earlier social theorists, such as Marx, Simmel and Veblen, as well as highlighting the work of recent thinkers such as Baudrillard, Bourdieu and de Certeau. To your good fortune, we have also made space for the brilliant insights of many other writers, whose inclusion in this Reader may well be unexpected, even for those who would consider themselves to be familiar with consumption studies. Indeed, we have aimed to provide a tantalizing *variety* of readings: readings that provide a fuller flavour of what consumption is all about; more abstract and demanding readings;

iconoclastic attempts to throw a spanner in the works and to undo conventional wisdom; and to mix more accessible and lucid pieces with much more challenging ones.

Needless to say, then, this is not a *closed* Reader. It does not contain everything you will ever need to know about consumption. Nor does it constitute an essential selection. Yet it is not simply an arbitrary collection of useful and insightful fragments, with a more or less convincing rationale for their selection and arrangement. We expect that this Reader will have opened your eyes to the value of rethinking the conditions and possibilities of life in a consumer society, and that you will have a renewed sense of the significance of innumerable everyday practices that have all too frequently failed to receive the attention that they deserve. We hope you enjoy consuming it – avidly, critically, reflectively, and in a way that prompts you, in turn, to write convincingly on the topic of consumption and to engage in consumption with a fuller appreciation of its significance and consequences for the kind of society in which we live.

PART ONE

History

INTRODUCTION TO PART ONE

By all accounts, the Middle Ages had no single word for what we now matter-of-factly refer to as 'consumption'. Only after being recognized as an opportunity for taxation did such formerly distinct activities as wearing clothes and shoes, eating and drinking, and dwelling in a house come to be thought of as having something fundamentally in common. It is difficult for those of us who live in a world where virtually anything and everything can be consumed to 'unthink' what we now take for granted. For us, it seems entirely obvious to say that all people, at all times, and in all places have consumed – they must have done so simply in order to survive. Yet while this is undoubtedly true, it hides the important fact that 'consumption' subsumes all kinds of disparate activities under one single heading, and presents this jumble as if it were self-evident, natural and unproblematic. It may be regarded as offensive to some people, for example, to suggest that religion, politics and morality are consumed. The term 'consumption' seems to lend itself more readily to chocolate bars, bottles of perfume, fast cars and toilet tissue. Yet while 'consumption' may be a more apt description of some activities than others, one way of defining a consumer society might be to say that it is a society where it increasingly *makes sense* to think of all kinds of incongruous activities as instances of 'consumption'. The history of consumption usefully 'defamiliarizes' our taken-for-granted assumptions. It reveals that, at one time, people did not so much 'consume' as eat, drink, dress, live, worship, listen, dance, and so on. No one yet thought of these diverse activities as mere variations on a single theme.

It is probably true to say that those of us living in the Western world are increasingly consumers *first and foremost*. We are citizens of consumer societies, and our principal duty is to consume: to take up everything that is endlessly produced; to exercise our freedom to choose; to act out our rational – and irrational – calculations; and to make demands on our economy, culture, society and polity. For a considerable part of human history, however, only a minority of people 'consumed' in anything like the sense we do today. While an elite once 'consumed', much of the populace merely 'subsisted'. Well before the dawn of modernity, for example, and stretching into the modern period, European consumption patterns were regulated by *sumptuary laws*, which governed who could and could not consume certain *luxury* items – materials such as silk or velvet, spices, and so on – in accordance with the doctrine of 'consumption by estates' (see **Hunt**). One of the most significant events in the history of consumption was the transformation of this situation into a state of affairs where an increasing proportion of the population was free to consume (though, clearly, those with a greater claim on resources had far greater freedom in this regard). Most of the extracts in this part of the Reader focus on the evolution of a *consumer society* in this particular sense: an evolution that was as much *qualitative* as *quantitative* – as much about *what* was consumed, *how* and *by whom*, as about simply *how much* was consumed. These extracts offer a range of perspectives on the way in which consumption gradually freed itself from being functionally tied to 'needs' and 'necessities' to assume its distinctive character and purpose – propelled, above all else, by its own intrinsic *pleasurability*. Accounting for the history of this fundamental change in consumption patterns, not to mention the very nature of consumption itself, is not surprisingly fraught with difficulties and invariably controversial. Nonetheless, the extracts we have selected at least have the following characteristics in common. First, they demonstrate that, with respect to consumption, things were once *otherwise* than they now are. They show that what we tend to regard unflinchingly as 'normal' is very much *historically* – and indeed geographically – *specific*. Second, they accept that certain fundamental changes in social conditions allowed the take-off of consumption as a significant force, which initiated not only changes in the nature of consumption but, ultimately, in the nature of society and social life itself. If the extracts share this much in common, however, there is much upon which they disagree. At the risk of caricaturing their carefully constructed arguments, the extracts contained in this part of the Reader are summarized below.

Joyce Appleby considers not only the important *material* changes associated with the growth in consumption originating in the seventeenth century, but also critical changes in the *discursive* construction of consumption. She focuses on the twists and turns of a series of broadly chronological representations of consumption, taking in the Restoration pamphleteers' observations on changing spending patterns and the expansion of trade; the Augustans' revival of classical wisdom on the dangers of luxury in the face of these changes; the Scottish Enlightenment's reformulation, as a potentially progressive force, of what classical republican thought regarded only as a harbinger of disorder; and Malthus' conservative strictures on the scarcity of resources in the face of burgeoning population growth – designed to curtail certain of the dangerous currents generated by the prevailing optimism about the new commercial basis of society. These discourses saw consumption as, respectively, heralding a new situation of abundance that would finally lift the mass of human society above the level of bare necessity; a dangerous relaxation of earlier sumptuary laws that would inexorably lead to corruption, licentiousness and the breakdown of the existing social order; an entirely new and positive force to reshape society according to the modern economic rationality of the market rather than the classical political principles of the republic; and an intractable social problem that needed to be confronted head-on, not least to quell the kind of dangerous passions recently unleashed by the French Revolution. Appleby ends by noting the irony of Malthus' concerns at a point in history where scarcity and famine finally seemed to have become yesterday's problems. In his wake, the 'dismal science' of economics was to become ever more narrow and abstract in its focus. What the history of consumption ultimately reveals, Appleby suggests, is that human society is the product of human efforts. The secularization of modern society, and Marx's revolutionary concerns with social justice in the face of an inequitable distribution of abundance, can both be seen more clearly in the light of the history of consumption.

In contrast with Appleby's wide-ranging account, the extract from **Neil McKendrick** locates a consumer revolution firmly in eighteenth-century England – on the face of it implying a far less drawn-out process than Appleby suggests. It should be noted, however, that McKendrick's work is not wholly contradictory with Appleby's version of events. McKendrick's piece initially gained fame as part of a volume that seemed to rearrange the conventional placement of the historical cart and horse – by suggesting that a consumer revolution *preceded* the Industrial Revolution in England. In some senses, the work of McKendrick and his colleagues may be regarded as having made more modest claims: that the consumer revolution set the *preconditions* for a subsequent industrial revolution, a revolution in production first heralded by changes in consumption mores, tastes and norms. Nonetheless, the specific mechanism that McKendrick held to account for the consumer revolution of eighteenth-century England has continued to attract considerable interest and criticism. That mechanism can be specified as *social emulation* (see **Veblen**). According to McKendrick, the key agents of this process were the new middle classes and their domestic servants, and the striving for what we would today think of in terms of 'upward social mobility' by those lower down the social ladder.

Arguably the most stringent critique of McKendrick's work is that by **Ben Fine and Ellen Leopold**, who seriously doubt that social emulation could ever serve as quite the kind of spur towards increased consumption implied by the notion of a consumer revolution, especially among domestic servants. Fine and Leopold challenge McKendrick on logical, empirical and political grounds – finding his account wanting on all counts. They argue effectively that emulation is assumed, not proven, by McKendrick. They note, for example, that domestic servants lacked the money and autonomy to emulate the tastes and consumption habits of their bourgeois employers. They also point out that emulation only makes sense for a limited range of goods. Most importantly, they argue against a single social pecking order – driven by an elite at the top, with everyone else attempting to climb the social ladder by emulating their social superiors. Instead, they stress the *class* dimension of society, suggesting that different classes not only experienced very different material circumstances, but also very different norms, discourses and practices of consumption. This tension between the singular and the multiple has arguably become one of the most crucial dimensions of consumption studies (see **Veblen, Bourdieu** and **Douglas**).

Colin Campbell's contribution is less damning of McKendrick than Fine and Leopold's, but Campbell evidently finds McKendrick's account irredeemably partial: social emulation, in Campbell's eyes, can hardly account for the dramatic changes in consumption that accompanied modernity. The title of Campbell's original book – *The Romantic Ethic and the Spirit of Modern Consumerism* – echoes Max Weber's famous work on the origins of capitalism: following in Weber's footsteps, Campbell explains the emergence of a specifically modern form of consumption in terms of a revolutionary shift in *attitudes*. Campbell is seeking a plausible answer to a long-standing enigma: how did the self-sacrifice and asceticism of the Protestants, whose work ethic defined the very spirit of capitalism and made the Industrial Revolution possible, unintentionally mutate into the kind of self-indulgent quest for gratification that underpins modern consumerism? For Campbell, the answer is that consumption is a quest for personal *pleasure*, not social status; desire, not recognition. Modernity, Campbell proposes, witnessed the emergence of a new kind of *hedonism*, which explains how consumers can have an endless desire for novel goods. Consumption involves not the desire for specific goods (which could be satisfied), but for pleasure itself (which is insatiable). Unlike traditional hedonism, which fixated on the satiable pleasures of the flesh, modern hedonism promotes the kind of 'non-specific pleasure-seeking' that can take literally *anything* as its object of desire. How does this square with Protestant asceticism? Protestantism demanded self-control, especially of the emotions: giving in to the vanities of the self could only lead one away from the perfection of God. Eighteenth-century Romanticism, however, served as a key agent of change: faith in perfection led inevitably to dissatisfaction with the world as it actually was. Displaying one's sensibilities by longing for an experience of perfection – for example, by daydreaming – suddenly came into vogue. People could increasingly hold out the hope that novelty would deliver on its promise to provide just such an experience and they could turn their ability to control their emotions towards this quest. Disappointment remains inevitable, however, since the real can never live up to the ideal: *insatiable* desire is the result. Campbell is keen to point out that the notion of a consumer ethic fuelled by vain hopes of self-fulfilment through consumption is not a universal theory. It is specifically meant to account for the 'spirit of modern (bourgeois) consumerism'.

Like Campbell, **Zygmunt Bauman** also focuses on the fundamental structural changes in society associated with modernity. Unlike Campbell, however, he does not attempt to account for these changes in terms of shifts in attitudes. Instead, his argument rests on the changing nature of social *practices* and especially their *unintended consequences*. Bauman is concerned specifically with the fundamental shift from a repressive disciplinary society to a seductive consumer society, a shift that takes place gradually, over the course of several centuries. According to Bauman, the trigger for this shift was a fundamental change in the organization of society in seventeenth-century Europe: from a patchwork of local communities that effectively governed themselves to a hierarchical society where those at the top instructed those at the bottom on how to live their lives. Unruly elements were brought under control by enclosure in places such as workhouses, prisons and asylums, where they were set to work: not principally for a productive contribution to the economy, but more or less as a form of punishment – as a way to instil *discipline*. Herein lie the origins of the factory system. The repetitive tasks of proto-industrialization fitted neatly into the repetitive routines of governing and disciplining institutionalized people. Needless to say, craft labour resisted these kinds of controls and the forms of discipline and subordination that they implied. Since they could not be forced to work under factory conditions, their compliance was effectively *bought*. The prospect of the freedom to consume compensated for the loss of control over their work practices – such was the historic exchange of *power* (autonomy in the sphere of production) for *purchasing power* (enhanced opportunities in the sphere of consumption). Interestingly, something akin to a producer society appears in Bauman's account as a 'vanishing mediator' between a world structured by disciplinary power to one structured by the seductive power of consumption. For Bauman, the purchasing power of consumerism – the so-called 'freedom to choose' – is basically a 'false power'. Real power would lie in the freedom of individuals to command *all* aspects of their lives. The 'freedom of choice' that underpins the consumer society detracts attention from other freedoms and ties people into a system that purports to be only in their interests. It manages to do so by providing no attractive alternative: those who are not admitted into the consumer society are increasingly excluded from feeling that they are members of society at all.

The last two pieces in this part of the Reader depart slightly from the concerns of the first five extracts (explaining the origins and emergence of a consumer society). Although they cover different ground, however, each offers insights into the complexities of the debates rehearsed in the previous extracts. **Alan Hunt**, for example, provides an account of how consumption has been regulated in different historical settings. He shows that while the sumptuary laws governing pre-modern consumption patterns sought initially to ensure that people knew their station in life and that they did not try to challenge it, the onset of modernity saw a series of transformations in the nature of sumptuary law. Yet the pre-existing ways of regulating consumption did not so much disappear as alter their form to fit the changing circumstances. Focusing on gender as well as social stratification, Hunt reveals how consumption has always been a site of struggle. Power and prestige have always been associated with particular kinds of consumption, not only in terms of specific objects but also in terms of *how* one consumes: gestures, mannerisms, etiquette, and so on. Yet again, this ties in with another important strand of consumption studies: consumption is not just a material process; it is also expressive, meaningful and communicative (see **Douglas**). Consumption is a language of identity and power.

When all is said and done, it is easy to imagine that consumerism is essentially a false freedom, a bogus power and a source of sham identity. For many, the notion of a consumer revolution is little more than a betrayal of what the world could have been like had the true interests of the working classes prevailed. It is also easy to imagine that the consumer society is the inevitable outcome of the movement of history. There seem to be fewer and fewer credible alternatives to the view that we are heading relentlessly towards a world of happy shopping. For many, the only glitch on the horizon is the likelihood that the natural environment will not be able to sustain this vision for very long – especially if it is to be exported to all areas of the world (see **Taylor**). **Martin Purvis'** study of the Consumer Co-operative Movement is instructive in this regard. Not only does it demonstrate the active struggle to effect positive social change from somewhere other than the workplace, it also shows the power of consumers when they act together as a self-conscious force for change: for what has been made by men and women can be made *otherwise*. Recognition of the fact that everyone was a consumer underpinned the nineteenth-century Co-operative Movement. It was potentially a much more *inclusive* social movement than the partiality and sectarianism of class politics rooted in the workplace. However, the sense in which consumption had already become detached from 'need' and put in the service of 'desire' – while the former situation remained the underlying legitimization of the Co-operative Movement – almost inevitably undermined this putative alternative to capitalism. The story of consumer co-operation, then, offers a powerful example of the politics of consumption.

Though we now know far more about the history of consumption than we once did, the study of the origins of the consumer society, and the nature and role of past consumption practices, remains essentially contested. Yet this is hardly cause for despair. It allows for the possibility of new insights as fresh perspectives on the past are generated and as new discoveries, from sources such as probate inventories, come to light. One final word about the limits of our selection: the extracts included here focus primarily on European and North American history – for the simple reason that the emergence of consumerism in these places is generally regarded as establishing the template for today's consumer society. This is undeniably a partial picture: it neglects the history of consumption in other geographical contexts, even if it does so for good reason. This deficiency is partially remedied in Part 2.

1

Consumption in Early Modern Social Thought

Joyce Appleby

My subject is consumption – the desiring, acquiring and enjoying of goods and services which one has purchased – and I will concentrate upon my historical predecessors' investigation of consumption, examining what they have said about these activities, but especially what they failed to say, for my search of relevant texts took me to a void, an emptiness – at best, a hiatus or lacuna. I can pose the puzzle of this silence in several ways: why is it that consumption has rarely been examined thoroughly or dispassionately despite its centrality to economic life? Why is consumption uniformly construed negatively even though there is abounding evidence that consuming is pleasurable and popular and brings rare moments of satisfaction? Why, in the floodtide of Enlightenment enthusiasms for freedom – free speech, free inquiry, free labour, free trade, free contract – was free consumption never articulated as a social goal? Or put another way, why has the opportunity to consume been made dependent morally upon the opportunity to produce, but functionally upon the opportunity to purchase? I can think of no other human predisposition so essential to economic growth which has been so perversely treated. Why is it, to put the question in more total terms, that consumption, which is the linchpin of our modern social system, has never been the linchpin of our theories explaining modernity?

These questions take us back to the initial efforts to understand the emerging commercial economy. English men and women in the middle of the seventeenth century did not know that they had crossed a barrier which divided them from their own past and from every other contemporary society. Yet they had. Somewhere around 1650 the English moved beyond the threat of famine. It is true that chronic malnutrition lingered on for the bottom 20 per cent of the population, not completely disappearing for another century, but famine was gone. In the future there would be food shortages, skyrocketing grain prices, distress and dearth, but never again would elevated grain prices go hand in hand with rising mortality rates. Agricultural productivity combined with the purchasing power to bring food from other places in times of shortage had eliminated one of the four horses of the apocalypse from England's shores. A powerful reason for maintaining strict social order had unobtrusively disappeared, leaving behind a set of social prescriptions whose obsolescence had to be discovered one by one in the course of the next two centuries. It would be hard to exaggerate the importance of freedom from famine just as it is exceedingly difficult to follow all of the ways this material circumstance influenced behaviour and belief.

A second feature I would draw your attention to was the population growth which started again in Europe in the middle decades of the eighteenth century. The world's population had expanded and contracted over three millennia, but with eighteenth-century population growth a vital revolution was in the making. Unlike the old accordian-like pattern that had characterized previous European population fluctuations, the increase in people this time laid a new basis for future growth with each augmented cohort forming a kind of springboard from which world population still continues to soar. Food supplies were to be severely strained but instead of shrinking they expanded to sustain new levels of population. The twenty million Frenchmen Louis XIV ruled in 1700 became the forty million Frenchmen who couldn't be wrong in 1914. English population grew at an even faster clip. And in

England's North American colonies – that catch basin of surplus people from northwestern Europe – the number of people doubled every twenty-five to twenty-six years.

Even more remarkable, the goods that people wanted grew apace – grew even faster than the number of people. A peculiar dynamic of the emerging world commerce had revealed itself most strikingly in England's first colony, that fragile outpost of European life established by the Virginia Company on the far side of the Atlantic. This settlement was explicitly tied to plans for extracting and producing vendible commodities. In 1617 John Rolfe successfully hybridized a tobacco strain which could compete with the much-esteemed Spanish orinoco. His leaf triggered a boom. Throughout the 1620s tobacco fetched between 1½ and 3 shillings a pound, a price high enough to encourage Virginia Company shareholders to pour money and men (and a few women) into their plantations. Cultivation spread along the tidal rivers emptying into the Chesapeake Bay. The volume of exports surged. When the inevitable bust of oversupply followed this boom of demand-driven expansion, prices dropped to as low as a penny a pound – a twenty-fourth of the price of good Virginia tobacco in the 1620s. However, at this cheap price a whole new crowd of consumers could and did begin to buy tobacco, or as we would say metaphorically, entered the market. Their demand in turn created an incentive to cut production costs in order to supply this larger body of consumers with cheap tobacco at a profit. Success at this endeavour sustained a slower expansion of tobacco cultivation for two centuries. A similar thing happened in 1634 with Dutch bulbs, only to be repeated over and over again with cutlery, calicos, printed pictures, blankets, pottery, pewter and pepper.[1]

When ordinary people joined their social superiors in the pursuit of the pleasures of consumption, their numbers changed the character of the enterprise. Retrospectively we can see that this boom and bust cycle unintentionally widened the market for new goods. Investors responded to the profits of the boom; ordinary people to the opportunity of the bust. This dynamic enabled commerce – a feature of human society as old as the Bible – to move out of the interstices of a traditional social order and impose its imperatives upon the culture as a whole. The enormous augmentation in the volume of goods when ordinary people became consumers meant enormous augmentations in the wealth and power of those nations and persons who participated successfully in supplying the new tastes.

We are of course used to hearing the litany of new products entering European markets from the sixteenth century onwards – first from the fabled East India trade, then from the homely shops of ingenious artisans, finally to be overtaken by the prodigious outputs of the marvellous machines of the factory age. Rattling off the names of new condiments, textiles and inventions has served as the incantation for summoning the spirits that presided over the rise of the west. These details of early modern enterprise have supplied the factual grist for the mill of material progress.

Told within the familiar narrative of the liberation of *homo faber*, man the maker, modern history presents no problems. There are no ruptures in the telling, if not the living, of this age so long as the stunning and devastating transformation of the world wrought by the cumulative revolutions of technology and human adaptations appear as the end-point of a plan which has design and meaning. But we, alas, live in a post-industrial era. We can't conceive of our own time as a mere coda. We've known civilization and its discontents too long to subscribe to the notion that the discontents are epiphenomenal. We have even begun to entertain doubts about the inevitability of the events in our past. We've lost faith that these transformations were either natural or evolutionary. Significantly, these doubts have enabled us to hear other voices from the past – the crazies who preferred occult mysteries to the plain and simple truths of nature; the atavists who harkened back to ancient prudence. Tuning into these alternative voices has unchained our imagination. We can begin to see that our history told as the history of progress might have served as the intellectual equivalent of whistling through the graveyard.

What was profoundly unsettling, even shattering about the cumulative gains in material culture which became manifest by the eighteenth century was that they made it evident that human beings were the makers of their world. There is no way to underestimate the reverberations of such a discovery; they resonate through every modern discourse. And if we are postmodern it is because we can now reflect upon these discourses in science, politics and literature from a perspective standing outside the engagement itself. We see how Hobbes's irreverences become the ingenious truths of Scottish moral philosophers, to be transmogrified once more into the social science disciplines of the first half of the twentieth century.

Here I am reminded of a passage in Louis Dumont's *From Mandeville to Marx*. After detailing the western conception of society as the interactions of rational, utility-maximizing, self-improving, materialistic individuals, Dumont commented that this was a radically aberrant world view shared by no other culture. Rather than ask why other people were taking so long to become like us, he suggested, we should turn our curiosity around and ask how 'this unique development that we call modern occurred at all'.[2] There has been a punitive arrogance in the West's refusal to see its cultural differences as differences and to characterize them instead as the end-point in a universal process. This grand explanation robbed the events of the indeterminacy essential to historical narrative and hence obscured the dynamics of change at work.

A peculiarly intense form of curiosity in Western culture drew the countries of Western Europe along the path of innovation which grew ever wider as the pathbreakers pushed against a comparatively weak attachment to customary practices. On this broad avenue of human inventiveness Europeans encountered themselves as the creators of their own social universe. But this discovery took place while the actual social arrangements of their world reflected traditional assumptions about divine punishments, fallen human nature and the inherent frailty of civil society. How was social order to be maintained when collective understandings were being undermined by the new Promethean powers at large in the world?

Consumption – the active seeking of personal gratification through material goods – was the force that had to be reckoned with. Like other social activities, consumption had first to be named before it could be discussed. I want to look at four responses to this new phenomenon, four sequential engagements with the idea of abundance and its social consequences: the Restoration pamphleteers on trade who first took note of new patterns of spending; the Augustans' revival of classical wisdom about luxury; the Scottish intellectuals' reaction to the classically inspired laments about corruption and finally Malthus's mordant rebuff to the enthusiasts of the French Revolution.

The first observers of England's material abundance had no trouble discerning the human impulse animating the lively round of goods that encompassed Europe and its colonies in a new trade system. I'll quote from a few:

> The Wants of the Mind are infinite, Man naturally Aspires, and as his Mind is elevated, his Senses grow more refined, and more capable of Delight; his Desires are inlarged, and his Wants increase with his Wishes, which is for everything that is rare, can gratifie his Senses, adorn his Body and promote the Ease, Pleasure and Pomp of Life.

From another:

> the main spur to Trade, or rather to Industry and Ingenuity, is the exorbitant Appetites of Men which they will take pains to gratifie, and so be disposed to work, when nothing else will incline them to it; for did Men content themselves with bare Necessaries, we should have a poor World.[3]

Research done within the last two decades has confirmed the assertions of contemporaries that it was domestic consumption, not foreign trade, that sustained England's manufacturing expansion in the eighteenth century.[4] Simon Schama has made a similar case for Dutch economic development in his *The Embarrassment of Riches*.[5] However, these early investigators of consumption, writing in the 1680s and 1690s, did not lay the foundation for a theory of commercial sociability. Rather, it was the critics of material abundance who seized the discursive high ground in England, appealing to classical republican texts to stigmatize novelty as the harbinger of social unrest. Using the essay form to inveigh against the new consuming tastes, these Augustan moralists read the goods they saw in haberdashery shops and food stalls as dangerous signs of corruption and degeneration. Against the delights of consumption, they pitted predictions of social disintegration. The only antidote: frugality and simple living for the people, austere civic virtue in their leaders. These alone could provide the social underpinnings for the Constitution, itself England's sole preserver from the terrors of history, that zone of irrational behaviour which made up the realm of *fortuna*.[6]

Consumption, as I have described it, figured in the political discourse of eighteenth-century England under the rubric of luxury. Luxury was not a thing, but a concept. As John Sekora has pointed out, the Greek view of luxury was a secular and rational complement to the Hebrew view. Luxury for the Hebrews represented a complex of evils moving from the personal and inveterate propensities of man to the ethical tendencies of the nation which collectively succumbs to temptation. The gravest feature of the repeated lapses recorded in the Old Testament was the evidence of disobedience. When a people ignore the law of necessity they undermine the established hierarchy between law-giver and subject. Necessity sets limits and happiness consists in having the rational capacity to abide by those restraints. Luxury brings disorder because it destroys harmony and prevents the human being from fulfilling his or her nature.

In both Christian and classical thought the central unworthiness of human beings stemmed from their desiring things that were unnecessary, that is from their desire to consume. The control of this endemic envy, vanity, gluttony and lust required draconian laws and God's redeeming grace. Essayists, political figures, novelists, journalists – all contributed to an unrelenting, unrelieved depiction of the horrors awaiting England if the nation did not mend its luxurious ways. Luxury was not a personal indulgence; it was a national calamity, as the account of the ravages of luxury offered in the books of Samuel and Kings so powerfully demonstrated.

Hebraic tradition, which gave English Puritans so rich a rhetorical resource for vivifying sin, identified luxury with desire and desire with disobedience. Eve indulged in luxury when she unnecessarily ate the fruit of the tree of knowledge. The Israelites persisted in the most serious of human errors in their yearnings for things that they did not need nor had the right to claim. If represented graphically luxury, of course, is a woman – sometimes a powerful evoker of desire carrying the comb and mirror of cupidity and self-love; at other times an abject naked woman under attack from toads and snakes.

Depicted as a constant psychological drive, the attraction to luxury can never be more than suppressed, and the act of suppressing it constitutes the reason and justification for the minute control of the status, duties and privileges of all members of society. When in the *Republic*, Glaucon asked why the state should not provide for the citizens' wants as well as their needs, Socrates describes the inevitable engorgement of people that would follow this abandonment of the limits of necessity:

> Now will the city have to fill and swell with a multitude of callings which are not required by any natural want; such as the whole tribe of actors, of whom one large class have to do with forms and colours; another will be votaries of music – poets and their attendant trains of rhapsodists, players, dancers, contactors; also makers of diverse kinds of articles, including women's dresses. And we shall want more servants. Will not tutors be also in request, and nurses wet and dry, tirewomen and barbers, as well as confectioners and cooks; and swineherds too?[7]

In other words Athens will be visited by economic development.

It fell to Aristotle to explain how authority and necessity were linked. I must say it's ingenious. As nature shows that the household is subject to the father so most persons must be subject to the dominion of the legislator. The rulers embody reason which teaches restraint and it is a sign of luxury for slaves, women, servants, tradesmen, artisans, mechanics, the immature, the illiterate and the weak to want what they do not need. In restraining them, the male leaders are demonstrating reason for the whole. From such an Aristotelian conception of order came the sumptuary laws common in Europe which elaborated specific standards of decorum and decoration under the doctrine of 'consumption by estates'. It was their obligation to maintain order among the predictably disorderly that saved the landowning elite of England from the sting of the criticism about its luxurious consumption. While technically as prone to sin as others, the elite supplied security to the whole society through its vigilance in controlling servants, young people and women – that trilogy of categorical unfitness. To incriminate the guardians was to weaken the only dyke against the floodtide of riotous consumption.

Both the sentiment and the metaphor are reflected in Henry Fielding's reference to a 'vast torrent of Luxury which of late Years hath poured itself into the nation . . . almost totally changed the Manners, Customs, and Habits of the People, more especially of the lower Sort'. A political evil, luxury has inspired in the poor, he went on to explain, a desire for things they may not and cannot have, hence their wickedness, profligacy, idleness and dishonesty. Daniel Defoe less dramatically spoke of the decline of the Great Law of Subordination.[8] Shops bulging with cheeses, sweetmeats, coffee, tea, table linens, dry goods, gadgets, pictures and prints gave the lie to Fielding's assertion that the lower sort desired things they could not have. It was exactly their increasing ability to buy what was being made available in ever cheaper forms that created the crisis of social leadership.

What Sekora so nicely captures is the way that the human desire for the sensual pleasures of eating, entertainment, adornment and comfort, made manifest in actual consumption, became evidence for the need for strictly enforced hierarchies of authority in the home, the shop, the street, the town hall and the church. However, the disjuncture between the jeremiads on luxury and the actually visible, even conspicuous, behaviour of ordinary people cried out for clarification. As Bernard de Mandeville had earlier pointed out, English moralists were not confronting the fact that they were preaching truths which, if followed, would bankrupt the nation and undermine its greatness. The private vices of personal indulgence, Mandeville warned, amounted to the public benefit of national prosperity. Vice, not virtue, stoked the engine of commerce. Mandeville's goal, however, was to point up the hypocrisy in the outcry against luxury, not to endorse the abandonment of society to the consuming impulses of the least discerning members of society.

Roy Porter, writing on the English Enlightenment, has pointed to the strain of eudemonism running through the century's public commentary. Indeed he has characterized the English Enlightenment by its mildness. Not forced to overthrow an oppressive old regime like their neighbours across the Channel, prosperous Englishmen settled down to enjoying the affability afforded them by urban life. Sipping coffee, displaying new forms of politeness, relishing the wit of Addison and Steele, Porter's 'affluent, articulate and ambitious' Londoners, along with their provincial imitators, bent their minds to considering ways to make the world safe for egoism. Because the English had dealt with political tyranny in a previous century, they could address the more fundamental modern problem – the one connected with a recognition that society is a human product – of how individuals could pursue life, liberty, wealth and happiness while maintaining the social solidarity and order agreed upon by all as essential.[9]

If the optimistic men of Porter's English Enlightenment preferred *belles lettres* to comprehensive philosophies, the same cannot be said for the Scotsmen – Adam Ferguson, Thomas Reid, John Millar, Dugald Stewart, David Hume and Adam Smith – who moved the discussions about luxury and egoism onto an entirely different plane.[10] Classical republicanism had taught that men – and it was just men and only men of independent means – realized their full human potential when they participated in civic affairs. Supported by a substructure of labouring men and all women, this idealized citizen realized moral autonomy because of his independence from the necessities imposed by nature and through the interaction of a community of peers. A highly artificial construct, classical citizenship elevated the citizen above the crass, mundane, earthy and vulgar, and tested his fitness by his capacity to be virtuous. Commerce reeked of all the proscribed qualities, linking men and women together in new systems of interdependence while trading on physical needs, worldly tastes, undisciplined wants and preposterous yearnings. Where classical republican thought utterly failed was in explaining the economic changes transforming society. Without abandoning a concern with the moral dimension of the new market society, the Scots directed their attention to analysing the new forces at work.

Following Hume, Smith saw that in the esteemed primitive societies where men and women retained the whole of their produce, there was material equality, but lives of misery and want. In commercial societies with their flagrantly unequal distribution of wealth, the labouring poor prospered as well. This apparent paradox led Smith to examine the secret spring of British abundance – the organization of labour through the division of productive tasks. Fed with ever-renewed freshets of capital, the modern commercial system would escape the cycle of luxury, corruption and decline, because it had enlisted the self-improving energies of most members of society.

Of course Smith's description of how nations grow wealthy through commerce – ingeniously detailed as it was – would not have answered the moral question posed by republicanism had he and Hume before him not considered human morality from a new perspective, that of the great sympathies and sociability enlisted in commercial society. Smith gave to all human beings the propensity to truck and barter, as well as the incessant drive to improve their condition. From these promptings men were drawn to each other's company. Here in the market place, not the political assembly of classical times, modern men developed the capacity to reflect upon themselves in society, to excel by emulating virtue and shunning dishonour. In the concourses of commerce, men acquired their notion of probity and justice. As Thomas Paine wrote in *The Rights of Man*, economic life drew upon the naturally sociable and co-operative aspects of human nature. Commerce works 'to cordialize mankind', Paine wrote, 'by rendering nations, as well as individuals, useful to each other'.[11]

It was also a feature of modern life that ordinary labourers were independent, feeding themselves through their wages and thereby participating in the system of natural liberty. By shifting investigations of human character from politics to economics, the Scots were including labouring people in their conceptual universe. Modern commerce had made it possible for all to be independent and thus cut the critical link in classical theory between independent citizens and the dependent, disenfranchised workers, leaving those categories to be redrawn on the basis of gender and race. Within the realm of independent men – wage-labourers, merchants, manufacturers and landlords – the natural operation of the invisible hand of the market could regulate affairs better than the legislator, thus adding to the freedom from servile dependency a freedom from overweening political authority. If commercial exchanges rather than government authority unified the nation, the talented few – the men of extraordinary virtue and rectitude – had no function which could justify their privileges. Indeed the whole concept of justification of privilege made its way into social discourse through the door marked utility.

Although it took him until Book IV to say it, Smith placed consumption at the heart of modern market society.

> Consumption is the sole end and purpose of all production and the interest of the producer ought to be attended to, only so far as it may be necessary for promoting that of the consumer. The maxim is so perfectly self-evident, that it would be absurd to attempt to prove it.[12]

Yet Smith was far from happy with the human propensity to consume, characterizing it variously as a fascination for 'baubles and trinkets', a passion for accumulating objects of 'frivolous utility' and, worse, a vehicle for deception with the false promise that wealth will bring happiness. Money will at best 'keep off the summer shower' he said, 'but not the winter storm', thus leaving humans more exposed than before to anxiety, fear and sorrow, disease, danger and death.[13]

Probing for the causes of the avidity so evident in his society in the last months before his death, Smith concluded that it was envy and admiration for the rich and powerful and fearful contempt of the poor that drove men to seek wealth. And since in modern society with its striking inequality of condition the prods from above and beneath were omnipresent, the material wants of man would be insatiable. In reasoning thus, Adam Smith anticipated at least a part of Max Weber's celebrated line that 'A man does not "by nature" wish to earn more and more money, but simply to live as he is accustomed to live and to earn as much as is necessary for that purpose'.[14]

It was one of the strengths of the Scottish moral philosophers to build upon human nature as they found it and to discern the springs of moral action from the close observation of men in their own society. In the 'uniform constant and uninterrupted effort of every man to better his condition' Smith found the greatest grounds for hope.[15] For this was the human disposition that prompted men to defer pleasure, to save, to compete and to shun prodigality.

Here the middle-class character of the Scottish ideal shows itself, but in fact no rigorous analysis of consumption was carried out. Rather it was sentimentalized. From the middle of the eighteenth century through to our own time a particular kind of consumption has been approved, that which was associated with respectable family life. In the eighteenth century the word 'comfort' began to figure as the happy mean between biting necessity and indulgent luxury. Working over a draft treaty sent to him from John Adams in 1787, Thomas Jefferson replaced the word 'necessities' with that of 'comforts'. The new American nation would establish commercial treaties on the basis of exchanging comforts, not necessities.[16] Mary Wollstonecraft elaborated the concept in her *Historical and Moral View of the Origin and Progress of the French Revolution* when she explained that the French people had never acquired an idea of that independent, comfortable situation in which contentment is sought rather than happiness, because the slaves of pleasure or power can be roused only by lively emotions and extravagant hopes. In fact she goes on to observe the French don't even have a word in their vocabulary to express comfort, 'that state of existence, in which reason renders serene and useful the days which passion would only cheat with flying dreams of happiness'.[17]

The urban conviviality which commercial prosperity introduced into the eighteenth-century Anglo-American world had narrowed to a family-based respectability in the nineteenth century. Increasingly the desire to better oneself became associated with the motive of providing for one's family. Novelists gave respectability a distinctly material embodiment in the cleanliness and cut of clothes, the privacy afforded in the home and the accoutrements required to support the round of domestic rituals. It is tempting to claim that the family was sentimentalized in order to supply the safe avenues for what otherwise might be riotous broadways of spending.

The passions which the French Revolution evoked challenged the benign optimism of those making their peace with Adam Smith's market society. Across the Channel it became apparent that competitive self-interest could translate quickly into violent clashes of interest. The discreet scepticism of David Hume flowered into the open irreverence of Thomas Paine, promulgated to ordinary people through mass printings. By making the economy rather than the polity the basic institution of the society, the Scots had left politics in something of a conceptual limbo. If labour created value instead of being God's curse on Adam, what was the position of the labourer? Even liberty and equality looked different when the economy rather than the polity became the pre-eminent social system. What need was there of the talented few whose extraordinary virtue and rectitude alone preserved the constitution if it was the economy that provided stability? And how firm was that stability? Commerce as the principal socializer lacked a certain disciplinary rigour.

These discursive speculations were shunted aside when economic commentators began groping for the certainties of science. The most striking reworking of consumption in modern social thought came from Thomas Malthus. Writing in the closing years of the eighteenth century, Malthus put forward a population theory which interpreted abundance as spurious and pernicious. He sidestepped the debate about human predispositions and socializing influences, arguing instead that human beings were ruled by a set of inexorable equations. Consumption was at the centre of his theory. Abundance created cheap food. In good times, men and women married early and had lots of children. Without the positive checks of war and disease (construed negatively in other discourses), human population would grow geometrically swiftly outpacing the incremental increases of harvests which brought forth the surplus births. Would these unequal potentialities come into actual collision? Malthus was unequivocal about the immediate relevance of his mathematical discovery. 'The period when the number of men surpass their means of subsistence has long since arrived, has existed ever since we have had any histories of mankind, does exist at present, and will for ever continue to exist', he wrote in the first essay which appeared in 1798.[18] Nor could deferred marriage and family limitations relieve this parlous human condition.

Malthus forestalled further speculation about the theoretical effects of material progress by consigning human beings to a new determinism, the one inflicted by nature. A proper understanding of the dynamics behind human procreation eliminated the troubling question of how to render social justice in an age of increasing abundance. Utopian dreamers like Condorcet and William Godwin could say that perverse social institutions accounted for the persistence of human misery in the presence of unparalleled wealth – Godwin had argued just this in his celebrated essay, *An Enquiry Concerning Political Justice* – but Malthus permanently reordered the debate. The crucial issue became whether men and women could regulate their numbers and thereby avoid the evils of population pressure. Malthus said, 'no', and for the next thirty-eight years he refined his explanation of Nature's great catch-22 about plenty and poverty.

The possibility of easy living demonstrated to Malthus that it was only biting necessity that got human beings to exert themselves. Thus while the fear of famine was evil, it was only a partial evil, because it acted for a greater good. And, he stressed, not enough people knew about it. Instead of forming correspondence clubs to circulate radical tracts, working men should be taught their true situation. Malthus's words bear quoting:

> the mere knowledge of these truths, even if they didn't operate sufficiently to produce any marked change in the prudential habits of the poor with regard to marriage, would still have a most beneficial effect on their conduct in a political light, making them on all occasions less disposed to insubordination and turbulence.

Although the lower classes were clearly the focus of Malthus's attention, his principle was universal: 'Want has not infrequently given wings to the imagination of the poet, pointed the flowing periods of the historian, and added acuteness to the researches of the philosopher.'[19]

Malthus's sober strictures on the inevitable tendency of abundance (that is, more food) gave economics its label as the dismal science. In the hands of Ricardo the dreadful implications of omnipresent scarcity were worked out in the famous iron law of rents and declining rate of profits. Much that had remained open-ended in Smith was now closed. Demand, the activity closest to consumption, re-entered the picture as marginal utility, a concept which permitted all the passions of motivation from frivolity, vanity and boredom to ambition, avarice and need to be weighed on the same scale.

In ensuing decades the Malthusian principle of scarcity moved from economics to biology and then returned to sociology with powerful reverberations through all educated discourses in the nineteenth century. Human beings were folded back into nature. Physiology replaced original sin as the source of suffering. A uniform human nature and the stinginess of the physical environment controlled human destiny. Only familiarity keeps us from enjoying the irony that at a time when human productive powers were about to explode, competition for scarce resources became the centrepiece of theorizing in both biology and economics. The range of choices open to people had never been greater and yet it was positivism not poetry that dominated social thought. Variety and abundance became a permanent feature of western society, revealing the fecundity of human inventiveness, the insatiability of human curiosity, the splendour of human talents and the inaccuracy of aristocratic assumptions about ordinary people's abilities. Yet the reigning social theories assumed that human beings invariably sought gain through the equally invariant invisible hand of the market. Scholarly light narrowed to a laser beam directed at the workings of rational choice, utility maximization and competition for scarce resources while the rich diversity of human personality found no place in social theory.

The most consequential intellectual response to abundance was the awareness that human society was the product of human effort. To a large extent this is what is meant by secularization. Enveloped within the story of progress, this fact holds no terrors and few problems. Our proleptic histories assume that people want to rush into the future to enjoy their share of progressive improvement. In reality this encounter with unmitigated social responsibility was very troubling because it threw into high relief the issue of social justice, or more simply, how abundance was to be distributed. The classical discourse on luxury held the ground for a while, but it offered no intellectual tools for analysing economic developments. One of the responses to dramatic changes in the material world was the desire to explore the dynamic behind economic development. As William Reddy has noted, this extraordinary effort to understand the exchange economy ended up with a doctrine of indifference.[20] The existence of system was perceived – no mean feat – but once perceived it was declared best left alone. Those who spoke of the delights of the new material culture and the prospect of a more just distribution of them were drowned out by the new social scientists who gave human beings a nature so invariant that its inexorable workings determined social existence.

NOTES

1 Sidney Mintz explores the complicated response to the popular consumption of New World commodities in *Sweetness and Power: The Place of Sugar in Modern History* (New York, 1985).

2 Louis Dumont, *From Mandeville to Marx* (Chicago, 1977), 6–7.

3 [Nicholas Barbon], *A Discourse of Trade* (London, 1690); [Dudley North], *Discourses upon Trade* (London, 1691), 14.

4 Neil McKendrick, 'Home demand and economic growth: a new view of the role of women and children in the Industrial Revolution', in idem (ed.), *Historical Perspectives: Studies in English Thought and Society in Honour of J. H. Plumb* (London, 1974).

5 Simon Schama, *The Embarrassment of Riches* (New York, 1987), 298–335.

6 I am indebted to John Sekora, *Luxury: The Concept in Western Thought, Eden to Smollett* (Baltimore, 1977) for this discussion of consumption considered under the rubric of luxury. See also J. G. A. Pocock, *The Machiavellian Moment: Florentine Political Thought and the Atlantic Republican Tradition* (Princeton, 1975).

7 Sekora, *Luxury*, 44.

8 ibid., 5, 299.

9 Roy Porter, 'The English Enlightenment', in Roy Porter and Mikuláš Teich (eds), *The Enlightenment in National Context* (Cambridge, 1981), 1–18.

10 For a particularly insightful discussion of the Smithian tradition see Keith Tribe, 'The "histories" of economic discourse', *Economy and Society*, vi (1977), 314–344. See also Isvan Hont and Michael Ignatieff, 'Needs and justice in the *Wealth of Nations*: an introductory essay' and Nicholas Phillipson, 'Adam Smith as civic moralist', in *Wealth and Virtue* (Cambridge, 1983). I am also indebted to the unpublished writing of Charles Nathanson.

11 Thomas Paine, *The Rights of Man* (London, 1791–2), 99.
12 Adam Smith, *An Inquiry into the Nature and Causes of the Wealth of Nations* (New York, 1937), 625.
13 Hont and Ignatieff, 'Needs and justice', 10.
14 Max Weber, *The Protestant Ethic and the Spirit of Capitalism* (New York, 1958), 60.
15 Smith, *Wealth of Nations*, 324–325.
16 Thomas Jefferson to John Adams, 27 November 1785, in Lester J. Cappon (ed.), *The Adams–Jefferson Letters: The Complete Correspondence between Thomas Jefferson and Abigail and John Adams*, vol. 1 (Chapel Hill, 1959), 103.
17 Mary Wollstonecraft, *An Historical and Moral View of the Origin and Progress of the French Revolution and the Effect it has Produced in Europe* (London, 1795), 511. I am indebted to Anne Mellor for this reference.
18 Thomas Robert Malthus, *An Essay on Population* (London, 1798), 54, as cited in Thomas Sowell, 'Malthus and the utilitarians', *Canadian Journal of Economics and Political Science*, xxviii (1962), 272.
19 Malthus, *Essay*, 2nd edn, vol. 2 (London, 1803), 200.
20 William Reddy, *Money and Liberty in Modern Europe: A Critique of Historical Understanding* (Cambridge, 1987), 78–82.

2

The Consumer Revolution of Eighteenth-century England

Neil McKendrick

There was a consumer boom in England in the eighteenth century. In the third quarter of the century that boom reached revolutionary proportions. Men, and in particular women, bought as never before. Even their children enjoyed access to a greater number of goods than ever before. In fact, the later eighteenth century saw such a convulsion of getting and spending, such an eruption of new prosperity, and such an explosion of new production and marketing techniques, that a greater proportion of the population than in any previous society in human history was able to enjoy the pleasures of buying consumer goods. They bought not only necessities, but decencies, and even luxuries. The roots of such a development reach back, of course, into previous centuries but the eighteenth century marked a major watershed. Whatever popular metaphor is preferred – whether revolution or take-off or lift-off or the achievement of critical mass – the same unmistakable breakthrough occurred in consumption as occurred in production. Just as the Industrial Revolution of the eighteenth century marks one of the great discontinuities in history, one of the great turning points in the history of human experience, so, in my view, does the matching revolution in consumption. For the consumer revolution was the necessary analogue to the Industrial Revolution, the necessary convulsion on the demand side of the equation to match the convulsion on the supply side.

We are only just beginning to realize how pervasive were the social and economic effects of that change, and how considerable were the pressures needed to bring it about. For the results were such as to bring about as great a change in the lifestyle of the population as was brought about by the neolithic revolution in agriculture which began some eight thousand years before the birth of Christ.

Changes of that order do not occur without comment, and contemporaries were eloquent in their descriptions and explanations of what Arthur Young in 1771 called this 'UNIVERSAL' luxury, and what Dibden in 1801 described as the prevailing 'opulence' of all classes.[1] Foreign commentators were astounded by what one called 'the inveterate national habit of luxury of the English', and almost invariably recorded their amazed reactions to it. The Göttingen professor Lichtenberg said of England in the 1770s that the luxury and extravagance of the lower and middling classes had 'risen to such a pitch as never before seen in the world';[2] the Russian writer Karamzin said of England in the 1780s: 'Everything presented an aspect of . . . plenty. Not one object from Dover to London reminded me of poverty';[3] the historian von Archenholz said of the 1790s: 'England surpasses all the other nations of Europe in . . . luxury . . . and the luxury is increasing daily!' 'All classes', he concluded, 'enjoy the accumulation of riches, luxury and pleasure.'[4] One does not have to take all such hyperbole literally to realize that something momentous was thought to be happening. By the end of the eighteenth century there was a deafening chorus of comment – full of wonder or, more often, complaint at the manifold signs of this great change in consumer behaviour.

The rich, of course, led the way. They indulged in an orgy of spending. Magnificent houses were built reaching a crescendo of building in the 1760s and 1770s when the brothers Adam designed so many memorable Georgian houses to replace Elizabethan and Jacobean mansions ruthlessly demolished to make space for them. Superlative furniture was commissioned from the published directories of Chippendale, Hepplewhite and Sheraton. Porcelain and pottery of a quality unparalleled in English history appeared, including Chelsea, Bow, Worcester and Derby for the few, and Wedgwood for the many. Silver ranged from the sturdy mastery of the early Huguenot silversmiths to the characteristic elegance of a Schofield candlestick to the feminine delicacy of Hester Bateman. Mirrors came from the great master Linnell; cutlery from the master smiths of Sheffield; 'toys' in protean variety, from the costly 'exclusives' of Matthew Boulton to the cheap buttons for the mass market. Wonderful new gardens were created with orangeries bursting with the latest 'exotiques' like pineapples and camellias. Whole estates were replanted for posterity with trees chosen from nurseries offering an unprecedented number and variety of species. Collections of aristocratic pets resembled menageries or private zoos. In all these areas there flourished the unmistakable signs of conspicuous consumption. In all a desire for novelty – so all-consuming that Dr Johnson complained that men were even 'to be hanged in a new way'. In fashion novelty became an irresistible drug. In possessions for the home, new fashions were insisted on – in pottery, furniture, fabrics, cutlery, even wallpaper. Even their animals must be new, and improved breeds of horses, cattle and sheep, dogs, fishes, birds and plants, were all deliberately pursued with a new intensity and a matching success. It is alarming to think of how many trees, flowers and animals which we take for granted they bred or imported for the first time.[5]

In imitation of the rich the middle ranks spent more frenziedly than ever before, and in imitation of them the rest of society joined in as best they might – and that best was unprecedented in the importance of its impact on aggregate demand. Spurred on by social emulation and class competition, men and women surrendered eagerly to the pursuit of novelty, the hypnotic effects of fashion, and the enticements of persuasive commercial propaganda. As a result many objects, once the prized possessions of the rich, reached further than ever before down the social scale.

Forster in 1767 neatly encapsulated all the features of the new demand stressed by modern historians when he wrote:

> In England the several ranks of men slide into each other almost imperceptibly and a spirit of equality runs through every part of their constitution. Hence arises a strong emulation in all the several stations and conditions to vie with each other; and the perpetual restless ambition in each of the inferior ranks to raise themselves to the level of those immediately above them. In such a state as this fashion must have uncontrolled sway. And a fashionable luxury must spread through it like a contagion.[6]

These characteristics – the closely stratified nature of English society, the striving for vertical social mobility, the emulative spending bred by social emulation, the compulsive power of fashion begotten by social competition – combined with the widespread ability to spend (offered by novel levels of prosperity) to produce an unprecedented propensity to consume: unprecedented in the depth to which it penetrated the lower reaches of society and unprecedented in its impact on the economy.

Both commercial activity and the consumer response to it were feverish. Uncontrolled by any sense of commercial decorum men advertised in unprecedented numbers – whole newspapers were taken over by advertisements, and a very large proportion of all newspapers was filled with advertising. And the customer had plenty to choose from. For, spurred on by rampant demand, designers produced both fashions of outrageous absurdity and styles of lasting elegance. Fashion in hats and hairstyles, dresses and shoes and wigs and such like, arguably reached even greater extremes than ever before and certainly changed more rapidly and influenced a greater proportion of society.[7]

NOTES

1 See in Neil McKendrick, 'Home Demand and Economic Growth. A New View of the Role of Women and Children in the Industrial Revolution', *Historical Perspectives. Studies in English Thought and Society*, edited by Neil McKendrick (London, 1974), p. 193.

2 G. C. Lichtenberg, *Lichtenberg's Visits to England*, translated and edited by M. L. Marc and W. H. Quarrel (1938).

3 N. M. Karamzin, *Letters of a Russian Traveller 1789–90*, translated and abridged by F. Jonas (1957), p. 261.

4 J. W. von Archenholz, *A Picture of England* (1791), pp. 75–83.

5 See J. H. Plumb, 'The Acceptance of Modernity', *The Birth of a Consumer Society: The Commercialization of Eighteenth-century England*, Neil McKendrick, J. Brewer and J. H. Plumb (London, 1982), pp. 316–34.

6 N. Forster, *An Enquiry into the Present High Price of Provisions* (1767), p. 41.

7 See J. H. Plumb, op. cit., pp. 60–3.

3

Consumerism and the Industrial Revolution

Ben Fine and Ellen Leopold

THE DEMAND FOR FASHION IN CLOTHES

McKendrick's article on the 'Commercialization of fashion' (in *The Birth of a Consumer Society*) illustrates some of the limitations of operating within the 'supply and demand' framework and, within that, of leaning too heavily on just one of its axes. In his article, social emulation, class composition and emulative spending in the second half of the eighteenth century become the prime movers behind the development of mass-based consumer demand fifty years later. Demand emerges as a necessary precondition and stimulant to the subsequent growth of mass production. In particular, the 'trickle-down' effects of the demand for luxury goods, i.e. the gradual percolation and diffusion of upper-class tastes through all strata of society, anticipate and expedite the arrival of mass markets. According to McKendrick, once the pursuit of luxury 'was made possible for an ever-widening proportion of the population, then its potential was released and it became an engine for growth, a motive power for mass production'.[1]

Demand, in this scheme, becomes an active transforming agent on its own, viewed independently of production; McKendrick's writing manages to attach a kind of dynamism to the force of demand by documenting the frenetic pace and manifestations of changes in taste. In the rapid displacement of one style, colour, eighteenth-century flavour-of-the-month by another, the course of fashion begins to acquire a self-sustained momentum of its own.

For McKendrick, this progressive emulation is manifested by the dress of the domestic servant which, in the second half of the eighteenth century, is often almost indistinguishable from that of the employer. Foreign fascination for the apparent blurring of social distinction that this implied is much quoted; the historian Von Archenolz 'complained that he could "not distinguish between guests and servants" when he visited the Duke of Newcastle, and was particularly thrown by the fact that the butler dressed like his master'.[2]

How this mirroring of costume between upper and lower classes came about is not addressed by McKendrick. For him, the transmission of taste, like that of demand, is a disembodied process, not explicable by reference to prevailing social or economic relations. There is no attempt to link the surge in demand in the late eighteenth century with the production and distribution of goods, nor is there any reference to the source and distribution of consumers' incomes unless it be by a minimal lip-service paid to the supply side of the extraordinary world of late eighteenth-century fashion in clothing. But his clear lack of interest in its dynamics leads him to make casual and misleading references to 'the sales of mass-produced cheap clothes' in an eighteenth-century context.[3] It may not be apparent to all his readers that no mass markets for cheap clothes (as opposed to fabrics and accessories) existed until almost a hundred years later, well after the arrival of the sewing machine in 1851. Even the systematic use of measuring tapes in the manufacture of clothes by size did not influence the cutting of fabric until the nineteenth century.[4]

Giving the impression that the supply underpinning mass markets already lay in readiness and simply awaited the catalyst of demand for luxury goods to spring into action is highly misleading. First of all, it conflates the idea of differentiation operating at the luxury end of the market with the standardization of goods that is the hallmark of mass production. The emphasis on the expansion of markets through rapid and continuous change in the products themselves cannot be used to explain the process by which goods are homogenized to meet the requirements of large-scale volume production. In fact, the former can be viewed as obstructing or at least delaying the arrival of the latter.

Having resolutely turned his back on both the producers and distributors to give centre stage to the role of demand, McKendrick is left without any means of explaining the transformation of a bespoke market in luxuries to a mass market of essential goods. It is simply observed, almost by a natural progression, in that 'luxuries' become 'decencies', and 'decencies' become 'necessities'.[5] At some stage, the frenzied proliferation and pace of fashion change just burns out leaving 'a greater social uniformity . . . which of course suited those producing and selling fashion'. He argues that 'demand could be controlled to suit their needs' but without postulating any mechanisms by which this might have been achieved.[6] The madness of compulsive differentiation in the consumption of hand-made luxury goods is presented as simply petering out, through some unexplained shift in the *zeitgeist*, transforming itself into a more even-keeled and sensible uniformity which could then be exploited by producers to consolidate and generalize the emerging features of modern-day consumer society.

Missing markets

One of the central flaws embedded in McKendrick's argument is his assumption that the expansion and broadening of interest in fashionable clothing during the second half of the eighteenth century in itself signified an expansion and broadening of the market demand for these goods. This is to confound consumption of goods with their exchange, i.e. to assume that the diffusion of fashion within society, from mistress to maidservant, demonstrated a progressive broadening of purchasing power. Social emulation, it is assumed, in itself begets emulative spending.

The argument is really a recasting of the optimist's position in the debate over changes in the standard of living caused by the Industrial Revolution. In the absence of any reliable information on the distribution of incomes during the period, the hypothesis of a closely stratified society which encouraged the idea of social mobility between groups would allow for the progressive redistribution of income as of fashion from the top down.

More specifically, McKendrick argues that 'The expansion in the market, revealed in the literary evidence, occurred first among the domestic-servant class, then among the industrial workers and finally among the agricultural workers.'[7] Though the argument depends upon emulative behaviour triggering emulative spending, it offers neither documentary evidence chronicling such behaviour nor any discussion of the social relations which might have influenced it.

In fact, neither spending nor any act of exchange appears in McKendrick's scheme of things at all. In the discussion of clothing, at least, the market remains an altogether shadowy not to say invisible figure. The acquisition of goods – as opposed to their display – is never addressed. Who purchases what, from whom and for what prices? What are the incomes or wealth of those making such purchases and what is their source? To answer those queries would be to detract too much from McKendrick's central concern which rests squarely, if narrowly, with the growth of consumer demand induced by luxury consumption. In an earlier article, McKendrick does attempt to address these issues; hypothesizing that the marked increase in the second half of the eighteenth century in the participation of women and children in wage labour explains the sudden surge in money wages that led to an expansion of demand.[8] Curiously, no mention is made here of the role played by social emulation and emulative spending. Women and children as industrial wage-earners are seen as contributing directly to an increased demand for household necessities and particularly for those goods formerly made by the women themselves rather than for luxuries. It is their wage income that pushes demand.

This at least grapples with the sources of income that are translated into effective demand for consumer goods. It also suggests that the process of transformation in demand and production occurred together over a much longer period than is suggested by McKendrick's later work. Moreover, it points to the contradiction between the entry of significant numbers of women and children into the labour force and the growth of unproductive domestic servants in the eighteenth and nineteenth centuries. On the other hand, it takes a positive view of this transition, arguing that the shift of women into wage labour represented a liberation from the oppression of 'cottage' industry, rather than the coercive deprivation of the means of home livelihood.

In short, there are problems of consistency across McKendrick's work. Either he sets aside the source of the income that provides for growing demand, simply relying upon the filtering down of tastes from the wealthy to the lower classes. Or, in a sideshow, wage income is generated by the proletarianization of women and children. Here, however, he would need to paint a much grander canvas of economic and social change over a longer period. Arguably, the destruction of domestic industry involves causal factors and associated effects whose own direct influence on demand is liable to be greater, through the wage income generated, than the indirect effect of the impact of fashion.

McKendrick's later article emphasizes the key role played by the unproductive domestic servant as a carrier rather than an active consumer of fashion – transmitting changes in taste both from the upper classes to the working classes and from London to the provinces. It makes no reference to the dampening role played by the servant class as a whole on the evolution of wage labour and its limiting effect on the extent of the market – of so much concern to Adam Smith. The perpetuation of almost feudal relations between servant and served could not be dismissed as insignificant when the number of men and women so employed was estimated at almost a million in 1806 – 800,000 female and 110,000 male, equivalent to about one in every eleven people in the population at large.[9]

Servants are withdrawn from other potentially productive activity and are paid a variable combination of wages, tips and goods in kind. Their exclusion from wage labour must have had an inhibiting effect on the development of both large-scale demand and mass markets, both key features of modern consumer society. In addition, their absorption into households other than their own held down demand for all those consumer goods which depended on the growth of individual households (everything from housing itself to sets of furniture and pottery) as well as on the growth of cash wages. In other words, the perpetuation of this relationship and its later extension to the middle classes can in itself be seen as moderating the growth of industrial capitalism.

Nowhere in McKendrick's article is the huge disparity of income between master and servant ever commented upon. Yet the annual income of most housemaids in the second half of the eighteenth century would have been insufficient to pay for the material for a single dress made up for her mistress (Barbara Johnson paid £7 15s 9d in the 1760s for material for a day dress, a 'negligee', at a time when her housemaid was probably earning a basic annual salary of about 7 guineas).[10] Clearly, domestic servants were not in a position to purchase newly made clothes in imitation of their employers. On their incomes, they could not possibly contribute to any growth in the effective demand for new fashion goods. Emulative spending emanating from below stairs appears highly improbable.

Those clothes worn by servants which echoed upper-class rather than working-class tastes came to them, for the most part, directly from their employers, unmediated by markets of any kind. No money changed hands. Most clothes were simply handed down from mistress to maid with increasing frequency as the turnover in fashion tastes increased (in an early demonstration of dynamic obsolescence). Extending the lifespan of still serviceable, though slightly outmoded garments could widen the general currency for fashion dress by raising its visibility and extending its use. In this, it functioned as a form of advertising. However, no discerning exercise of taste on the part of domestic servants was required.

In fact, the strength of emulation as a transforming mechanism can be even more comprehensively questioned. Employers often selected and purchased clothes for their servants to wear, in pursuit of a kind of 'vicarious consumption' which McKendrick mentions in connection with exotically dressed black page boys who adorned some eighteenth-century households. (There is much evidence of this practice, from Parson Woodforde to Anne Lister.) Lavish dress provided by the employer, which cut across accepted social distinctions, reflected his or her taste, not that of the servant. Furthermore, clothes purchased in this way often remained the property of the employer.

Extravagant dress for female servants in lieu of uniforms served as a kind of livery comparable to the more formal outfits worn by coachmen. Though male servants wore livery from an early date, uniforms were not adopted for female servants until the second half of the nineteenth century. Until then, their dress had always echoed the style and taste of their mistresses. For both male and female servants, clothes were a highly visible sign of their employers' wealth and status

– much more so than their living quarters or diets, which remained completely hidden from the visitor's gaze and hence were less extravagantly catered for.[11]

The more exposed the servant to the public life of the house, the more he or she contributed to setting the overall social tone of the establishment, forming an integral part of the domestic display. Upstairs maids were clearly better dressed than those working exclusively below stairs. The coachman's livery was often the most elaborate of all, because he carried the employer's reputation directly into the public realm. Even when servants purchased their own wardrobe, its contents would be carefully vetted by an existing or potential employer to guarantee that it was in acceptable taste. In this respect, emulation – whether reflecting genuine preference or not – might be perceived as enhancing job prospects.

Many servants were also bequeathed dresses upon the death of their mistresses. For instance, Sarah, Duchess of Marlborough stipulated in her will that her wardrobe be divided between her lady's maid and two other maidservants.[12] The acceptance of the bequest in no way signified a preference for whatever fashion might be reflected in the clothes passed on. As an alternative to extending a housemaid's own wardrobe, inherited dresses might be sold on to the second-hand clothing market, a substantial and lucrative trade in the eighteenth century. So established was the market in used, reconditioned or repaired clothes that many servants were granted the rights to their masters' cast-offs as an agreed condition of their contract of employment. The preference for cash (from selling hand-me-downs) over the pleasures to be had from keeping and wearing the master's old clothes suggests a definite limit to the allure of emulation.

The persistence of a strong second-hand market does lend weight to the influence of the trickle-down effects of fashion. But equally powerful would be the attraction of slightly used goods set at low prices, relative to their cost when new. If cost is the primary factor in making such a choice – as it must be where a pair of boots could account for more than a month's wage – the contribution made by fashion or style is liable to be negligible.

Perhaps the clearest evidence of servants' apparel as an accoutrement of the household is in their wearing of mourning dress upon the death of an employer's relative. Clothes for this purpose were supplied by the employer. Following the death of her uncle in 1817, Anne Lister noted in her diary:

> The mantua-maker came & bombasines & stuffs were sent over from Milne's & bombasines from Butters. Chose mourning for the 2 women servants from the former place, & from the latter, 50 yds (a whole piece) bombasine at 4/9 for ourselves. The servants had each 8 yds, ⅝ wide twilled stuff at 2/4, and 3½ yards of the same for a petticoat.[13]

It is interesting to note that material for the servants' mourning dress was half the price of that worn by the mistress.

Servants had no choice in the matter, any more than would children in the household. They were required to assume – literally – the mantle of a grief which they were unlikely to feel, particularly since mourning dress was worn in memory of relatives outside the immediate household as well as for those within it. Mourning dress is simply the sharpest example of the control exercised by the employer over a servant's attire. None of the other means by which servants acquired their masters' or mistresses' clothes – through loans for the duration of service, hand-me-downs, bequests – demonstrates any preference for their employers' taste. The element of choice or preference expressed through monetary or any other form of exchange simply does not come into it.

McKendrick concentrates on the surge in demand for fashionable clothing in the late eighteenth century, i.e. before the advent of significant advances in the technology of clothes production (all clothing remained hand made). This allows him to raise the importance of demand as a pre-condition of, and catalyst for, the revolution in production that was presumed to follow. But, as suggested above, the omission of any evidence relating to the development, source or exercise of purchasing power weakens credibility in this concept of demand, and the emulation on which it is founded is also unproven. Moreover, by the time capitalist production of clothing, as opposed to fabric, is finally under way – in the second half of the nineteenth century – the structure of purchasing power is very different from that prevailing fifty years earlier. By the middle of the nineteenth century, a distinct pattern of demand emerges reflecting the rising income, and numbers, of industrial workers.

This coincides with a significant shift of population growth away from the south to the Midlands and the north, where the new centres of industry were located. While in 1801 only five towns in England apart from London boasted populations of more than

50,000 (Liverpool, Manchester, Bristol, Birmingham and Leeds), by 1851, their number had risen to 24. The same period showed the proportion of the labour force engaged in manufacturing, mining and industry rising from 29.7 per cent to 42.9 per cent. This considerable growth in cash wages as a source of demand was, therefore, based on a completely different pattern of social relations governing consumption (as well as those governing production). The influence of London fashion and the importance of domestic servants were both much weaker in the industrial hinterlands and largely contained with the squirearchy maintaining active links with the capital. There is something more than a little absurd in reducing these regional patterns of, and stimuli to, industrialization to the dictates of London fashion emanating from the London upper classes.

It seems highly probable that the emergence of demand as of consumption norms among the industrial working class took place outside McKendrick's social nexus altogether, i.e. that the influence of the dynamic he describes was both self-contained and historically specific. McKendrick has no way of bridging the gap between the luxury trade he describes and the evolution of mass markets, just as he supplies no connecting thread to link the bespoke fashion trade with the evolution of the machine-based clothing industry.

One of several more plausible explanations for the change in taste is the spread of the Protestant ethic in dress.[14] This emphasized modesty and conformity in place of worldly showiness and individuality, and was indisputably a middle-class phenomenon. The growth of the professions also contributed to the increasing sobriety of middle-class dress – clergymen, doctors and lawyers all donned working clothes that were uniformly black and unadorned. The gradual adoption of these habits by the aristocracy is in part a reflection of the decline in the influence of the court within the economic and social life of the country and of a rise in the secular culture of work. Emulation *upwards* signifies the court's (and the aristocracy's) decline as an arbiter of taste and style. The wearing of court dress, with its expensive and impractical finery, became increasingly rare, limited to gala royal occasions and grand social events. As the influence of the rising bourgeoisie grew, so their taste's emphasis on working dress began to take hold across a wider spectrum of society.

Equally overlooked by McKendrick was a more naked emulation by the upper classes of the dress of labourers. The introduction of the workman's frock-coat into the higher reaches of society is a clear example of this countervailing trend (repeated in the twentieth century by the fashion history of jeans). According to the Cunningtons' history of costume, the 'frock' had become the common wear of the urban man by 1700. Designed for comfort (and therefore loose-fitting) from cheap materials, its distinguishing characteristic was a flat turned-down collar. By the middle of the eighteenth century, the 'gentleman' had taken it up as a comfortable alternative to his traditional heavy stiff coat. In the following decades, it became established ordinary wear for gentlemen. By the end of the century it annexed the classier word 'coat', becoming the 'frock-coat' 'with a long career ahead as a symbol of Class; a truly remarkable garment to have climbed the whole social scale from the farmyard to the Royal Enclosure'.[15] Nor were those who took up the wearing of the frock-coat unmindful of its political implications. With its reputation as a more 'democratic' mode of dress, it became popular during the period of the French Revolution among those wishing to show their opposition to the Ancien Régime.[16]

Both the examples of the frock-coat and that of the Puritan ethic suggest that the trend towards equality in dress emanated from below, pushing its way upwards towards the wealthier classes, as well as from above and trickling down. This two-way influence, however, deprives luxury spending of its progressive attributes. The lower orders are no longer simply passive beneficiaries and transforming multipliers of consuming habits imposed from above, and spending by the upper classes is no longer required to serve as a catalyst in the transformation of goods from luxuries into basic necessities. The diffusion of the frock-coat represents an opposing tendency – the mutation of a humble garment into an essential component of the upper-class wardrobe. It also demonstrates that emulation need not necessarily lead to a cheapening of the final product through either increases in productivity or cheapening of materials. In other words, the widening of demand, on its own, cannot provide an impetus for progressive technical change.

NOTES

1 N. McKendrick, 'The commercialization of fashion', in N. McKendrick, J. Brewer and J. H. Plumb (eds), *The Birth*

of a Consumer Society: The Commercialization of Eighteenth-Century England (1982), 66.

2 *Ibid.*, 58.

3 *Ibid.*, 53.

4 J. Tozer and S. Levitt, *Fabric of Society: A Century of People and their Clothes, 1770–1870* (1983).

5 McKendrick, *op. cit.*, 1.

6 *Ibid.*, 56.

7 *Ibid.*, 60.

8 N. McKendrick, 'Home demand and economic growth: a new view of the role of women and children in the Industrial Revolution' in N. McKendrick (ed.), *Historical Perspectives: Studies in English Thought and Society in Honour of J. H. Plumb* (1974).

9 J. Hecht, *The Domestic Servant Class in Eighteenth-Century England* (1956), 34.

10 *Ibid.*, and B. Johnson, *Barbara Johnson's Album of Fashions and Fabrics* (1987).

11 Hecht, *op. cit.*, 120.

12 *Ibid.*, 116.

13 A. Lister, *I Know My Own Heart: The Diaries of Anne Lister (1791–1840)* (1988).

14 A. Ribeiro, *Dress in Eighteenth-Century Europe 1715–1789* (1984).

15 C. and P. Cunnington, *Handbook of English Costume in the Eighteenth Century* (Boston, 1972), 18.

16 Tozer and Levitt, *op. cit.*, 55.

4

Traditional and Modern Hedonism

Colin Campbell

Traditional hedonism involves a concern with 'pleasures' rather than with 'pleasure', there being a world of difference between valuing an experience because (among other things) it yields pleasure, and valuing the pleasure which experiences can bring. The former is the ancient pattern, and human beings in all cultures seem to agree on a basic list of activities which are 'pleasures' in this sense, such as eating, drinking, sexual intercourse, socializing, singing, dancing and playing games. But since 'pleasure' is a quality of experience, it can, at least in principle, be judged to be present in all sensations. Hence the pursuit of pleasure in the abstract is potentially an ever-present possibility, provided that the individual's attention is directed to the skilful manipulation of sensation rather than to the conventionally identified sources of enjoyment.

These two orientations involve contrasting strategies. In the former, the basic concern is with increasing the number of times one is able to enjoy life's 'pleasures'; thus the traditional hedonist tries to spend more and more time eating, drinking, having sex and dancing. The hedonistic index here is the incidence of pleasures per unit of life. In the latter, the primary object is to squeeze as much of the quality of pleasure as one can from all those sensations which one actually experiences during the course of the process of living. All acts are potential 'pleasures' from this perspective, if only they can be approached or undertaken in the right manner; the hedonistic index here is the extent to which one is actually able to extract the fundamental pleasure which 'exists' in life itself. To pursue this aim, however, it is necessary not only for the individual to possess special psychological skills, but for society itself to have evolved a distinctive culture.

THE GROWTH OF MODERN HEDONISM

The key to the development of modern hedonism lies in the shift of primary concern from sensations to emotions, for it is only through the medium of the latter that powerful and prolonged stimulation can be combined with any significant degree of autonomous control, something which arises directly from the fact that an emotion links mental images with physical stimuli. Before the full potential of emotionally mediated hedonism can be realized, however, various critical psycho-cultural developments have to have taken place.

That emotions have the potential to serve as immensely powerful sources of pleasure follows directly from their being states of high arousal; intense joy or fear, for example, produces a range of physiological changes in human beings which for sheer stimulative power generally exceed anything generated by sensory experience alone. This is true no matter what the content of the emotion. It is certainly not the case that some emotions, such as gratitude or love, are pleasant, while others, such as grief or fear, are not, for there are no emotions from which pleasure cannot be obtained.[1] Indeed, since the so-called 'negative' emotions often evoke stronger feelings than the others, they actually provide a greater potential for pleasure. The question, therefore, is not which emotions can supply most pleasure but what are the circumstances which must prevail before any emotion can be employed for hedonistic purposes.

An emotion may be represented as an event which is characteristically 'outside' an individual's control (or, at least, this is true biographically and historically, if subsequent developments are ignored). It is, in that sense, a behavioural storm which is endured, rather than an activity which is directed. Under the influence of very intense emotions, the behaviour of people is frequently so extreme and chaotic that they are said to be 'out of their minds', or 'to have taken leave of their senses', even, to be 'possessed'. Individuals may laugh or cry uncontrollably, dance, or run wildly about, even beat themselves or pull out their hair. Clearly experiences of this kind inundate the individual with such an excess of stimulation that there can be little possibility of enjoying it. What is more, as the examples suggest, such emotional arousal is merely part of a larger directive behavioural complex, involving overt motor activity, in the way in which fear is linked to flight or anger to aggression.[2] Thus not only is the individual's capacity to 'appreciate' his aroused state negated by his being subjected to a form of sensory overload, but he also has his attention directed away from any introspective appreciation of the subjective dimension of his experience by the preparation and implementation of action. Before any emotion can possibly be 'enjoyed', therefore, it must become subject to willed control, adjustable in its intensity, and separated from its association with involuntary overt behaviour.

This form of emotional control is not to be confused with that ordering and regulation of affective responses which must necessarily be a feature of all social life. That process is primarily concerned with the co-ordination of patterns of emotional restraint and display, and is primarily achieved through common socialization experiences. It is obvious that all cultures require individuals to learn both when and how to suppress, as well as express, emotions – a process which consists, in essence, of learning which situations are associated with what emotions. Control rarely extends, however, beyond the exercise of restraint in circumstances where no expressive response is permitted. In other words, it does not embrace a process of self-determination with regard to emotional experience, yet it is precisely in the degree to which an individual comes to possess the ability to decide the nature and strength of his own feelings that the secret of modern hedonism lies.

Such self-regulative control is clearly more than a mere capacity to suppress, although this is the connotation most usually associated with the expression, 'controlling one's emotions'. Obviously this is a necessary part of such a skill, and a soldier who endeavours to subdue his fear when in battle can indeed be said to be trying to 'control' both his state of arousal and its manifestation in observable actions. If he succeeds his fear will not be translated into the action of fleeing from the battlefield, and perhaps, in time, a certain diminution in the tendency to experience the emotion may occur. This ability, however, is one of limited behavioural, rather than full emotional, control; power being exerted over overt action rather than the psychophysiological dimension of emotional experience itself.[3] The term 'self-control' or 'self-discipline' is appropriate to describe success in this respect.[4] A more crucial part of the capacity for emotional control concerns the deliberate cultivation of an emotion, especially in the absence of any 'naturally occurring' stimulus, and although this is, in part, a corollary of the power to suppress feeling, it also transcends it.[5] The attainment of emotional 'self-control' in the negative sense is hence both a precursor and a prerequisite of the development of full voluntaristic emotional control, for, while it is perhaps natural that problems presented by the presence of unwanted emotions should be more pressing than those created by the absence of desired ones, efforts directed at suppressing emotion succeed in breaking the intimate association between feeling and overt behaviour. By thus separating anger from aggression, or fear from flight, a start is made on the process by which emotion becomes defined as a largely interiorized facet of human experience.[6]

Of course, if an individual is to determine his own emotional state, then it is necessary to be 'insulated' in some way from those inevitable exigencies of life which typically prompt such responses. To the extent, therefore, that advances in knowledge, wealth and power reduce a person's exposure to the threats of famine, disease, war or disasters in general, one might anticipate an increased possibility of emotional control. Although this is true, the development of cultural resources to provide such 'insulation' would seem to be an occurrence of far greater significance, for this process allows considerable latitude in the way in which any situation is defined. Thus, to take an example, a clergyman may organize the frightened passengers of a sinking ship into a religious congregation joined in prayer, and in this way counter the environmentally stimulated fear and panic with ritually stimulated hope and calm. Alternatively, a commander-in-chief, like Henry V, might employ a rhetoric rich in powerful and suggestive images to instil courage and determination into his exhausted and demoralized troops. In this fashion the symbolic resources of a culture can be employed to re-define the situations in which groups find themselves and thus bring about changes of mood, a process which extends beyond mere self-control to embrace the substitution of one emotion for another. The trouble, of course, with using symbolically triggered emotions in this way to combat environmentally prompted ones, is that the individual may simply be exchanging one form of external determination for another. Only if the individual is himself in control of the employment of symbolic resources can true emotional self-determinism emerge. For this reason, a decline in the importance of the collective symbolic manipulation of emotion is important. Literacy, in conjunction with individualism, would seem to be the key development in this respect, for this grants the individual a form and degree of symbolic manipulation which was previously restricted to groups.

The central point to be emphasized in this context is that only in modern times have emotions come to be located 'within' individuals as opposed to 'in' the world. Thus, while in the contemporary world it is taken for granted that emotions 'arise' within people and act as agencies propelling them into action, it is typically the case that in pre-modern cultures emotions are seen as inherent in aspects of reality, from whence they exert their influence over humans. Thus Barfield has pointed out how, in the Middle Ages, words like 'fear' and 'merry' did not denote a feeling located within a person, but attributes of external events; 'fear' referring to a sudden and unexpected happening, and 'merry' being a characteristic of such things as the day or the occasion.[7] The attitude and emotion of 'awe' is another good example of an aspect of experience which was regarded as primarily a characteristic of God rather than of a man's typical reaction to his presence. These examples show how the main sources of agency in the world were viewed as existing outside of man, from whence they not only 'forced' him into actions but also 'filled' him with those distinctively aroused states called emotions.[8]

This view of man and his relationship to the world was to change dramatically as a consequence of the process which Weber called 'disenchantment'; that is, the collapse of the general assumption that independent agents or 'spirits' were operative in nature.[9] The origins of this development can be traced back as far as ancient Judaism but it was accelerated by the Reformation, attaining its most complete expression in the Enlightenment. A significant corollary of disenchantment was the accompanying process of de-emotionalization such that the environment was no longer seen as the primary source of feelings but as a 'neutral' sphere governed by impersonal laws, which, while they controlled natural events, did not, in themselves, determine feelings. A natural consequence of this fundamental shift in world-view was that emotions were re-located 'within' individuals, as states which emanated from some internal source, and although these were not always 'spiritualized', there is a sense in which the disenchantment of the external world required as a parallel process some 'enchantment' of the psychic inner world.[10] A new set of terms was required in order to describe this transition, and to this end old words were pressed into fresh uses. Examples would be 'character', 'disposition' and 'temperament', all words which had originally referred to some feature of the external world and which now came to stand for a subjective influence upon behaviour.[11]

This increasing separation of man from the constraining influence of external agencies, this disenchantment of the world, and the consequent introjection of the power of agency and emotion into the being of man, was closely linked to the growth of self-consciousness. Such a uniquely modern ability is itself a product of these processes, as, in becoming aware of the 'object-ness' of the world and the 'subject-ness' of himself, man becomes aware of his own awareness poised between the two. The new internal psychic world in which agency and emotion are relocated is that of the 'self', and this world is, in its turn, also increasingly subject to the cool, dispassionate and inquiring gaze which disenchanted the outer, with the result that consciousness of 'the world' as an object separate from man the observer, was matched by a growing consciousness of 'the self' as an object in its own right. This is revealed by the spread of words prefixed with 'self' in a hyphenated fashion, words such as 'self-conceit', 'self-confidence' and 'self-pity', which began to appear in the English language in the sixteenth and seventeenth centuries, and became widely adopted in the eighteenth; 'self consciousness' itself apparently being first employed by Coleridge.[12]

Associated with this development were attempts to understand the laws which link the inner and the outer worlds: to grasp how exactly certain features of each are connected. In part, this meant examining the way in which aspects of externality tend to prompt particular emotional responses from within. Hence the proliferation of words which relate to the effects which objects can have upon people, words like 'amusing', 'charming', 'diverting', 'pathetic' and 'sentimental', while the effects which the 'self' has on the environment are summarized by the terms, 'character', 'disposition', 'taste' and the like mentioned above.

Of crucial significance for this discussion is the fact that the growth of self-consciousness had, as one of its many consequences, the effect of severing any remaining necessary connection between man's place in the world and his reaction to it. Objective reality and subjective response were now mediated through consciousness in such a way that the individual had a wide degree of choice concerning exactly how to connect them. Beliefs, actions, aesthetic preferences and emotional responses were no longer automatically dictated by circumstances but 'willed' by individuals. Such a contrast is, of course, exaggerated, but, in so far as individuals gained control over their own tendency towards impulsiveness, and could, on the other hand, manipulate the symbolic meanings of events, then it is indeed reasonable to speak of the growth of an autonomous control of emotional expression.

The first major historical expression of success in this direction was apparent in Protestantism, and it is natural that one should automatically think of the Puritan ethic when discussing the issue of emotional control, as the success obtained by the Puritan 'saints' in suppressing all manifestation of unwanted emotion was formidable indeed. But it would be wrong to envisage such control in the purely negative form of suppression, for once this power had been attained then some controlled expression also increasingly became possible. In fact, not even the Puritan ethic prohibited the expression of emotion upon all occasions.

To stress the crucial part played by Puritanism in the evolution of modern hedonism may seem, at first sight, to be somewhat strange, and yet as far as the emergence of sentimental hedonism is concerned, Protestant religion, and especially that harsh and rigorous form of it which is known as Puritanism, must be recognized as the primary source. This is precisely because as a movement it adopted a position of such outright hostility to the 'natural' expression of emotion, and consequently helped to bring about just that split between feeling and action which hedonism requires. In addition to this, however, it also contributed greatly to the development of an individualistic ability to manipulate the meaning of objects and events, and hence towards the self-determination of emotional experience.

Religion is the most important of all areas of culture as far as the evolution of an ability to cultivate emotion is concerned. This is because such intensely fateful issues as one's state of sin (or grace) and one's hopes for salvation, together with the extremely powerful emotions which they can arouse, are coupled with the necessity of presenting invisible divine agencies by means of symbols. Naturally enough, the potential to arouse these feelings then becomes attached to the symbols themselves. This is in marked contrast to the powerful emotions aroused by such real events as a battle or a shipwreck, where the emotion arises from the experienced reality rather than a 'symbol'. In fact, as has been noted, religious symbols can serve to counteract such experientially induced emotion, just as, more significantly, they can serve to induce emotion in the absence of any discernible environmental stimulus.

That individualism was carried to unprecedented lengths in Protestantism is particularly significant in relation to this last point, for while in Roman Catholicism symbols also served to arouse (and allay) powerful emotions, their control was kept firmly in the hands of the priesthood, and hence situationally located in communal ritual. In Protestantism, by contrast, not only was there no one to act as mediator between the individual and the divine, but both 'magical' ritual and the use of idols was proscribed. The consequence of this was that those symbols which did serve to arouse religious emotion were of an abstract and general character. Death and mortality, for example, which were commonly regarded as evidence of man's inherently sinful state, could be represented by a very wide range of objects and events in the world, from coffins, graves, churchyards and yew trees, to sickness, worms and church bells, with any one of these acting as the 'trigger' for emotional experience. Such a situation clearly gives the individual considerable scope to decide when and where he will choose to undergo a particular emotion. It is, however, the religious beliefs which ultimately underpin these emotions, and hence, as long as the beliefs are accepted as true, then this ability to manipulate symbolically the occasion of their expression is of comparatively little relevance; but when such beliefs begin to atrophy, a significant change can occur.

Clearly belief dependent emotionality is a rather different phenomenon from that which is event dependent in so far as the potentiality exists for the individual to gain control over his own emotions without first having to obtain mastery over the real world. As long as the validity of the beliefs is taken for granted, however, there is little obvious difference between an individual's terror on encountering the devil and that on meeting a lion. But the waning of conviction inevitably affects the intensity of the emotion even if it still occurs: of more significance, however, is its probable effect on the emotion's genuineness. For as doubts about the truth of beliefs crystallize, the likely initial consequence is to remove the basis for the emotion rather than the emotion itself, which has, over time, become habitually associated with the given symbols. There thus remains a tendency for it to occur even though the individual knows that it is not entirely necessary. It is under these circumstances that the real possibility of gaining pleasure from emotion can arise.

This is best illustrated by reference to the fate of beliefs concerning hell, eternal damnation, the Devil and sin in the late seventeenth and early eighteenth centuries when they gradually faded in the face of the scepticism and optimistic rationalism of the Enlightenment. As they did not disappear altogether, the powerful emotional resonances which such beliefs created remained in the minds of many, and their conventional symbols became employed as a means of gaining emotional pleasure. Thus out of a background of real religious terror there developed such artistic genres as graveyard poetry and the Gothic novel both of which catered for the 'thrill' of being frightened.

In order, therefore, to possess that degree of emotional self-determination which permits emotions to be employed to secure pleasure, it is necessary for individuals to attain that level of self-consciousness which permits the 'willing suspension of disbelief';[13] disbelief robs symbols of their automatic power, while the suspension of such an attitude restores it, but only to the extent to which one wishes that to be the case. Hence through the process of manipulating belief, and thus granting or denying symbols their power, an individual can successfully adjust the nature and intensity of his emotional experience; something which requires a skilful use of the faculty of imagination.

While it is possible to employ the power of imagination to summon up physical sensations such as the feel of the sun on one's back or the taste of grapes, this is an exceptionally difficult exercise. To that extent it is almost impossible to gain real pleasure from directly imagined sensations. By contrast, it is comparatively easy (at least for modern man) to use imagination to conjure up realistic images of situations or events which produce an emotion in the imaginer; an emotion, which, if controlled, can itself supply all the stimulation necessary for a pleasurable experience. This is an ability which it is all too easy to take for granted, forgetting that it is a comparatively recent addition to mankind's repertoire of experiences.[14]

Modern hedonism presents all individuals with the possibility of being their own despot, exercising total control over the stimuli they experience, and hence the pleasure they receive. Unlike traditional hedonism, however, this is not gained solely, or even primarily, through the manipulation of objects and events in the world, but through a degree of control over their meaning. In addition, the modern hedonist possesses the very special power to conjure up stimuli in the absence of any externally generated sensations. This control is achieved through the power of imagination, and provides infinitely greater possibilities for the maximization of pleasurable experiences than was available under traditional, realistic hedonism to even the most powerful of potentates. This derives not merely from the fact that there are virtually no restrictions upon the faculty of imagination, but also from the fact that it is completely within the hedonist's own control. It is this highly rationalized form of self-illusory hedonism which characterizes modern pleasure-seeking.

NOTES

1 While it is true that some emotions may be more commonly accompanied by feelings of pleasantness than others, this does not mean that such emotions as anger, fear or sorrow cannot become sources of pleasure under the right conditions.

2 It is pertinent in this context to note the common stem to the words 'emotion' and 'motion'.

3 Strictly speaking, all aspects of emotional arousal could be considered to constitute 'behaviour', even that which is subcutaneous. In the context of social interaction, however, overt manifestations of arousal clearly have a far greater significance than any covert indications.

4 Where the suppression of emotion extends beyond the control of overt actions to embrace all subjective dimensions of the experience, as it does in the stoical ideal of *apathea* or emotionlessness, it would be more correct to speak of emotional extinction than control. This path clearly leads directly away from modern hedonism.

5 In one sense, the power to inhibit the expression of emotion logically implies an ability to express it when

desired. All that would seem to be required is the cessation of the effort of suppression. Individuals do not normally live, however, in a state of permanent suppressed emotional excitement, and hence there will be occasions when something more than the abandonment of inhibition will be necessary. In addition, full emotional control implies the ability to choose the emotion desired.

6 As long as emotion remains locked into a behavioural complex which includes motor activity of an extravagant and dramatic kind, there can be little hope of gaining control of the subjective dimension of that experience. All that one can do is to try and adjust one's real experiences, or alternatively try to shut out information about the world, as children often do by hiding their eyes (or covering their ears) when frightened.

7 Owen Barfield, *History in English Words*, new edn (London: Faber & Faber, 1954), pp. 169–170.

8 Contemporary linguistic usage suggests the continuing influence of the idea that emotion is an attribute of situations rather than of persons as references to 'sad' or 'happy' occasions imply. Usually, however, this is interpreted to mean that individuals feel themselves under an obligation to experience the required emotion, and hence is not really comparable with situations in which there is no voluntaristic ingredient present.

9 Max Weber, *The Sociology of Religion*, trans. Ephraim Fischoff (London: Methuen, 1965), chapters 2, 3, *passim*.

10 Those thinkers and artists who were part of the Romantic Movement in Europe in the eighteenth and nineteenth centuries can be said to have been well to the fore in accomplishing this process, with the 'self' and 'genius' twin foci of their 'spiritualizing' endeavours.

11 Barfield, *History in English Words*, pp. 165–169.

12 Ibid., p. 165.

13 The phrase is, of course, Coleridge's. The full quote is 'That willing suspension of disbelief for the moment, /which constitutes poetic faith' (see Coleridge, *Biographia Literia, or Biographical Sketches of my Literary Life*, 2 vols, first edn repr. (London: Rest Fenner, 1817), vol. 2, p. 6).

14 It might be objected that the ancient profession of acting had always presupposed possession of the ability to employ imagination to successfully manifest a chosen emotion at will. While this seems a plausible claim, it would be anachronistic to assume that acting, either in Graeco-Roman, Renaissance or Elizabethan times, necessarily involved the ability to actually experience chosen emotions at will. For the fact that a character was undergoing a given emotion was typically conveyed by stylized gestures and expressions (in addition to utterances), and it was not necessary to the performance for either the actor to actually experience, or for the audience to believe that he really was experiencing, that emotion for the portrayal to be accepted. Thus, although acting has always encompassed imitation and mimickry, it is only in modern times that it has also typically embraced the ability to voluntarily 'become' another person in the sense of taking over their experience of reality; something which is perhaps most associated with the ideas of Konstantin Stanislavski.

5

Industrialism, Consumerism and Power

Zygmunt Bauman

I

For all we know of past societies, the disciplinary, capillary power-through-surveillance, which Foucault considers the mark of the modern era, could not be an invention of the seventeenth century; it could be only its discovery. A strong case can be made for the assertion that power of exactly the kind described by Foucault was the major tool of social control, integration, and reproduction of society well before the dawn of modernity.

Indeed, for most of their history people lived under close and permanent scrutiny of the collectivities they were born a part of. Village communities were not prominent for their tolerance and normative ambivalence; nor were, for this matter, craftsmen guilds or noblemen courts. The spate of historical studies triggered off by Phillippe Ariès has convincingly demonstrated the absence of borderlines between private and public lives which characterised pre-modern communities; the openness of family homes, the street as the principal sites of social intercourse and child education, the absence of enclosed territories reserved for the invited or otherwise selected persons at selected times, all contributed to the high probability of being seen, gazed at, evaluated and censured. Being seen, or being aware (even if only subliminally) of the likelihood of being seen, was a constant factor of life and certainly a crucial determinant of the stable regularity of conduct. Universality of surveillance was reflected in the universality of enforced uniformity. Indeed, if the kind of diffuse communal power responsible for the behavioural routine of the members of the pre-modern community differed from disciplinary power typical of industrial society, it did so above all by its ubiquity and comprehensiveness no later power would be able to match.

One can say that the social reproduction of the mode of life which constituted the pre-modern (traditional, pre-industrial, etc.) society was accomplished by a kind of power which in most important respects answered the description reserved by Foucault for the modern society only. Did, therefore, the nature of power remain unchanged? Did anything at all happen in this field in the seventeenth century? If something did, what?

Three things did.

First, communal control, however wide-ranging and uncompromising, was exercised in a diffused way, matter-of-factly, with the participation of few if any explicitly specialised agencies, and hence unobtrusively. In Heidegger's words, we become aware what the world is when something goes wrong. As long as the capacity of communal surveillance roughly corresponded with the prerequisites of the smooth reproduction of social order – the awareness of the presence and exact nature of the means with which this balance was achieved had little chance to appear. Disciplinary power, though present and active, was not noticed; still less was it problematised, forged into an issue, a task, a matter of conscious design. It dissolved in the totality of daily routine, as an integral attribute of the life-world distinguishable only through theoretical analysis armed with retrospective wisdom.

The idea of secular power, under the circumstances, could be easily monopolised by its 'sovereign' variety: the power of the King, the Noble, the Bishop to appropriate a part of other people's surplus in the form of a levy, a tax, or a tithe. Unlike the disciplinary power, the sovereign power was visible. It arrived periodically from outside the realm of daily life to execute its demands, poorly incorporated in the rhythm of quotidianity. Even given this handicap, however, the sovereign power would remain hidden, 'unseen', if not for the periodic disturbance of the habitual rules of the game. Time and again, more and more often the closer the dawn of the modern age, people had to defend their *Rechtsgewohnenheiten* against the rising appetites of their lords and princes. It was this contest which brought the phenomenon of power into relief and made its articulation, as a concept, possible.

No wonder the attention of the enlightened opinion of the time was focused undivided on the 'sovereign power'. It was then that the idea of power as an object which can be appropriated and expropriated, transferred from King to people, etc., and mostly as a negative faculty, i.e. the ability to prevent others from doing what they would wish to do, or to force them to do what they would not wish, was conceptualised in the emerging political science, where it was destined to remain virtually unchanged for two centuries or more.

From the seventeenth century on, however, the most pronounced and threatening disturbances of the social order evolved around the regulation and reproduction of quotidianity – functions with which the 'sovereign power' was never intended, nor armed, to cope. The power crisis resulted from inadequacy of the communal framework for the deployment of 'disciplinary power'. For once local powers were progressively eroded with the growth of the centralised state. More importantly, demographic expansion, which the existing mode of production could not absorb, resulted in a large, excessive, economically redundant population, which exploded the finite potential of communal tutorship. This process reached its peak, throughout Western Europe, towards the end of the eighteenth century. It was perceived from the contemporary perspective as, first and foremost, a problem of law and order: the emergence of dangerous classes, rabble, riff-raff, the crowd, the mob – uncontrolled, unpredictable, subject to no rules, supervised by none of the existing authorities. Legally and politically, these proto-masses were defined as vagrants or vagabonds; this definition grasped the roots of the threat the 'dangerous classes' presented: they were unattached, belonged nowhere, were mobile, and hence they escaped the pervasive surveillance – thus far exercised only within, and by, the local community. The 'dangerous classes' were dangerous because they were the first to live under conditions where probability of 'being seen' was low. The communal principle of the deployment of 'disciplinary power' had an in-built limit of application and could not be extended to absorb the new phenomenon (though the legislators for many decades tried in vain to achieve just that, decreeing for instance a forceful return of the vagabonds to their native parishes and declaring vagabondage itself a crime). The sovereign power, armed only with coercive agencies specialised in preventing resistance against the expropriation of surplus, was organically unfit for positive action, i.e. generation of a comprehensive totality of behaviour. The result was a power void.

This was to be filled by a two-fold process of redeployment of disciplinary power and a radical extension of the scope of the sovereign one.

What did happen in the seventeenth century, in other words, was not the birth of disciplinary power, as Foucault's analysis would imply, but 'problematisation' of this power, both in political and social practices and in their intellectual reflections. Formerly an aspect of communal life blended fully into the totality of life process, it had now been brought into the field of vision as a method in search of its institutional tools. What used to be accomplished matter-of-factly without ever surfacing at the level of conscious political practice had now become a matter of conscious design, planning, and legislation. In its institutional redeployment, the state had to play the decisive role.

Second, the transfer of disciplinary power from its former communal setting to deliberately constructed institutions brought to life and supported (or at least launched) by the state, meant the separation of bodily control from the other aspects of the traditional mode of life with which it was previously blended. Bodily drill which disciplinary power necessarily entailed was now not only seen as such and perceived as a consciously constructed condition, but also as one uncompensated for by, say, reciprocity of rights and duties or, in general, communal, parish, or manorial tutelage. This circumstance made salient the externality of discipline, and served as an epistemological premiss for the articulation of the autonomy/heteronomy opposition. It also exposed the essential asymmetricality of disciplinary power and thus cast it as an object of contest and as a negative aspect of the human condition which is not naturally balanced by tied benefits.

Third, from the seventeenth century on, disciplinary power moved into the space of inter-class relations. For the first time reproduction of the social order at the grass-roots level became the concern, and then responsibility, of the dominant classes. In the traditional society, they rarely, if ever, interfered with the daily occupations of their subjects. In particular, they did not consider the production of surplus their responsibility, fully satisfied as they were with the enforcement of its repartition. They left folk culture to its own resources and its own developmental logic; when they noticed differences between their own way of life and those of the lower ranks, they would hardly consider it as a problem calling for action. Interaction between coexisting class cultures, located on different levels of the social hierarchy, was not one of *Kulturkampf*, cultural crusade, civilising, educating, humanising, or whatever the name capturing the gist of later one-way action coinciding with the modern era. In the traditional society, the dominant class ruled, but they did not teach or proselytise.

Now the situation changed radically. The dominant classes stepped into the role of collective teacher. The subordinate classes, 'the people', were cast in the roles of pupils, and it became the duties of the dominant classes not just to punish their mischiefs and so prevent departures from the 'natural' way of life proper for them; or rather to prompt 'the people' to learn and to observe the one way of life considered properly human or 'civilised'.

The traditional imagery of the continuous 'chain of being', in which for everybody there is an inter-connected, but nevertheless separate and different, type, was now gradually replaced by another, dichotomous and dynamic image of one truly human, or civilised, or rational pattern of life. It confronted many imperfect forms of life which it was called to extirpate, amend or banish: the ignorant, the superstitious, the emotion-ridden, the beastly. Alternative ways of life, and above all the popular cultures, were now redefined as cruel, crude, beastly, inhuman, a target for incessant and radical proselytising crusade.

'Being human', instead of remaining a natural condition enjoyed by all though in many alternative forms, became now a skill to be learned, an end to a tortuous effort, which everyone had the duty to undertake, but few only were able to accomplish unassisted. The necessary complement of the dichotomous imagery and proselytising practice was, therefore, a 'tutelage complex' (Donzelot's term) which simultaneously interpreted society in the light of a school metaphor, and asserted the permanence of asymmetrical school roles.

The result was a total re-definition of the relationship between the top and lower regions of the social hierarchy. If the dominant function of the traditional hierarchy was, so to speak, to assure the upward flow of the surplus resulting from an essentially autonomous productive activity, the function of the new hierarchy was more than anything else to assure the reproduction of a form of life compatible with the continuation of hierarchical order. This 'disciplinary power' was accordingly deployed in the service of this function.

What has happened from the seventeenth century on, in other words, was the shift of disciplinary power from the area of community reproduction into that of the reproduction of class hierarchy.

II

The reproduction of social hierarchy (perceived, at that time as at any other time, as the preservation of social order) was first and most directly threatened by the appearance of 'dangerous classes' exempt for the time being from any institutionalised form of surveillance and hence bodily control. It was to the refractory and scattered gangs of vagabonds and itinerant poor that the state-operated disciplinary power was first applied. In the effort to contain the overflow of social structure, to erect artificial dams and dykes where the ditches in the care of parishes and manors proved too shallow to hold the gathering waves – new, consciously designed techniques of supervision were developed. These developments took place simultaneously in a number of areas, apparently distant from each other and unconnected; and yet they showed amazing similarity of means and purpose. There was, as Michel Foucault convincingly demonstrated, an essential unity of method and effects between hospitals, mental clinics, prisons, poorhouses, workhouses, military barracks which all appeared on a grand, unprecendented scale approximately at the same time in all areas of Western Europe. All of them were means of social control; all of them exercised control through deployment of surveillance power; all of them made the deployment of surveillance viable by confining their charges to a limited, easily supervised space – by transforming the charges into inmates (consider, for instance, the nearly total abolition of outdoor assistance to the poor); and they used surveillance to regiment the totality of inmates' conduct in its most minute detail, to impose well-nigh complete heteronomy of the body. In one sense the massive total institutions were merely about reconstructing the conditions of deploying disciplinary power, practised on a large scale by local communities, for the categories of population which evaded communal control. In another sense, however, the fact that disciplinary power in the total institutions was now a matter of a functionally differentiated design, that it was prised off from other customarily connected dimensions of social interaction, and that it entailed a permanent and unremediable asymmetricality of power positions, made the newly developed technique of bodily control an entirely new phenomenon. For once, this technique had a potential of almost infinite extension which its diffuse communal proto-form demonstrably did not.

The category of disciplinary power, and restructuring of social control at the dawn of the modern era which this category helps to bring into the open, can therefore serve as a pivot in the new model of modern (industrial) society and its origins, which Foucault's analysis suggests: a model called to replace the understanding of modernity as, above all, a novel attitudinal syndrome formed first among the extant or aspiring elites and then trickling down, through education, ideological pressure and a complex of positive and negative sanctions, to the rest of the population, particularly to the pre-industrial poor, now to be transformed into factory labour. The new model, built around the redeployment of disciplinary power and the rearrangement of social control in general, substitutes an essentially political concern with public order and the task of containing 'dangerous classes' for the Puritan concern with earthly success as the symptom of spiritual salvation. This model also assigns centrality to the political, rather than economic, interests of the elites, at least in the triggering off of the chain of transformations which led to the establishment of the 'mature' industrial society.

The story of the first factories can be then reread as just one chapter, or rather one theme, of a much larger history of the containment of the new poor, uncontrolled and hence redefined as dangerous classes. The possibility (indeed, the plausibility) of such an interpretation has been overlooked presumably because of the different structure of research relevancies induced by the theories of economy-led or attitude-led changes.

There is more than sufficient evidence (I have indicated some, related mostly to Britain, in my *Memories of Class*), that the first factories were perceived by the contemporaries as another variety of poor- or workhouses, and their owners as sui-generis agents of authorities, making the communal task of the care for material and spiritual welfare of the poor their responsibility, and thereby simultaneously relieving the local taxpayer from an excessive financial burden and promising to secure the sought-after control over the bodies of potential rebels as well as morally regenerate their souls. Labourers of the first factories (in most cases women and children) were more often than not delivered to the willing entrepreneurs direct from the parish-supported poorhouses, and kept there by force by the same guardians of order whose task it was to chase and capture runaway inmates of workhouses. Gains in productivity (indeed, the productivity activity itself) made possible thanks to subjecting a large number of labourers to the uniform rhythm of bodily action were, in public view at least, secondary to the direct gain in the efficiency of control achieved thanks to the close supervision and minute regimentation of life-processes in the factories and attached dormitories. Attentive reading of the factory rules of this period shows an extremely tenuous link between most of the prohibitions and stipulations, and demands of economic effectivity. Pernickety, often contradictory, rules, prohibiting whistling or singing, setting penalties for having dirty hands but also for washing them, etc., make sense only when seen as an aid to drilling factory labourers into complete submission to their supervisors, and to extirpating the last shreds of the autonomy of the labourer's body.

In other words, the emergence of the industrial system followed a political logic, rather than the logic of technology. Rather than training the pre-industrial poor into the new habits of regular work required by new technology, entrepeneurs were prompted to accept technological contraptions because the use of them had now become possible thanks to the availability of large numbers of well-drilled potential factory hands and of the means to exercise an effective control over their bodies.

The factory hands meeting these conditions most fully came from the ranks of the homeless and the destitute already processed through the exacting regimes of poorhouses. By the nature of their drill they were fit to operate only the simplest technology, which called for no autonomy of judgement and no individual discretion. Such technology was quite early available for the least skilled aspects of production.

Long into the industrial era such technology was not available to the craftsman type of work; man-based, rather than machine-based, skills for many decades remained an indispensable complement of the industrial process. Factory regime was made possible thanks to the effectivity of surveillance power applied to an unskilled labour force crowded together inside mill enclosures; but factory type of production could not function unless the co-operation of skilled workers was enlisted and assured on a regular basis.

Such skilled workers were not to be found within the walls of poorhouses. They could be drawn only from the artisan workshops, where the tradition of autonomy and disciplinary power, both tightly wrapped into a dense tissue of multidimensional reciprocity, was only relatively slightly dented by social translocations elsewhere. Set against this tradition, bare bodily control and unconcealed asymmetricality of power identified with the new factory system must have appeared to the craftsmen strange and repulsive. Expectedly, the advances of the factory system met with the fierce resistance of the craftsmen. This resistance against factory order constituted the major plank of all workers' protest movements, conspiracies, and their ideological platforms throughout the first half of the nineteenth century (from a much later and distorted perspective, these movements will go down in interpreted history as the utopian, petty-bourgeois and otherwise immature, pre-historical stage of the 'true' working-class movement).

Partly because of the freshness of their autonomous tradition and the resulting strength of their resistance, partly because of the nature of the functions they were called to perform, the skilled workers could not be fastened to the factory system with the same bolts of naked disciplinary power as their unskilled predecessors.

The first factories applied, therefore, double standards. Unadorned disciplinary power was deemed both appropriate and sufficient for exacting unskilled labour. With the small, but indispensable contingent of skilled workers the entrepreneurs had to settle for less. For several decades, until 'skilled' machines sapped the indispensability of skilled labour, the skilled part of the factory labour retained most of the autonomy and self-rule typical of the craftsman workshop. In a great number of cases the craftsmen were drawn into the productive process in the role of subcontractors, controlled more by the rules of market exchange than by any direct supervision of their work.

Compliance of such workers with the demands of factory production had to be bargained for. It had to be bought tit for tat, more money for more discipline. Within the frame of reference of factory labour, skilled craftsmen appeared as a 'labour aristocracy'. What set them apart from the rest of labour was not, however, merely the difference of income. More than anything else, the peculiarity of their position was grounded on their freedom, for the time being, from the most obtrusive excesses of disciplinary power. Their gradual surrender of one aspect of the craftsman's tradition – self-management – had to be obtained through the boosting of another aspect: market orientation and self-interest.

In other words, the conflict over *control,* triggered by the attempt to extend over the skilled part of factory labour the disciplinary forces developed in dealing with the unskilled part (or the fear of such an extension), was displaced and shifted into the sphere of surplus distribution. Legitimation of the new structure of power and control was obtained through the delegitimation of the division of the surplus – the one thing which the 'sovereign power' in pre-industrial society sought to keep clear from contest.

This fateful process of displacement I propose to call 'economisation of power conflict'. From the perspective of its ultimate consequences, it may be depicted as the trade-off between the acceptance of the stable asymmetry of power and heteronomy inside the productive activity, and the rendering of the share in surplus open to contest. Money becomes a makeshift power substituted for the one surrendered in the sphere of production; while the experience of unfreedom generated by the conditions in the workplace is re-projected upon the universe of commodities. Correspondingly, the search for freedom is reinterpreted as the effort to satisfy consumer needs through appropriation of marketable goods.

As the latter is only a displaced effect of the experience independent from, and unaffected by, its progress – it is unlikely that the emancipatory urge, originated and perpetually refuelled by the heteronomy of productive activity, will be ever quenched by a success, however spectacular, of its surrogate form. It seems, on the other hand, that the unsatisfied need of autonomy puts constant pressure on the consumer urge, as successively higher levels of consumption become disqualified and discredited for not bringing the hoped-for alleviation of stress. This stress is a virtual time-bomb lodged in industrial society from its very inception; and the inescapable, though unanticipated, effect of disciplinary power deployed to control the producer's body. It is perhaps the major fly-wheel of industrial society and the source of its irresistible dynamism: it makes industrial society the only one in human history which experiences stability as crisis rather than bliss. By the same token, it is the major structural fault generative of an ever increasing scale of contradictions which ultimately this kind of society is incapable of solving.

III

The most powerful minds of the early modern era (Adam Smith, Jeremy Bentham, James and John Stuart Mill, W. Nassau Senior among others) never counted 'economic growth' among the attributes of industrial economy. They welcomed advancing capitalism as the way to improve the general welfare of society. They praised free trade for its tendency to decrease the toil necessary for decent life and bring into balance the needs and the productive capacity of man. They would not trust the market with accomplishing this task alone, though; if production, in their view, was ruled by natural laws, and all interference with its course could bring more harm than benefit (a fair reflection of the 'naturalisation' of productive relations perpetrated by the growing heteronomy of producers), distribution called for some corrective interference by the state (again a fair anticipation of the needs arising from the 'opening up' of the division of surplus to non-economic forces). This corrective interference, as they envisaged it, would be confined, however, within the total volume of goods the 'natural laws' would make possible to produce. This volume, they believed, was finite and fairly limited. Social welfare would arrive, in their view, within the 'stationary state' of the economy, which was just round the corner.

It was not to be as they believed it would. Incessant economic growth proved to be if not the undetachable attribute of industrial society, then certainly a *sine qua non* condition of its survival. Being the condition of its survival, but not a feature guaranteed by the inner logic of industrial economy, economic growth – as a postulate more than as the reality of economic life – turned gradually into a major factor in shaping the system, its contradictions, and the way of coping with contradictions, of a society based on an industrial mode of production.

Ever growing consumer pressure renders the balancing of industrial output impossible; it can be satisfied only on condition of an equally intense increase in industrial output. With satisfaction being the function of growing appropriation of consumer goods, production must grow – only to keep satisfaction of its present level and prevent the feeling of 'outraged justice' (Barrington Moore Jr.) from emerging. Moreover, virtual demand for consumer goods tends to outstrip what economists call the 'effective demand' (demand backed by monetary resources and hence able to clear the existing market supply simply as an effect of buyers' decisions); the gap between the two cannot be bridged by 'natural laws' of the economy, and hence becomes a political issue, which leads to the growing functionality of state intervention within the system of industrial society. Ultimately, the state is forced to switch its attention from 'correcting the distribution' to stimulating, if not directing, the production of surplus itself.

In the long run, this new role of the state entails the likelihood of a 'legitimation crisis' (Habermas). Perpetuation of economic growth becomes now the main measure of a state's performance, by which the governments and parties are judged, win, or suffer defeat. Given, however, that gains in the share of surplus are, for industrial society, the only token with which the asymmetry of power can be retained and compliance bought – the state must go about the task of continuing economic growth, and for this purpose apply means which, at least in the longer run, must undermine the chances of the task being performed. There seems to be an insoluble contradiction between the need to compensate for the lasting asymmetry of power with a visibility of constantly improving welfare defined in consumption terms, and the ability of the productive process, operated by disciplinary power, to turn out a surplus large enough to balance the intensity of tension this power continues to generate. The final outcome of the original sin of economisation of the essentially political conflict is politicisation of economics.

The early economisation of power conflict gave the initial push to the reorientation of life interests from the reproduction of subsistence in its traditional form to the improvement of living standards, i.e. to the increasing consumption. This new and unprecedented attitude, gestated and continuously reinforced by the blocked power relations in the productive sphere, was destined to bring in the end an enormous pressure upon production which the system of power responsible for its original success would not be able to sustain. First, however, the reorientation towards consumption and life improvement in general opened up, before industry, unheard of prospects of expansion. It did not take long for merchandising firms to start to compete with each other for the greater share of the generalised 'readiness to buy'. Without visible help from the experts in psychology, the competitors soon realised that with the field of productive expression tightly fenced and unencroachable, the potential customers seek in the commodity market the means to mend their somewhat incomplete or scarred identity and to make up for their injured self-dignity. The aggressive advertising, which took off towards the end of the nineteenth century, set about linking this urge with specific goods the market had to offer; giving a tangible and purchasable content to compensatory dreams. This concentrated effort helped to harness the energy of the power conflict, already channelled into the contest over distribution, to the commodity market. Consumerism was born as a twice-removed offshoot of the frustrated resistance against disciplinary power which penetrated, and finally conquered the field of productive activity.

Disciplinary power, as we remember, was first and foremost about bodily control. It was the human body which for the first time in history was made, on such a massive scale, an object of drill and regimentation. Later consumerism was a product of failed resistance to such a drill and regimentation. But what was negated could not but determine the substance and the form of its negation.

The origin left three salient birthmarks on its offspring.

First, disciplinary power in the form in which it was redeployed from the seventeenth century on, produced the body as an object of conscious attention, as a receptacle of potential powers which, to materialise, must be selectively developed and properly channelled – as, in short, a thing incomplete in itself, under-determined, needing cultivation towards an ideal which would not be attained without a conscious and consistent effort. This power produced a body consisting of two parts of unequal value and different relation to the ideal; a part to be suppressed, tamed, hidden, and preferably eradicated, and a part to be tended to, cared for, brought into full fruition.

Much in the same way as Rod Steiger's *pawnbroker* manifested his liberation from concentration camp slavery in *choosing* the non-interference with evil into which he was *forced* by his camp guards, consumerism as a compensatory reaction to *heteronomous* bodily drill selects an *autonomous* bodily drill as its principal target. Consumerism is not about the emancipation of the body from control; it is about the joy of controlling the body of one's own will, with the help of sophisticated products of technology which offer all the visibility of the formidable power of one's controlling agency. The body is subjected to, say, the uncompromising drill of slimming or jogging discipline, applied against the body's natural drives and wants, but this time administered by the body's own master. Consumerist freedom drags behind it a huge shadow of its slave origin. To satisfy itself it does not need to break the manacles. It satisfies itself by locking the manacles with its own key.

Second, the bodily training associated with the disciplinary power in its production setting, failed to develop some attributes of the body which under conditions of consumerism acquire particular significance. Growing consumption is a novel task, to which the body must be prepared as it was prepared in the past to accomplish tasks. It must be *made fit* to absorb an ever growing number of sensations the commodities offer or promise. Once again the body is to be trained, but this time its capacity as a 'receptacle of sensations' is the training target. It is a condition *sine qua non* of consumerism that the body becomes richer, and life is fuller, depending on the ubiquity and comprehensiveness of its training. Often there is little to distinguish between developing a capacity to absorb new (or more) sensations, and the more traditional forms of bodily drill. The objective, for instance, is not to train the capacity to enjoy music, but to make the body capable to withstand a permanent exposure to the flow of sometimes deafening, sometimes barely discernible, sounds. The objective is not to teach appreciation of dramatic plot, but to drill the body into a need to be exposed again and again to unfamiliar and familiar plots alike, in slow motion or reverse order – a need into which the contraption of a video-recorder fits like hand into glove. The objective is not to develop the suppressed capacity of erotic sensuality, but to drill a body into the capacity of going over and over again the codified routine of sexual acts. Orgasm is just one of the many promised prizes, calculated to prompt willing embracement of 'do-it-yourself' or 'teach-yourself' drill. They retain their attraction in as far as the elusive bodily sensation is translated into 'objectified' indices of the observable routine, and hence the relevance is shifted from the outcome to the bodily drill itself. On the whole, it is a condition of consumerism that the body is trained into a capacity to will and absorb more marketable goods, and that routines are instilled, through a self-inflicted drill, which make possible just that.

Third, the body, this object of loving care, has retained its central defining features as articulated and imputed during the dawn of disciplinary power: it has remained first and foremost the paramount source of evil and suffering, and as such it cannot be left unattended: the care of the body as the crucial time- and money-consuming activity of the denizens of consumer society is an uneasy, poorly balanced, mixture of love and horror (which renders the body not unlike the divine objects of religious fervour of the past). As before, the body is charged with the responsibility for success and failure in earthly endeavours, and the urge 'to do something about my life' is most eagerly translated into a precept 'to do something about my body'. For many years now the two kinds of books most likely to win a place on the bestseller list in the USA have been cookbooks, including the most eerie and exotic ones (the idea that the body could miss any of the experiences available to other bodies is terrifying), and diet books, prohibiting the consumption of practically everything the first category of books recommends. Cookbooks offer the images of things which diet books order to deprive the body of: the body must remain an object of constraint and drill however much it consumes, surrendering to the barrage of marketing allurements; and so there is a need of a constant supply of new, applied or imagined, instruments of torture and punishment.

BIBLIOGRAPHY

Barthes, R. (1967) *Système de la mode*, Paris.

Bauman, Z. (1982) *Memories of Class*, London: Routledge & Kegan Paul.

Donzelot, J. (1979) *The Policing of Families*, London: Hutchinson.

Elias, N. (1978) *The Civilizing Process*, Oxford: Basil Blackwell.

Foucault, M. (1980) *Power/Knowledge: Selected Interviews and Other Writings 1972–77*, edited by C. Gordon, Brighton: Harvester Press.

Habermas, J. (1976) *Legitimation Crisis*, London: Heinemann.

Muchembled, R. (1978) *Culture populaire et culture des élites dans la France moderne (XV–XVIII siècles)*, Paris: Flammarion.

Yeo, E. and Yeo, S. (1981) *Popular Culture and Class Conflict: Explorations in the History of Labour and Leisure*, Brighton: Harvester Press.

6

The Governance of Consumption

Sumptuary Laws and Shifting Forms of Regulation

Alan Hunt

The overwhelming response of commentators throughout the long life of sumptuary regulation has been to view sumptuary laws as doomed to failure. It is insisted that such laws were widely evaded and that little effort was made to enforce them. Sumptuary law has been regarded as an immature or unsophisticated stage of legal development. I will challenge this dismissive view. It seems improbable that a form of social regulation, bearing highly charged economic, political and cultural manifestations and found in virtually all major civilizations and which persisted until, at the very least, the dawn of modernity, should be regarded as a failure such that it is not worth studying, save perhaps for antiquarian interest.

To challenge this overwhelming and pervasive dismissal of the significance of sumptuary law creates the risk that I might fall into the trap of overcompensating and of lapsing into a simple reversal. It is not the intention to argue that sumptuary governance had some intrinsic importance or that it had some profound impact on societies that experienced it. My project is more modest: it is to advance the contention that sumptuary law was significantly associated with the major stages in the rise of modernity. The study of sumptuary law yields significant insight into the characteristics of the changing forms of governance that are associated with the passage from premodernity to modernity. It will be further argued that, far from disappearing without trace, projects of sumptuary rule left their imprint on modernity. In particular, I will argue that sumptuary laws have lived on in the ubiquitous quest for moral regulation that found expression in the social purity and prohibition movements of the nineteenth and early twentieth centuries and is today alive and well in the contemporary purity movements and projects of moral regulation.

The argument that I develop is that sumptuary law, while present in many different societies, can best be viewed as existing on the borders of modernity; it can be understood as a first response to modernity. It is this historical location astride an old world and a new world that leads me to focus attention on the way in which sumptuary projects face in two directions at the same time. Sumptuary law was a response to at least three of the most distinctive features of modernity. These I take to be: urbanization, the emergence of class as the pervasive form of social relations and the construction of gender relations in these 'new' conditions. As such, sumptuary law connected 'backwards' to the medieval world by building on the critique of luxury and a general moralization of social relations embodied in the primacy of religious discourses, but it also points 'forward' to a shift of preoccupation with the 'economy', and thus with the role of the state, which involves a radically different way of thinking and understanding the way in which social relations are constructed and acted upon by the multiple projects of governance. The sumptuary ethic was strongly implicated in some of the most crucial phases of the advance of urbanization and of the transition from mercantile to manufacturing capitalism. With respect to the experiential impact of urbanization, it brought with it the problem of how it is possible to live in close physical proximity with others and sustain relations of mutual dependence with strangers. I will advance and defend a 'recognizability thesis' which views sumptuary law as one significant component in the solution to this potentially traumatic dilemma.

The existence of sumptuary law at the gateway to modernity also invites an inquiry that focuses on the attempt to understand the anxieties and tensions that accompanied the experience of modernity. Sumptuary laws speak to us, usually indirectly, about the impact of plagues, famines and crises of all kinds that were such a significant feature of late medieval and early modern life. These concerns went hand in hand with the ubiquitous preoccupation with a 'world turned upside-down' which might well be treated as the most pervasive social-psychological response to the onset of modernity. It is not just the perceived experiential anxieties that should be of concern; these provide a starting point, a point of articulation (rather than of causation), but this methodological strategy requires exploration of the deeper level of the structural tensions, the forms of class, gender and status relations that constitute the bedrock on which sumptuary legislation rested and of which it can best be regarded as a barometer of both popular and governmental anxiety.

Sumptuary laws were a form of symbolic politics. This draws attention to the importance of symbolic struggles that surround the recognition of appropriateness of dress, appearance or demeanour, captured in the rhetorical question 'Who does s/he think s/he is?' It is precisely the capacity to assert and deny social recognition that lies at the heart of all those forms of regulation, including sumptuary laws, that have addressed this relation between being and seeming, with all its capacity for misreading, misrepresentation, dissimulation and pretence.

The character of sumptuary projects changed over time. They provide a case study of the 'functional transformation' of legal norms whereby the same legal framework is deployed in changed circumstances and in the service of different projects (Renner 1972). With the rise of modernity, sumptuary laws did not so much die out, but rather spread and changed their form.

Studying the trajectory of sumptuary law also provides an opportunity to study the construction of gender during the period of the transition to modernity. Dress is central to the construction of gender. Sumptuary regulation centrally revolved around the practices that regulate the gendered ordering of appearance; they emerged in a period in which sexual dimorphism in dress became entrenched and in which there was a deepening anxiety about the ordering of gender. Gender relations are central to the concerns over a 'world turned upside-down', and they manifested themselves in a specific anxiety about cross-dressing.

The historical record of dress is extensive. Dress has the distinctive characteristic of being, at one and the same time, both important and trivial. As such, it has the capacity to communicate in ways that are not only highly self-conscious, but also in ways quite outside the realm of intention. Dress is a form of consumed material culture and as such involves a complex system of communication.

A SHORT HISTORY OF SUMPTUARY LAW

I offer an overview of the history of sumptuary law because this is a slice of social history that has rarely been told and will be unfamiliar to many readers. My account follows, with occasional departures, a chronological approach. I focus on 'shifts', the implication being that the identification of such shifts stimulates questions about what was happening and why it happened in the way it did.

In medieval Europe the presence of sumptuary regulation is at first uncertain and irregular. The first traces are found in Charlemagne's extension of the range of governmental action to secure the general conditions of feudal relations. Symbolism came to the fore in the provision that peasants were required to walk and not ride to church on Sunday, and to carry a cattle prod, suggesting that peasants should know their place and be reminded of it. More regular sumptuary regulation begins to appear by the end of the twelfth century. By the second half of the fourteenth century sumptuary laws are present in significant numbers over most of Europe.

Some patterns are detectable in this early phase. Continuity with ancient societies is evident in the widely imposed restrictions on conspicuous consumption in connection with funerals and weddings and, to a lesser degree, other public celebrations of private rites of passage (births, christenings, betrothals, etc.). The regulation of dress was present from the beginning, and became increasingly important; it became the typical target of medieval sumptuary law. Dress rules took two distinctive forms: the first imposed expenditure limits as part of an economic discourse which counter-posed luxury and productivity. The second form was to reserve particularly significant types of clothing categories; this was done either by granting a privilege to a social category (e.g. only nobles may wear ermine) or by negative prohibition (e.g. no female servant may wear a train on her robe). Both types of rule reserved privileges associated with the internal regulation of social hierarchy limiting or channelling competition within the dominant classes. The negative prohibition was associated with the imposition of an imagined social order on real or imagined challenges from below; typical of this category are rules that relate to merchants, apprentices, servants and prostitutes.

There was a complex articulation alongside hierarchical considerations in the use of sumptuary dress provisions to regulate gender relations. In its simplest form a sumptuary dress code provided a codification of the dress restrictions for males and for females in each designated social category. While many ordinances were gender neutral, others differentially imposed restraints, sometimes it was men who were targeted and at other times it was women; those statutes that target women frequently employed strongly misogynistic sentiments. However, on balance, medieval sumptuary law was directed against men more often than it was against women who became the target of sumptuary regulation in the early modern period. However, there is abundant evidence that throughout enforcement was directed more rigorously against women than it was against men.

As European feudalism waned, mercantile capital achieved economic power and, in some cases, political power. If Europe stood on the brink of modernity then it did so with sumptuary regulation at its most active and extensive. This should be sufficient to dispel the tendency to equate sumptuary projects with the kind of social hierarchy conceived of as typically feudal or pre-modern. Sumptuary laws straddle the often imperceptible but fundamental divide between the pre-modern and the modern. In the early modern period in which commercial capital had come to the fore, from the sixteenth into the eighteenth century, there was a great volume of sumptuary laws directed against conspicuous consumption in matters of dress and ornamentation. Dress codes become more, rather than less, preoccupied with the preservation of hierarchy.

My purpose here is to review the major trends in sumptuary law that manifest themselves in the transition to modernity. The sumptuary regulation characteristic of feudalism came to expand its range and its volume; it increased rather than decreased with the decline of feudalism. Sumptuary projects become a standard feature of governmental activity in the early modern period and only declined when capitalism had triumphed. Four major trends are observable. First is the gradual disappearance of funeral and wedding regulations. The second is that dress became the central target and dress codes became more comprehensive and increasingly codified in form. The third feature is that the legislation became more focused upon the preservation of the external symbols of class hierarchy. As dress became the major regulatory object, attempts at the regulation of food, whether in the form of general expenditure limits or the reserving of privileged items for the upper classes, declined. The fourth trend saw a marked turn towards protectionist discourses. An initial bullionist concern with the outflow of gold and silver from the import of luxuries was succeeded by considerations about the balance of trade, and finally became expressed within increasingly protectionist discourses in which the projects became those of advancing some particular domestic trade or industry. An example of this distinctive admixture and supersession was the official sponsorship of luxury industries in France in the second half of the seventeenth century and the elaboration of protectionism into a national economic policy, an economic discourse still familiar in the late twentieth century.

The legitimatory motives for sumptuary rules reveal a number of 'shifts' in economic discourses. In the early phases sumptuary laws appealed to the associated phenomena of 'dearth' and 'ruin', but gradually gave way before two other loosely connected considerations, the first being that 'extravagance' and 'luxury' became the central discursive features focused on the diversion of economic resources away from productive investment; this form was succeeded by the persisting positive protectionism that seems to be outlasting the long reign of national capitalisms. Sumptuary laws were found in virtually every type of political system in medieval and early modern Europe. They were as prevalent in highly centralized nation-states already well on the way to absolutism as they were in the relatively democratic cities and communes.

SUMPTUARY LAW, REGULATION AND GOVERNANCE

My argument has been directed against the widely held view that sumptuary laws were an unconscionable constraint on the liberty of the individual in the marketplace of consumer choice. Such views evince a flavour of the evolutionist assumption of the unquestioned superiority of the modern over the pre-modern. This view was exemplified by the nineteenth-century commentator who argued that 'the Law becomes tyrannous when it prescribes what shall be eaten and what shall be drunk; all such interference is a relapse towards the barbarism of the Middle Ages' (Pike 1968: II: 585). This rejection of the very idea of sumptuary regulation with its assertion of the autonomy of a field of private consumption is buttressed by the historical generalization that sumptuary laws were in any case 'doomed to failure'. Since, from this standpoint, such laws played no part in the emergence of modern law and government they have simply been dropped from the record or, at best, relegated to a footnote. The result has been a lack of scholarly attention to sumptuary law and to its treatment as a quaint reminder of a time now sufficiently distant to be looked back on with amusement rather than rancour.[1]

The internal history of sumptuary laws shows that the shift from one form to another was almost always associated with the emergence of a new linkage or 'discursive coupling' in which the immediate rationale of the sumptuary project became associated with some new discursive context. The most important of these shifts was the uneven and faltering transition from the theological discourses against luxury to the economic discourses of protectionism. The first stage of this transformation involved the theological discourses surrounding the sin of luxury, as a species of the major sin of pride. This gave way before distinctively secular discourses that still retained the central figure of luxury, but inflected it towards the economic wrong of 'extravagance'. Extravagance was conceived as the wasting of resources that could be more gainfully employed. In the next stage the sumptuary element came to be displaced or relegated to a minor role in a discursive formation organized around the figures of the economy and the nation; it expressed itself most distinctively in economic protectionism. The result of this process was that it was no longer such forms of consumption as the wearing of silk and furs that were the target, but it was the wearing, for example, of 'Italian silk' or 'imported furs' that the regulation sought to control. What is distinctive is that the shift from the discourses of luxury to protectionism moved imperceptibly: it is only at the end of the process that it is apparent that the sumptuary content has entirely disappeared from the legal norm to be replaced by discourses of economic and national interest.

Alongside such discursive shifts (decouplings and recouplings) there were persistent combinations with other separate discourses. The most important of these linkages were those with the discourses of gender, generation, class and population.[2] It is in the context of teasing out the couplings between these different discourses that it becomes possible to get beneath their surface and to approach explanation. To move from meaning to explanation requires an inquiry that starts by seeking to identify the anxieties and preoccupations that underlie the many and varied regulatory projects. For example, the central part played by gender in sumptuary projects reveals the profound social ambiguities and contradictions about the self-presentation of women in an historical period in which women were subject to the processes of privatization and vicarious consumption, placed on public display but without public role. These changes and their location in the expanding urban world generated the specific forms within which patriarchal authorities directed sumptuary laws to govern the appearance of women. Central to this was the distinctively urban problem of securing and stabilizing the distinction between 'good girls' and 'bad girls', between respectable women and prostitutes. More generally, the construction and regulation of gender is situated in the dense and practical field of the political economy of marriage. This focus is exemplified in the connection established between sumptuary laws and attempts to regulate the dowry system. Any factors that led to an increase in the size of dowries made it more difficult for fathers to secure marriages for their daughters. The complexity of the political economy of marriage is illustrated by the fact that bigger dowries, perhaps paradoxically, weakened the position of husbands in the household as they became increasingly dependent upon the capital brought in by their wives.

Sumptuary law addressed inter-generational relations. Frequent regulatory targets were (male) apprentices and (female) servants. Provisions sought to provide a visible distinction between them and their masters and mistresses and to repress a youthful enthusiasm for current fashion. Such rules reveal not only general hierarchical concerns, but more importantly attest to the distinctively urban problem of the growing influx of unattached young men and women now beyond parental control that has proved such a potent stimulus to projects of moral regulation such as the spate of anti-seduction laws at the end of the nineteenth century.

The urban sumptuary legislation of the early modern period that was endemic in the city states from the Mediterranean to Northern Europe can best be understood as a response to the quest for recognizability. The modern subject was located within a widening gulf between the individual as private being and their public social roles. The former is marked by processes of individualization which are manifest, to modify Bourdieu's terms, in private cultural capital deployed in the search for 'distinction' and 'taste' as a means of securing both self-identity and class-identity (Bourdieu 1984). The quest for individuality through the medium of dress and self-presentation is one dimension of this process. The other is the process of 'individuation' whose distinctive form of cultural capital is exemplified by public credentials (qualifications, skills, references and testimonials) and of the disposal of cultural capital (taste, distinction, refinement) that increasingly depends on a non-genetic sense of 'breeding'.

The attempt to render social status legible can be seen in the common form of sumptuary rules reserving the wearing of silks and furs to designated status categories. The problem of the social visibility of rank, status and occupation was a distinctive problem generated by urban life. The urban social landscape as a 'world of strangers' generated a concern to 'know' strangers through their appearance and amplified anxieties about the capacity of appearance for dissimulation (Lofland 1973).

The imposition of recognizability also took more coercive forms. One widespread expression was in the attempts to impose distinguishing marks on particular categories of persons. Prostitutes were an important target; sumptuary regimes sought to impose 'marks of infamy'. They have variously been required to wear coloured patches (yellow for some reason being the most widely employed), special hats, bells and other means of providing visible and audible evidence. The sumptuary regulation of prostitution manifests evidence of another strategy that fluctuated between two tactics. The first played on the distinction between the respectable and the whore by denying fashionable or luxurious dress to prostitutes as a sign of their lack of social respectability while 'rewarding' the virtuous by granting them access to fashionable attire. The second tactic was diametrically opposed; it allowed fashion and luxury to prostitutes in the hope that respectable women would be discouraged from emulating their sinful sisters in order to confirm their own respectability.

The attempt to prescribe visibility finds its most dramatic expression in the use of branding of the flesh with an authoritative and irreversible imprint of its crimes. Branding allows considerable potential for the symbolization of wrongs. The Japanese harnessed the techniques of tattooing to provide a highly legible code in which the nature of the offence could be read from the body or facial tattoo. Medieval and early modern Europe wielded the branding iron to imprint signifying letters into the flesh of those judged to merit such designation. In England branding was employed against vagabonds. It appeared in English legislation in the late fourteenth century in an attempt to control the movement of labour that had become a serious concern of the landed classes. A statute of 1361 required that fugitives, those leaving employers without leave, be branded with 'F' on the forehead for 'falsity'.[3] A statute of 1547 provided that sturdy beggars should be branded with 'V' for vagabond on the chest and an 'S' for a second offence.[4] That distinguishing marks also served an evidential role is clear from the widespread use of 'badging' to designate authorized beggars from vagrants within many municipalities.

One consequence of the widening separation between the private and the public is that it becomes both less possible and less necessary to seek to stabilize recognizability through sumptuary law directed towards external appearance. The desire for a legible ordering of appearance never entirely disappeared but it became seriously attenuated and resurfaced only in periodic and episodic form. While sumptuary laws were concerned to secure a stable connection between appearance and entitlement, the social conditions in which it did so were everyday making the achievement of its goals ever more unrealizable. The history of women's veils attests to the instability of external appearance. On the one hand, the veil exhibits piety and modesty, but, on the other hand, it signifies allure by facilitating concealment of identity. In Italy there were tensions between clerical and secular sumptuary activity over the wearing of veils. While the Church favoured head covering as symbolizing religious piety and sexual modesty, the secular authorities were concerned that veils could also encourage conspicuous consumption. In addition, veils allowed women a problematic anonymity. In Sienna officials were empowered to demand of a veiled woman the name of her father or husband (Hughes 1983: 79). What was at stake was the degree of freedom veils allowed for assignations that breached patriarchal control and facilitated some degree of personal and sexual licence.

This insight provides a more adequate understanding of the 'failure' of sumptuary laws. It was not so much that they were doomed to failure, but rather that they should be understood as attempts to grapple with a set of social and economic changes that would sooner or later swamp the idea that to secure social recognizability and stabilize it through law was an appropriate object of governance. Yet the persistence of calls for such a project echoed long after the demise of active sumptuary legislation. The sumptuary ethic lived on long past its active legislative existence as a component of a cultural nostalgia for a time when people knew their allotted social place which was recognizable through a semiotics of appearance.

The long-run shift of sumptuary projects traversed the field of moral regulation to the field of economic policy; this process ran powerfully throughout the early modern period without ever being completed. While the targets of regulation continued to shift and their discursive expression changed accordingly, a deeply embedded and resilient preoccupation formed around a persistent moralization of appearance constructed around the two key parameters of class and gender. Class moralization involved two features: a degradation of the labouring classes through the almost universal preoccupation with their oft asserted idleness and, second, the concern with imitation that speaks of a concern to render social difference visible and thus legible. Gender moralization revolved around the dichotomy of respectability and dishonour and its close associate, eroticism. The general sumptuary phenomenon and its concrete manifestations thus occupied the central social field bounded by the three axes of urbanization, class differentiation and gender polarization.

NOTES

1 There are a number of still useful, but dated studies of local sumptuary regimes (Baldwin 1926; Boehn 1932–1935; Greenfield 1918; Hooper 1915; Newett 1907; Vincent 1935). More recent scholarship, aside from Harte's study of the economic history of English sumptuary law (Harte 1976), has focused on the abundant evidence from Renaissance Florence (Hughes 1983; Kovesi-Killerby 1993; Rainey 1985).

2 This mode of procedure which seeks to trace shifts in specific forms of social regulation is effectively employed by Arthur Stinchcombe's study of moral categories in eighteenth-century France (1982) and by Nicola Beisel in her study of the nineteenth-century censorship campaigns of Anthony Comstock (1992).

3 34 Edw. III, c.10; SR 1, 367. The legislation, by providing for the postponement of the branding provision for first offenders, suggests that it was intended as a threat.

4 1 Edw. VI, c.3; SR 4(2), 5–8.

REFERENCES

Baldwin, Frances (1926) *Sumptuary Legislation and Personal Regulation in England*, Baltimore, MD: Johns Hopkins University Press.

Beisel, Nicola (1992) 'Constructing a shifting moral boundary: literature and obscenity in nineteenth century America', in Michèle Lamont and Marcel Fournier (eds) *Cultivating Differences: Symbolic Boundaries and the Making of Inequality*, Chicago, IL: University of Chicago Press, pp. 104–128.

Boehn, Max von (1932–1935) *Modes and Manner* (4 vols), London: Harrap.

Bourdieu, Pierre (1984) *Distinction: A Social Critique of the Judgement of Taste*, London: Routledge.

Greenfield, Kent R. (1918) *Sumptuary Law in Nürnberg: A Study in Paternal Government*, Johns Hopkins University Studies in Historical and Political Science (Vol. XXXVI), Baltimore, MD: Johns Hopkins University Press.

Harte, N. B. (1976) 'State control of dress and social change in pre-industrial England', in D. C. Coleman and A. H. John (eds) *Trade, Government and Economy in Pre-Industrial England*, London: Weidenfeld & Nicolson, pp. 132–165.

Hooper, Wilfred (1915) 'The Tudor sumptuary laws', *English Historical Review* 30: 433–449.

Hughes, Diane (1983) 'Sumptuary laws and social relations in Renaissance Italy', in John Bossy (ed.) *Disputes and Settlements: Law and Human Relations in the West*, Cambridge: Cambridge University Press.

Kovesi-Killerby, Catherine (1993) 'Practical problems in the enforcement of Italian sumptuary law, 1200–1500', in K. Lowe (ed.) *Crime and Disorder in Renaissance Italy*, Cambridge: Cambridge University Press.

Lofland, Lyn (1973) *The World of Strangers: Order and Action in Urban Public Space*, New York: Basic Books.

Newett, M. Margaret (1907) 'The sumptuary laws of Venice in the fourteenth and fifteenth centuries', in T. F. Tout and J. Tait (eds) *Historical Essays*, Manchester: Manchester University Press, pp. 245–278.

Pick, Daniel (1989) *Faces of Degeneration: A European Disorder, c.1848–c.1918*, Cambridge: Cambridge University Press.

Pike, Luke (1968) *A History of Crime in England* [1873–76] (2 vols), Montclair, NJ: Patterson Smith.

Rainey, Ronald (1985) 'Sumptuary legislation in Renaissance Florence', PhD thesis, Columbia University, New York.

Renner, Karl (1972) *The Institutions of Private Law and their Social Functions* [1949] (ed. Otto Kahn-Freund), London: Routledge & Kegan Paul.

Stinchcombe, Arthur L. (1982) 'The deep structure of moral categories: eighteenth-century French stratification, and the revolution', in Eno Rossi (ed.) *Structural Sociology*, New York: Columbia University Press.

Vincent, John (1935) *Costume and Conduct in the Laws of Basel, Bern and Zurich 1370–1800*, Baltimore, MD: Johns Hopkins University Press.

7

Societies of Consumers and Consumer Societies

Co-operation, Consumption and Politics in Britain and Continental Europe c. 1850–1920

Martin Purvis

The growth of industrial capitalism which often gave a particular stamp to the creation of a resource of collectivity among workers also eroded the relevance of some older traditions of association. The ideal of independence for small artisanal groups through co-operative production was marginalized amidst recognition of the increasing imbalance between the meagre resources of workers, even collectively, and the growing concentration of industrial capital in large-scale productive units. Denied direct entry to the sphere of production and self-employment workers looked to other means to improve their circumstances. An association of consumers promised, and in part provided, a degree of empowerment to working people denied other outlets for political participation and leadership. It also offered a service of especial utility to those growing elements of the European population reliant upon commercial markets for their food supply.

Urban and industrial consumers, in particular, experienced dietary changes with increasing use of imported, processed and packaged foods during the later nineteenth century. These developments heightened the impact of their increasing distance, both functional and spatial, from traditional forms of agricultural production and locally focused systems of self-provisioning still current in many rural areas throughout the long nineteenth century.[1] The potential vulnerability of urban and industrial consumers increased the appeal of this form of co-operation. Societies were in part a reaction against the worst abuses of private shopkeepers, including adulteration, short-weight, punitive terms of credit and high prices. Hence, contemporary commentators noted their utility and success in settlements where previous lack of retail provision and competition had produced profiteering by private grocers and bakers.[2] However, co-operation could also be proclaimed as a means of effecting change at a systemic level, rather than merely localized amelioration of capitalist exploitation. While older means of securing morality in trade, such as food riots, declined in relevance (although not to the point of total extinction) co-operation promoted a new vision of the potential power created by the association of the masses as consumers.[3] Moreover, from mid-century there were signs that the initiatives taken by capital to control the masses within the productive sphere were extending into the arena of consumption. Co-operatives can thus also be seen as a means of defence by consumers against the efforts of increasingly powerful individual interests among distributors and producers to create and 'discipline' mass markets.[4]

TOWARDS THE 'CONSUMERS' COMMONWEALTH': CONTESTING THE NATURE OF CONSUMPTION

In principle, consumers' co-operation sought to distance acquisition of the necessities of life from the sphere of competitive capitalism by establishing the people's own distributive and productive operations. In so equating the producer and the consumer, the trader and the customer, they sought a new equity for all, eschewing the beguiling display and advertising that increasingly characterized capitalist commerce.[5] Yet in practice, co-operation was itself shot through with contradictions. Just as retail societies could not fully separate themselves from the world of competitive commerce (and some, in truth, did not try very hard to effect such a separation) so co-operative members could not be wholly apart from evolving consumerism and the desire to possess more and different goods. However, integral to implementation of the co-operative project was a promotion of morality in trade and a concomitant rejection of conspicuous display and the creation of unnecessary demand for luxuries. This reflected not simply the modest incomes of most co-operators but, for some at least, a conscious assertion of freedom from the power of capitalism to create an infinitely expansive and thus unsatisfiable desire to consume.[6]

Thus, co-operative principles stood opposed to competitive capitalism, critical not merely of its worst excesses, but of the system as a whole in its control over consumers.[7] To strengthen that opposition, however, co-operation had to project an alternative, not just to the prevailing system of distribution, but ultimately to the economy as a whole. In France, Italy and Belgium during the final decades of the nineteenth century, elements of co-operation became explicitly identified with socialist programmes for societal transformation. Such links reflected an appreciation of the potential of consumption to generate both funds and publicity for political causes and, through daily interaction in the acquisition of the necessities of life, to consolidate a collective identity. Elsewhere, co-operation was regarded as a project in its own right, distinguished from both capitalism and socialism in its pursuit of the 'consumers' commonwealth'. This new world to be created through the association of the masses as consumers emphasized democracy and mutualism in the ownership, control and divisions of reward from business as against the increasing concentration of productive and distributive capital in the hands of a minority. From a foundation in retailing it was envisaged that the territory controlled by the masses would grow as co-operators appropriated and reinvested the economic and social resources necessary to expand their operations. The productive capacity of industry could thus pass into the control of consumers, ensuring that it was harnessed to meet true needs, rather than creating illusory desires which led only to profit for a minority of producers and traders.

Such visions rested on a belief in the power of association among consumers. Analysis of both the contemporary and historical experience of consumption under competitive capitalism has tended to emphasize one of its characteristics as an inherent individualism. Consumption is often deemed to be collective but atomizing in that it sustains and is sustained by the myth of individual satisfaction and the creation and defence of individual status through the acquisition of particular material possessions. An activity undertaken in common does not therefore generate a collective consciousness or solidarity and consumers do not possess the means to challenge capitalism in a manner analogous to associative labour.[8] Many co-operative ideologues, however, subscribed to a different reading of the nature and power of consumption. Indeed, they saw consumption as *the* experience that *all* individuals shared and through which they might unite consciously to take common cause against competitive capitalism.[9] By comparison, perhaps in part reflecting changes in the composition of the labour force in the more mature industrial economies which increasingly marginalized women and juveniles, the world of work was deemed to be predominantly the preserve of the adult male.[10] Moreover, notions of a unity of class flowing from the experience of labour were seen as being over-ridden by the sectionalism of the individual workplace, craft or trade association.[11] Thus, for some, it was consumers who were invested with considerable collective power and moral authority:

> Since wealth arises from exchange, it follows that those who exchange, as we all do by spending money, have it in their power to end this system of calculated impoverishment. Without revolutionary violence or the employment of legal force they can detach support from capitalism and build up a new system of socialised wealth. . . . All labour and effort, all material resources, must be consciously devoted to the good of the whole people as consumers; and all owners and business organisers, all workers, ceasing to be masters and exercisers of absolute rights, must become the agent and stewards of the community and of the world order of communities.[12]

Thus consumption could be seen as a vehicle for collective redemption rather than individual gratification, a means not of entrapment by competitive capitalism, but of escape. The transformation was theoretically total; some envisaged both a new way of life and a new geography of settlement and economy, naïvely predicting the concomitant withering away of the pre-existing system. That such visions proved to be illusory does not undermine the central point that the rise of modern consumer society was not an uncontested process. Co-operators thus contributed to wider debate on the nature and morality of consumption in nineteenth-century Europe.[13] Furlough has explored French interest in the development of a 'co-operative republic'[14] and a similar project of socioeconomic transformation is evident in the constitution of an Italian society which aimed to create:

> a regime in which production would be organized for consumers collectively and not for profit, and to acquire gradually the means of production and sale for the associated consumers, so that they may have in future the surplus riches which they have created.[15]

Although there were attempts to establish consumers' co-operatives in many parts of western and central Europe from the early nineteenth century, it was Britain that saw the first sustained expansion of societies during the late 1850s and early 1860s.[16] By comparison the foundations of modern co-operation in France, Italy and Belgium were laid in the 1880s.[17] German consumers' co-operation also entered an important phase of growth during the late 1880s as workers began increasingly to organize societies themselves, rejecting the influence of middle-class patrons who had initially promoted co-operation as a source of social quiescence.[18] In Russia and Russian-ruled Finland and Ukraine, development of independent working-class co-operatives was an early twentieth-century phenomenon.[19] Given this chronology of development it is unsurprising that the British example was often invoked by European co-operators.

As early as 1862, Holyoake's history of the Rochdale Pioneers,[20] which confirmed that society as an inspiration to British co-operators, was serialized in the French journal *Le Progrès* of Lyon.[21] However, Furlough considers that the English example made a greater impact in France during the final quarter of the nineteenth century, as a source of ideas (particularly the accumulation of surpluses to augment collective capital and as a direct benefit to members through dividend on purchases) which were 'crucial for the clarification of a set of practices that supported collective democratic and social values' in opposition to an increasingly powerful capitalist commerce.[22] Elsewhere, including Belgium, Germany, Italy, Denmark, Sweden and Russia, there is further evidence that contacts with Britain influenced the development of consumers' co-operation. Paradoxically, however, the key individuals identified as the carriers of information, particularly the potent 'myth' of the Rochdale Pioneers, were often intellectuals and middle-class patrons whose personal vision of co-operation as an agency of social stability was at odds with the collective assertion of independence and authority by working-class consumers.[23]

Explanation of this chronology and geography cannot, however, rest simplistically on notions of the diffusion of co-operation from Britain. In part this reflects the importance of alternative foci of co-operative inspiration. Nicolas Ballin, for example, a pioneer of Russian consumers' co-operation in the late 1860s and 1870s, visited Germany and France as well as Britain in search of inspiration.[24] The German cooperative movement from the 1860s assumed increasing influence in central Europe and its organizational structures, incorporating credit and consumer's societies, found echoes in neighbouring states, including Austria and Switzerland. The writings of the leading German advocate of co-operation Hermann Schulze-Delitzsch circulated in translation in the Francophone countries, which also had an indigenous resource of associative writing, including the earlier work of Fourier and St Simon, on which to draw.[25]

Examination of the pattern of co-operative development cannot, of course, attend only to the circulation of ideas. We should also consider the more concrete circumstances which influenced the extent to which such ideas appeared relevant to working people and which permitted, or prevented, their translation into practice. Full discussion of the context of the evolution of co-operation in nineteenth-century Europe is beyond the scope of the present paper. Rather the aim is to sketch some key aspects of socioeconomic and political change which interacted with the development of consumers co-operation. Attention has already been drawn to appreciation of the relevance of the co-operative critique in the broad context of urban and industrial development, and concomitant innovation in practices of retailing and consumption. Brief discussion is also required of the political context of co-operative development. This will be defined both narrowly in terms of the process of definition of the legal status of societies, and with wider reference to the politics of consumption, and association between consumption and political projects working to maintain or transform prevailing socio-economic systems.

CREATING AND REGULATING A POLITICAL SPACE FOR CO-OPERATION

The development of consumers' co-operation reflected in part the changing balance between the collective resources of working people and the exercise of control by the state and elements of the socio-economic elite. Co-operatives were a product of popular perceptions of advantage through association, but also of the willingness of establishments to permit both association *per se* and the application of collective initiative to particular ends. Political elites have often been grudging in conceding such freedoms. Nineteenth-century Britain saw an important change from overt repression to control through regulation.[26] Legislation, initially covering friendly societies and after 1852 a series of Industrial and Provident Societies Acts, offered legal 'protection' to co-operatives, allowing them to safeguard their financial resources, defend against fraud and recover debts.[27] But legislation exacted a price: a greater transparency in the activities and finances of societies, which were required to submit to registration and the annual presentation of their accounts. Moreover, legislation involved, at least implicitly, the definition of that which was illegal in the extension of popular initiatives. The state thus contributed to the fracturing of the earlier concept of comprehensive 'union' among workers which sought to establish new rights as producers, consumers, social and political actors, into separate, and less threatening, legally defined entities including friendly societies, trades unions and co-operatives.[28] Such fragmentation did not extinguish support for grand projects for the total recasting of society, economy and polity, but the retreat to a series of institutions with prescribed and restricted functions constrained popular initiatives.

There were also specific restrictions on the operation of co-operatives; the initial legislation of 1852 prevented the funding of education, ownership of land and property other than retail premises and investment in the operation of another society.[29] Such provisions were ostensibly intended to defend the financial security of individual local co-operatives, but they also countered co-operative aspirations to extend their own activity at the expense of competitive capitalism into production and wholesale distribution. Subsequent reform of Industrial and Provident Societies legislation from 1862 onwards removed some of these restrictions but within a legal framework that emphasized the commercial rather than social or political nature of co-operation.

The legal curbs placed on popular institutions by many continental regimes were more severe than those in Britain, with a concomitant impact on co-operative development. In France the 1834 law on associations:

> reflected the underlying conviction of the social and political elite that most kinds of association were fundamentally unwholesome. . . . Like most of their middle-class constituents, the government viewed workers' associations as little more than the organized extension of seditious gatherings.[30]

The declaration of the Second Republic in 1848 created a temporary political space for popular association. While most attention has been paid to the producers' associations of the period, there were also some important consumers' co-operatives, perhaps numbering 40–50 in total, in and around centres including Paris, Lille, Lyon, Vienne, Rheims, Sedan and Nantes. However, the projected social order of fraternal harmony was choked by increasingly conservative republican governments and finally extinguished with the imperial *coup d'état* of 1851.[31] The following years saw some co-operative foundations; one estimate suggests a national total of only 37 between 1855 and 1869,[32] although another source records around 120 co-operative stores and bakeries in France by the latter date.[33] At this stage, however, most new societies avoided association with programmes of socio-economic or political transformation, concentrating instead on the provision of cheap foodstuffs.[34]

The 1860s saw efforts by increasingly beleaguered French governments to broaden their support by yielding some ground to workers. In 1867, co-operatives were granted the concession of limited liability that was essential if societies were to recruit a substantial membership. However, proposals for a special law for co-operatives were rejected, with the changes being incorporated into the *Code Commercial*.[35] Other European regimes, including Italy and Belgium, similarly countenanced an official status for co-operation only under commercial law, thus identifying societies primarily as trading organizations.[36] In Germany the political upheavals of 1848 also created a temporary space for the workers' own consumer co-operatives, of which there were several in Berlin.[37] But the state was normally much less tolerant of popular association. Indeed the denial of limited liability until 1889 and constraints on working-class initiatives imposed by the Anti-Socialist Law of 1878–1890 curbed co-operative growth.[38] Even after these restrictions had been removed, the state attempted to prohibit involvement by public servants, including postal workers and railwaymen, in co-operatives considered too closely identified with the socialist cause.[39]

Restrictions placed on the involvement of government employees with co-operation also reflected a desire to placate private traders who protested against what they interpreted as state sanction for forces opposed to the interests of capitalist commerce.[40] Moreover, the political influence of the *petit bourgeoisie* in parts of continental Europe helped to fuel controversy regarding the taxation of co-operatives.[41] The distinction drawn by states, including France, Germany and Britain, between co-operatives and conventional retailers in assessing liability for commercial licence fees and the taxation of trading surpluses was interpreted by private traders as discrimination in favour of co-operation. This issue attracted particular attention from the 1880s onwards, partly reflecting co-operative expansion, but also as a result of the increasing vulnerability of smaller shopkeepers in the face of general economic downturn, overcrowding and inefficiency in key sectors of retail trade (including foodstuffs) and the challenge, faced particularly in some of the larger urban centres, of substantial new private stores and multiple retailers.[42] Some changes did follow from these protests, including the extension of the payment of the *patente* or commercial licence to all French consumer cooperatives in 1905.[43] However, original governmental assessments of liability for taxation were designed not so much to advantage co-operatives as to define their extent and operations. Exemptions from licence fees and other forms of commercial taxation were often granted on condition that co-operatives did not compete with private traders for general trade but dealt only with their own members. Thus the official definition of a status for co-operation again involved the imposition of constraints upon its development. Such measures were reinforced, at least in theory, as German legislation of 1889 and 1896 made clear, by significant financial penalties.[44]

In practice, the taxation issue perhaps serves chiefly to reinforce perceptions of the ambiguity of the status of consumer co-operatives as organisations within, but not wholly of, the legal and capitalist commercial framework of the Europe of the long nineteenth century. In some instances, indeed, this ambiguity seems to have had a particular impact on the geography of co-operative development. Denmark was one of the few European states where consumers' co-operatives trading in foodstuffs and household goods, as opposed to agricultural supply associations, initially grew most strongly outside urban centres. Indeed rural societies accounted for 92 per cent of recorded national membership and 93 per cent of sales in 1910.[43] This partly reflects the strong links between consumers' societies and agricultural producer co-operatives and, indeed, the importance of agriculture within the wider Danish economy. However, it was also an unintended consequence of legal restrictions on the operation of private traders in the rural hinterlands of the main towns, which were not rescinded until 1920. Intended as a measure to protect the competitive position of urban traders, in practice the law allowed, even encouraged, the development of co-operative societies which, again by virtue of doing business with only their own members, were exempt from this element of commercial law.[46]

Overall, however, consumers' co-operation was subject to the efforts of political establishments, aimed not so much at its elimination, but at negating its potential as an agency of socio-economic transformation. Encouragement was given for the incorporation of societies and their membership within an expansive capitalism whose worst abuses were thus ameliorated. Such aspirations are apparent in statements of support for co-operation from its middle- and upper-class champions.[47] In Britain, Ludlow and Jones' characterization of co-operation contains little suggestion of a sustained critique of capitalism. Rather the stores 'were meant chiefly as a defence against the inroads of the distributing classes on the working-man's pocket; and also as a means of promoting ready-money dealings, and the prudence in expenditure which usually accompanies such dealings'.[48] Lord Reay, in his inaugural address to the Fourteenth Co-operative Congress in 1882, even more pointedly asserted that:

English co-operators have never boasted that they were going to renovate English society, and the consequence is, undoubtedly, that by their aid English society has been spared a good deal of the friction which we see elsewhere. . . . Co-operation has been the best friend of capital, and is therefore the strongest ally of the middle-classes. . . . The co-operative stores have increased the security of the small capitalist, and have contributed materially to our escape from the great peril of French society – the dislike of the 'bourgeois' by the 'ouvrier'.[49]

NOTES

1 W. H. Fraser, *The Coming of the Mass Market, 1850–1914* (London 1981); R. Price, *The Modernization of Rural France. Communications Networks and Agricultural Market Structures in Nineteenth-Century France*. See also working-class budgets in Mrs A. Sidgwick, Household budgets abroad. I Germany, *Cornhill Magazine* 17 (1904), 98–101; *Royal Commission on Labour. Foreign Reports Vol. 5. Report on the Labour Question in Germany*, P.P. 1893–1894 XXXIX Part II, 139–142; L. Villari, Household budgets abroad. IV Italy, *Cornhill Magazine* 17 (1904), 479–480.

2 M. Purvis, *Nineteenth-century Co-operative Retailing in England and Wales: A Geographical Approach* (unpublished D. Phil thesis, University of Oxford 1987), 371–385; A. Roulliet, *Des Associations Coopératives de Consommation* (Paris 1876).

3 P. R. Hanson, The 'Vie Chère' riots of 1911: traditional protest in modern garb, *Journal of Social History* 21 (1987–1988), 463–482; R. Price, *A Social History of Nineteenth-Century France* (London 1987), 182–186.

4 J. Benson and G. Shaw (Eds), *The Evolution of Retail Systems, c.1800–1914* (Leicester 1992); Fraser, *op. cit.* (Note 1); cf. J. Baudrillard, Consumer society, in M. Poster (Ed.), *Jean Baudrillard: Selected Writings* (Cambridge 1988) especially pp. 30–32, 49–50.

5 A. Faure, The grocery trade in nineteenth-century Paris: a fragmented corporation, in G. J. Crossick and H-G. Haupt (Eds), *Shopkeepers and Master Artisans in Nineteenth-Century Europe* (London 1984), 155–174; P. Mathias, *A History of Multiple Retailing in the Food Trades based upon the Allied Suppliers Group of Companies* (London 1967); P. C. Nord, *Paris Shopkeepers and the Politics of Resentment* (Princeton 1986); T. Richards, *The Commodity Culture of Victorian England: Advertising and Spectacle, 1851–1914* (London 1990); R. H. Williams, *Dream Worlds: Mass Consumption in Late Nineteenth-Century France* (Berkeley, California 1982); M. J. Winstanley, *The Shopkeeper's World 1830–1914* (Manchester 1983).

6 Baudrillard, *op. cit.* (Note 4), 29–55; N. Xenos, *Scarcity and Modernity* (London 1989).

7 P. Gurney, *Co-operative Culture and the Politics of Consumption in England, 1870–1930* (Manchester 1996); T. R. Tholfsen, *Working-Class Radicalism in Mid-Victorian England* (London 1976), 253.

8 D. B. Clarke and M. Purvis, Dialectics, difference, and the geographies of consumption, *Environment & Planning A* 26 (1994) 1091–1109; D. Miller, Consumption as the vanguard of history: a polemic by way of an introduction, in D. Miller, (Ed.) *Acknowledging Consumption. A Review of New Studies* (London 1995), especially pp. 23–28.

9 Gurney, *op. cit.* (Note 7).

10 E. Jordan, Female unemployment in England and Wales 1851–1911, *Social History* 13 (1988), 175–190; E. Jordan, The exclusion of women from industry in nineteenth century Britain, *Comparative Studies in Society and History* 31 (1989), 273–296. Even in France, where female participation in the paid workforce increased during the later nineteenth century, the rhetoric of organized labour proclaimed a domestic rather than industrial role for women: R. Magraw, *A History of the French Working Class, Volume 2: Workers and the Bourgeois Republic, 1871–1939* (Oxford 1992), 8, 60–61.

11 Cf. T. B. Greig, *The Consumer in Revolt* (London c. 1912?); P. Redfern, *The Consumers' Place in Society* (Manchester 1920), 67.

12 Redfern, *op. cit.* (Note 11), 39, 92.

13 Cf. W. G. Breckman, Disciplining consumption: the debate about luxury in Wilhelmine Germany, 1890–1914, *Journal of Social History* 24 (1990–1991), 485–505.

14 E. Furlough, *The Politics of Consumption: The Consumer Co-operative Movement in France, 1834–1930* (unpublished Ph.D. thesis, Brown University 1987).

15 Objectives of the Alleánza Cooperativa Milanese, quoted in D. Coffey, *The Cooperative Movement in Jugoslavia, Rumania and North Italy During and After the World War* (New York 1922), 76.

16 T. E. Coles, *The Evolution of Urban Retail Systems in Germany, 1848 to 1914: An Historical Geographical Perspective* (unpublished Ph.D. thesis, University of Exeter 1996), 149–150, 162–165; L. F. Dvorak, *The Co-operative Movement in Czechoslovakia* (Prague 1924); C. R. Fay, *Co-operation at Home and Abroad: A Description and Analysis* (London 1908); C. Gide, *Consumers' Co-operative Societies* (Manchester 1921); Purvis, *Nineteenth-century co-operative retailing* (Note 2); M. Purvis, The development of

co-operative retailing in England and Wales, 1851–1901: a geographical study, *Journal of Historical Geography* 16 (1990) 314–331; *Reports of Her Majesty's Representatives Abroad, on the System of Co-operation in Foreign Countries*, P.P. 1886 LXVII, 429–570; L. Pizzamiglio, *Distributing Co-operative Societies: An Essay in Social Economy* (London 1891) Roulliet, *op. cit.* (Note 2); M. L. Stewart-McDougall, *The Artisan Republic. Revolution, Reaction and Resistance in Lyon, 1848–1851* (Kingston, Ontario 1984)

17 Fay, *op. cit.* (Note 16), 298–309; Furlough, *op. cit.* (Note 14); Gide, *op. cit.* (Note 16); P.P. 1886 LXVII (Note 16), 433–438, 515–516, 535–538; V. Serwy, *La Coopération en Belgique: II La Formation de la Coopération 1880–1914* (Bruxelles 1942).

18 T. Cassau, *The Consumers' Co-operative Movement in Germany* (Manchester 1925); Coles, *op.cit.* (Note 16), 149–153.

19 J. V. Bubnoff, *The Co-operative Movement in Russia: Its History, Significance and Character* (Manchester 1917); Y. Imai, *Nicolas P. Ballin, A Pioneer of[the] Russo-Ukrainian Co-operative Movement and his Letters to English Co-operators 1871–1888* (Tokyo 1992: manuscript in Co-operative Union Library, Manchester); *International Co-operative Bulletin* February 1909 and March 1909.

20 G. J. Holyoake, *Self-Help by the People: The History of the Rochdale Pioneers* (London 1858).

21 Letter from A. Talandier 5 October 1862, Item 1444, Holyoake Collection, Co-operative Union Library, Manchester.

22 Furlough, *op. cit.* (Note 14), 45.

23 The Belgian example stands out as one in which the element of bourgeois paternalism was least apparent. An influential pioneer of co-operation, Edouard Anseele, had encountered consumers' co-operation while working as a London dock labourer: M. K. Z. Anafu, *The Co-operative Movement of Reggio-Emilia, 1889–1914* (unpublished Ph.D. thesis, University of Cambridge 1980); W. M. Childs, *Sweden: The Middle Way* (New Haven 1938); Coles, *op. cit.* (Note 16), 153–154; J. Earle, *The Italian Co-operative Movement. A Portrait of the Lega Nazionale della Cooperative e Mutue* (London 1986), 10–19; Furlough, *op. cit.* (Note 14); 35–36, 46–47; Gurney, *op. cit.* (Note 7), 89–92; Imai, *op. cit.* (Note 19); P.P. 1886 LXVII (Note 16), 553–554; U. Rabbeno, *La Cooperazione in Inghilterra* (Milan 1888); U; Serwy, *op. cit.* (Note 17); L. Smith-Gordon and C. O'Brien, *Co-operation in Many Lands* Volume 1 (Manchester 1919), 53.

24 Imai, *op. cit.* (Note 19).

25 For example, H. Schulze-Delitzsch, *Cours d'Economie Politique à l'Usage des Ouvriers et des Artisans* (Paris 1874).

26 P. Corrigan and D. Sayer, *The Great Arch: English State Formation as Cultural Revolution* (Oxford 1991), 114–165.

27 A. Bonner, *British Co-operation* (Manchester 1970), 66–67.

28 J. F. C. Harrison, *Robert Owen and the Owenites in Britain and America: The Quest for the New Moral World* (London 1969).

29 Bonner, *op. cit.* (Note 27), 66–67.

30 J. M. Merriman, *The Agony of the Republic: The Repression of the Left in Revolutionary France 1848–1851* (New Haven 1978), 52–53.

31 Furlough, *op. cit.* (Note 14), 15–20; R. Magraw, *A History of the French Working Class, Volume 1: The Age of Artisan Revolution 1815–1871* (Oxford 1992), 158, 165; Merriman, *op. cit.* (Note 30), 68–78; P. P. 1886 LXVII (Note 16), 434–435; Stewart-McDougall, *op. cit.* (Note 16), 116–122.

32 Furlough, *op. cit.* (Note 14), 62.

33 P. P. 1886 LXVII (Note 16), 438.

34 Magraw, *op. cit.* (Note 31), 208; P. P. 1886 LXVII (Note 16), 436; Roulliet, *op. cit.* (Note 2).

35 Fay, *op. cit.* (Note 16), 379–380; Furlough, *op. cit.* (Note 14), 40–43.

36 Fay, *op. cit.* (Note 16), 355–378; P. P. 1886 LXVII (Note 16), 515–516.

37 Cassau, *op. cit.* (Note 18), 6.

38 C. Eisenberg, Artisans' socialization at work: workshop life in early nineteenth-century England and Germany, *Journal of Social History* 24 (1990–1991), 507–520; Fay, *op. cit.* (Note 16), 287; G. Shaw, Large-scale retailing in Germany and the development of new retail organisations, in Benson and Shaw, *op. cit.* (Note 4), 168–173.

39 Cassau, *op. cit.* (Note 18), 176; *International Co-operative Bulletin* March 1909.

40 Gide, *op. cit.* (Note 16), 144–145.

41 G. Crossick and H-G. Haupt, Shopkeeper, master artisan and the historian: the petit bourgeoisie in comparative focus, in G. Crossick and H-G. Haupt (Eds) *Shopkeepers and Artisans in Nineteenth-Century Europe* (London 1984), 3–34; R. Gellately, *The Politics of Economic Despair: Shopkeepers and German Politics 1890–1914* (London 1974); B. King, Co-operation in Italy, in *CWS Annual* (Manchester 1902), 181–183; P. Nord, The small shopkeepers' movement and politics in France, 1888–1914, in Crossick and Haupt *ibid.*, 175–194.

42 Multiple retailers probably grew more slowly in France and Germany than in Britain; *The Report of an Enquiry by the Board of Trade into Working Class Rents, Housing and Retail Prices Together With the Rates of Wages in Certain Occupations in the Principal Industrial Towns of France*, P.P. 1909 XCI, notes their limited progress in France and suggests that multiples were of importance only in a small number of urban centres including Lyon, Toulouse and

Saint-Etienne. The *Report of an Enquiry by the Board of Trade into Working Class Rents, Housing and Retail Prices, Together With the Rates of Wages in Certain Occupations in the Principal Industrial Towns of the German Empire*, P.P. 1908 CVIII, concluded that multiples had 'made so far but little progress in Germany', although such stores were recorded in individual centres such as Düsseldorf. Nord, *op. cit.* (Note 5); Shaw, *op. cit.* (Note 38); G. Shaw, The evolution and impact of large-scale retailing in Britain, in Benson and Shaw, *op. cit.* (Note 4), 135–165; Winstanley, *op. cit.* (Note 5), 75–93.

43 Gide, *op. cit.* (Note 16), 175–176.

44 Gide, *op. cit.* (Note 16), 50.

45 International Labour Office, The consumers' co-operative societies in 1919, *Studies and Documents Series H* **1** (1920), 3–4.

46 Childs, *op. cit.* (Note 23), 133–141; International Labour Office, *op. cit.* (Note 45), 3; Smith-Gordon and O'Brien, *op. cit.* (Note 23), 52.

47 Gurney, *op. cit.* (Note 7), 143–168.

48 J. M. Ludlow and L. Jones, *The Progress of the Working Class 1832–1867* (London 1867).

49 Co-operative Union, *Proceedings of the Annual Congress* (Manchester 1882), 3.

PART TWO

Geography

INTRODUCTION TO PART TWO

'Consumer society' is a term that is often bandied about with little if any consideration of *where* consumer societies are supposed to be. Although Part 1 considered *when* the consumer society is generally held to have originated and the changes it has undergone over time – revealing that these issues are far more problematic than is typically assumed – it is noteworthy that many of the extracts also took the *place* of consumption seriously. Places such as London, England; Paris, France; America; the British Empire; Europe; and the West figure prominently in such accounts. Indeed, it seems highly unlikely that one could compose an effective *history* of consumption and the consumer society without it being a historical *geography*. Part 1 provided an initial step in composing such a historical geography. The extracts in Part Two of the Reader take this further by concentrating on the specifically geographical aspects of consumption. In so doing, they raise a whole series of questions concerning such things as the precise *global reach* of consumerism; the significance of particular *spaces* of consumption; the way in which consumption *transforms* space and place, *displacing* and *dislocating* established practices; the possibility of *places* and *spaces* themselves being consumed; and the role of the *geographical imagination* in shaping practices of consumption.

The key theme that returns over and over again within this part of the Reader is the interrelationship between consumption and geography: the impact of consumption on particular places and spaces, and the impact of particular places and spaces on consumption. Not all the extracts focus on both sides of this relationship with equal weight. Some abstract from one the better to understand the other. Taken together, however, they enhance considerably our appreciation of the geography and historical geography of the consumer society.

It is a truism of geographical analysis that things always become more complex when one allows for the fact that not everything is the same everywhere – and that not everything occurs in the same way everywhere. Many of our taken-for-granted concepts – such as 'city', 'country', 'Third World', 'Western World' – essentially boil down to the fact that the world is *uneven* and *differentiated.* Cities contain more people, buildings, cars, and so on, than the countryside. The countryside contains more grass, cows, trees and tractors than do cities. The West contains more affluence, the global South more famine. These characterizations are obviously stereotyped, and risk appearing banal and trite. Yet they also point up the fact that the world is differentiated in ways that are often allowed to slip past our reflection on how society works. Once one pauses for thought, however, one begins to recognize that this unevenness implies not only a basic *contrast* between different places but also a *mutually constitutive* spatial relationship. For example, it is not *simply* the case that much of the world is poor while some of the world is affluent: much of the world is poor precisely *because* some of the world is affluent. At first glance, this might not seem to work for the other spatial contrast mentioned above – the contrast between urban and rural areas. Yet as urban historians and archaeologists have discovered from studying the earliest cities, in the beginning there were urban areas *because* there were rural areas capable of producing a surplus that could support and sustain an urban civilization. One of the surprising things about cities is that they have always been, in one sense or another, 'spaces of consumption' – a point that is picked up by a number of the extracts in this part of the Reader.

One other fundamentally geographical idea, which intersects with this sense of unevenness and differentiation, is the idea of *scale.* Occasionally, one encounters the notion that 'space' refers to one –

very broad – scale, and that 'place' refers to another – much more intimate – scale. This is, however, a misleading and inaccurate way of thinking. Different territorial areas do not simply map on to 'space' (at one end of the scale) and 'place' (at the other). It is perfectly meaningful, for example, to speak of 'domestic *space*' and 'a *place* like Los Angeles' – even though a living room or kitchen is clearly much smaller than LA. Space usually implies something more *abstract*; place something (or, rather, somewhere) more *specific* – without implying anything in particular about *scale*. If we speak of 'spaces of consumption', therefore, we do not usually imply a certain scale; we typically imply a certain *kind* of space – the kind of space given over to consumption activities. Nonetheless, scale is of fundamental importance when considering consumption. The extracts below range across a variety of spatial scales: global, international, national, regional, local, and the (for the human world) 'micro-geographical' scale of individual buildings and bodies. Consumption is seen to have an influence on each and every scale.

The contrast between space and place alluded to above also marks another crucial geographical distinction. Some of the extracts in this part of the Reader present more abstract considerations of 'space' and 'spaces', others more concrete considerations of particular 'places' – though sometimes the distinction seems to break down. This distinction might break down simply because our existing vocabulary is imprecise (one might speak of 'this kind of *place*' – when referring to a shopping mall – when 'this kind of *space*' might be a more accurate phrase). Or it might break down because the distinction between 'space' and 'place' has *really* been undermined – in certain circumstance, at least. For example, tourism – which is about the consumption of *place* and often depends on the *uniqueness* of places such as Paris, Thailand and Stonehenge – frequently stands accused of rendering places 'placeless' and turning 'places' into 'non-places'. Whether consumption as a whole tends to promote a monotonous cultural homogeneity is a moot point, but much of the evidence suggests that this is too simplistic a picture (even if it contains an element of truth). As certain of the extracts below suggest, consumption tends to *reconfigure* space and place, often disrupting, undermining and *displacing* consumption activities that were once thought of as being related to specific places (think of 'Italian' food, 'exotic' fruit, or even the humble potato). The complexities of geography tend to undermine all simple all-or-nothing generalizations, not least when it comes to consumption. The geography of consumption frequently seems to pull in two different directions at once – setting up contrasts between spaces that are *spectacular* and *seductive*, on the one hand, and spaces that are *ordinary* and *mundane*, on the other; creating – paradoxically – *homogeneity* and *heterogeneity* at the same time; promoting both *mobility* and *fixity* without contradiction; and ensuring that *space* and *place* sit alongside each other in a way that challenges unreflexive assumptions about the way the world works. Above all, the transformative force of consumption becomes evident as soon as one begins to take geography seriously.

As well as pointing forward to subsequent parts of the Reader, many of the extracts included here resonate with Part One of the Reader. As we noted at the outset of this Introduction, history and geography need to be considered together in order to provide an adequate framework within which to set out the social processes and practices of consumption. In our first offering, **Wolfgang Schivelbusch** considers the way in which the distinctive spaces of consumption we know today first developed, providing a useful historical contextualization of the growing body of work dealing with *spectacular* sites of consumption such as shopping malls and theme parks. His account tackles the issue obliquely, via a consideration of the transformative power of *light*. He nonetheless succeeds in showing that light was absolutely fundamental to the development of modern spaces of consumption.

As we discovered from Part One, consumers are never merely users of things (see **Barthes** and **Baudrillard**). While the world of the user is supposed to be ruled by cold calculation, with objects being evaluated on the basis of their usefulness and ability to satisfy needs, the world of the consumer is *seductive* and *enchanted*. But there is a paradox here. How can the consumer's world be seductive and enchanted, when everybody knows that science, reason and technology have *disenchanted* the modern world by disclosing the *objective* functioning of things? Schivelbusch answers this question by tracing the emergence of two very different *regimes of light* in the West. On the one hand, the development of various forms of street lighting over the past three centuries has given the ruling authorities, and especially the police, an increasing ability to bring public space under control. This regime of light has produced the harsh light of *surveillance*, which one finds in current perfection in technologies such as closed-circuit television systems and consumer-related databases for targeted marketing and credit referencing (see **Goss**). Little wonder, then, that violent resistance often accompanied the spread of street lighting, particularly when it was installed as a mechanism to facilitate the extension of the working day as well as the suppression of deviant behaviour. On the other hand, light has not always been either harsh or in the service of surveillance. Living flames have long been intimately related to occasions of *festivity* and *pleasure*. One thinks, for example, of the hearth, bonfires and fireworks. In addition, the *play of light* has always been associated with enchantment and spectacle. Consider in this regard chandeliers, sparkling crystal, reflections, and innumerable optical devices such as the camera obscura, phantasmagoria and cinematographé. Taken together, these two regimes of light provided the context within which something that is of massive importance for consumer societies gradually emerged: 'night life'. At first, those who were not compelled to work for a living extended the pleasures of conspicuous consumption (see **Veblen**) later and later into the night. In the nineteenth century, the pleasures of the night became available to the vast majority of the population. The hours after nightfall were, of course, never *entirely* colonized by the lighting of order and festivity. Nonetheless, while the lighting of order has gradually brought about a securing of the consumer's world, the emphasis has been displaced increasingly on to the commercial development of enchanted forms of commodified display. Indeed, in the trade-off between security and seduction, the former invariably defers to the latter (see **Abelson**).

Elegant shop interiors were first developed to cater to customers who were members of the aristocracy. The rich furnishings of such courtly shop interiors were initially at odds with the Puritan morality of the bourgeois merchants and traders. Yet as social life became increasingly anonymous and individualistic – and traders sought to appeal to a wider public than their established aristocratic clientele – the enchanted light of courtly shop interiors could no longer be contained. From the late eighteenth century the seductive light of enchantment spilled out into the streets of European capital cities, and eventually spread through the shopping arcades and department stores. By the end of the nineteenth century Western cities were filled with the light of festivity and seduction, and virtually all the spaces of consumption were bathed in an enchanting glow. Of particular importance was the invention of large plate-glass windows in the mid-eighteenth century which enabled the window to become a stage, the street a theatre, and passers-by an audience. Our own consumer society is still subject to these two regimes of light – both for seduction and surveillance (see **Bauman**). Taken together, they secure the enchanted dreamworlds that compose the principal spaces and places of consumption (see **Kiaer**).

While Schivelbusch is able to explain how the enchanted spaces of the consumer society could emerge alongside the disenchantment usually associated with modernity, **Peter Taylor** considers the historical geography of modernity from a very different direction. Instead of dwelling on the spaces and places of consumption, Taylor enquires into the fundamental logic of the modern world-system – a system that has been forged increasingly around consumerism. He is keen to stress the *ordinary* side of modernity – the way in which modernity becomes a normalized and naturalized aspect of everyday life (see **de Certeau**). In line with this concern, Taylor asks: 'Can we identify a single concept, like rationality or trust, to capture the essence of ordinary modernity?' Perhaps surprisingly, his answer is '*comfort*'. So begins Taylor's 'geological' account of the emergence and consolidation of the modern world in terms of the successive layering of different kinds of comfort. The seventeenth-century rise of mercantilism enabled the Dutch to create a private domestic space for family life. Nineteenth-century industrialization let the British transform this private space into a comfortable home, furnished with carpets, ornaments, armchairs, and so on. Finally, the emergence of consumer capitalism in the twentieth century meant that the Americans could refine comfort through the adoption of labour-saving devices and the associated spread of *suburbanization* (see **Miller**). On each occasion, the new form of comfort remade the world in its own image. As do McKendrick and Veblen, Taylor argues that this is achieved through *emulation* rather than coercion. While McKendrick and Veblen focused on emulation with respect to individuals and classes, Taylor considers it in relation to nation-states. Through emulation on a worldwide scale, the rising standards of comfort accomplished by these different historical developments have tended to become *hegemonic*: they have set the standards to which all other nations aspire. Ultimately, Taylor argues that even the antagonism between capitalism and communism was played out and resolved according to the logic of the struggle for comfort in everyday life. The fatal failure of communism was not ideological, but material: communist states were unable to create comfortable lives for their citizens. Yet if comfort provides the key to unlocking the secrets of the modern world as a consumer society, and the American Dream serves as the culmination of history, Taylor is at pains to point out that the world is not big enough to support the realization of this dream. As many environmentalists have noted, it would probably take six or seven planet Earths to sustain the expansion of consumerism to all four corners of the globe. For consumerism, then, the world is not enough. While this may mean that the end of the world is nigh, Taylor hopes that it is only the end of world *capitalism* that is nigh. In the meantime, the world will remain divided between a large 'zone of comfort' and a vast 'zone of struggle'. Furthermore, many would argue that on the one hand the zone of struggle is no longer confined to the so-called Third World, but is expanding its presence everywhere – even into the heart of the most affluent cities on Earth – and on the other hand that in an age of anxiety the zone of comfort can never be comfortable enough.

Whereas Taylor concentrates on the hegemonic powers of the capitalist world-system in general terms, **Jean Comaroff** considers one particular aspect of this situation. Comaroff focuses specifically on the impact of West European values on the Tsawna people of Southern Africa during the age of imperialism, and the *colonial* refashioning of local modes of dress. The 'civilizing' zeal of Protestant missionaries involved a sustained effort to clothe the (to Western eyes) 'naked' bodies of the indigenous people in subdued, dark-coloured garments in order to inculcate a Western sense of *decency*; to render Western clothing *ordinary*; and to 'awaken the desire for property and self-enhancement, for a life of righteous getting and spending'. It also involved an effort to reconfigure the existing *gendering* of sartorial matters: where indigenous clothing manufacture used animal hides, and was thus an extension of animal husbandry (and hence a male concern), the wives of the Nonconformist missionaries set up women's needlework circles, attempting to *domesticate* African women along European lines. Even so, the shortfall in the local supply of suitable garments necessitated, at first, a supply of *charitable* donations of 'cast-offs', which were secured readily from the Christian citizens of the colonial power. The situation ensured that the colonial frontier would eventually be brought within the sphere of *commercial* exchange: the civilizing mission proceeded from a moralistic starting point, yet it fitted hand in glove with the economic imperatives of Western production, commerce and trade. Nonetheless, the transformation of the colonial frontier was not without its contradictions and complexities.

The indigenous élite initially *resisted* the attempts to refashion their identities by reasserting *traditional* forms of dress. At a later stage, however, the contradictions between the conspicuousness of Western fashion and the ascetic values of Protestantism asserted themselves; not least as the 'enhanced language' of Western fashion combined with traditional clothing to produce a promiscuous kind of '*bricoleur* tailoring' – a hybrid combination of clothing styles that exasperated the missionaries by turning their best efforts against them. The refashioning process set in train by the missionaries was, however, completed ultimately by the *urbanization* that accompanied the development of a colonial labour force. While the efforts of the missionaries saw the development of a 'folk style' based on 'missionary-approved' clothing in the countryside, the public space of the city witnessed the widespread commodification of clothing, as 'Kaffir storekeepers' responded to the increasing demands of the ever-growing numbers of migrant labourers. The refashioning of the colonial subject, therefore, involved a complex interplay of missionary work, commercial expansion and local custom. It eventually gave rise to the adoption of a modified system of Western clothing, as the changing social structure of the local populace learnt to speak the language of Western fashion and to adapt it to their own ends (see **de Certeau** and **Simmel**).

Comaroff's contribution offers one example of the way in which people *learn* to consume – and not always in ways their mentors intended. **Michael Wildt** offers a very different example – West Germany in the 1950s – and neatly demonstrates that Comaroff's consideration of colonial Africa has a far wider bearing. As both Wildt and Comaroff show, the way in which *modernity* encroached on traditional forms of consumption was far more complex than is often assumed. With specific reference to food, Wildt shows how consumption involves a substantial amount of work on the part of consumers (see **de Certeau**), even when the things consumed promise to deliver 'convenience'. In line with a number of authors, Wildt emphasizes the *gendering* of such 'productive' consumption (see **Part 3**), particularly with regard to the rise of *domestic technologies* (see **Redfern**). There is a constant tension between 'efficiency' on the one hand and the loving expenditure (or 'waste') of time in the household economy on the other, as well as between 'convenience' foods and the superiority and relative cheapness of 'home-made' foods. Yet as Wildt's discussion of condensed milk shows, the overall trend has been away from practical considerations such as convenience and efficiency towards a more *aestheticized* rhetoric of consumption that operates through the 'invocation of desire' (see **Baudrillard** and **Falk**).

The shift towards a more aestheticized form of consumption and a broadening of tastes also involved a broadening of the consumer's horizons – in a way that drew explicitly on the *geographical* associations of food. The post-Second World War period saw geographically and pecuniary limited foods and tastes gradually give way to more *exotic* and *extravagant* ones, marking the beginnings of a transition in Western food consumption that would eventually lead to the situation described by **Cook** and **Crang**. (However, one should note that Western Europe has not always had limited food horizons. For example, in the Middle Ages pepper and spices were not only exotic products from the Orient, and some of the very first status symbols, they were literally considered to be tastes of Paradise.) The broadening of culinary horizons after the Second World War was given a very particular inflexion in the German case, in that it expressed the desire of the German people to re-enter the 'family of nations'. The shift in tastes also saw a transition away from traditional (or 'anti-modern') notions of a wholesome diet towards new views of healthy living based on a more 'scientific' nutritional knowledge – an awareness of calories, vitamins, and so on. New retail forms, such as *self-service* shops, reinforced these overall changes, as goods began to 'speak' directly to consumers without the intervention of sales assistants. The enchanted form of *commodified display* that was pioneered in department stores (see **Schivelbusch, Abelson, Bowlby** and **Schwartz**) was then extended to grocery retailing. However, the qualitative changes in consumption practices, as people learned to consume in an age of affluence, did not lead to any simple cultural homogenization – to an undifferentiated 'mass consumption society'. On the contrary, consumers learned to use the proliferation of goods to mark finer and finer distinctions between themselves and others (see **Bourdieu, Douglas, Falk** and **Veblen**).

Bringing the trends discussed by Wildt fully up to date, **Ian Cook** and **Phil Crang** offer a wholesale reconceptualization of the geography of consumption (and the consumption of geography) by considering the claim that London's cosmopolitan restaurant scene offers consumers 'the world on a plate'. Given Britain's colonial history, not only is Britain's 'national' cuisine effectively a matrix of worldwide combinations and permutations; the identities of other national cuisines are also drawn out of this abstract matrix. So, what counts as 'Indian' or 'Italian' food in Britain need bear little relation to the cuisine enjoyed in those countries – which in turn may or may not have escaped the maelstrom of globalization. Indeed, claims to 'authenticity' appear increasingly anachronistic in an age where the world is given to us on a plate, unless they are delivered as a rhetorical device. In an attempt to get to grips with this situation, Cook and Crang review some of the most common ways of representing the relationship between geography and food. Their aim is to reconcile two distinctive spatial figures or tropes through which cultural geography has commonly been articulated: the long-established trope of the '*cultural mosaic*' (whereby national, regional and local cuisines are figured as intimately related to place); and the more recent spatial trope of '*cultural flows and networks*' (which captures a much more fluid situation). These two figures have been *articulated* in two principal ways – neither of which adequately grasps the reality of the situation for Cook and Crang. The first – usually characterized as 'McDonaldization' – sees culinary cultures, embedded within locally meaningful contexts, as being increasingly *swamped* by the invasive flows of standardization and globalization. The second articulation offers a more optimistic view than the outcome of simple cultural homogenization allows: local cuisines do not submit passively to the overarching forces of standardization and globalization but resist this onslaught actively. Typically, for instance, changes that seem to 'descend from above' on particular places are woven into their traditional cuisines, producing a heterogeneous, hybrid or creolized form of cuisine. Cook and Crang suggest that both of these articulations amount to faulty conceptions. Culinary cultures have never been related strictly to place in a pristine, uncontaminated way – only to be infected at some later stage by external influences. It may help, therefore, to refigure the whole debate in terms of processes of *displacement* and *transformation*, whereby mobility and fixity in space are constantly renegotiated; where the senses of 'local', 'regional', 'national' and 'global' are appealed to without ever representing a 'pure' or 'authentic' situation; and where an ever-changing system of flows, intersections and connections positions both actors and places differentially in relation to the overall system. Cultural geographies of food, therefore, both actively constitute and are constituted by processes of displacement and transformation. Three 'interrelated geographies that constitute food-consuming worlds' are brought out particularly clearly by this conceptualization: local geographies of food consumption (which relate to spaces and practices of identity); extensive (often global) systems of food provision; and geographical knowledges (upon which consumers draw in managing their identities and determining their consumption habits). For Cook and Crang, this emphasis on displacement and connectedness – which implies the constant and ongoing reconstruction of culinary cultures – is far more capable of handling the complexities of the geography of food consumption (or any other kind of consumption, for that matter). In terms of food, then, postcolonial 'national' cuisines are no longer reducible to the countries and regions that bear their names. This situation carries important lessons for all other geographies of consumption.

While the geographical imagination is evidently stimulated, nourished and sustained by food consumption, tourism arguably represents a more literal 'consumption of geography'. **John Urry** offers a discussion of the peculiarities of consuming place, steering his way through a variety of economic conceptions to develop a more nuanced sociological view. Many debates on tourism have commented on its self-negating or self-defeating consequences: large numbers of tourists flock to particular places to 'get away from it all', to see an 'unspoilt' area of the world – and end up spoiling it, along with crowds of like-minded individuals. Urry points out that such a critical view is rather partial: increasing numbers of tourists might create 'bads' (such as congestion) but they also create 'goods' (such as better facilities). The consumption of place does, however, have the character of what Fred Hirsch calls '*positional*' goods: unlike the 'non-rival' consumption of, say, a radio broadcast (where my listening to a programme does not affect your listening to it as well), positional goods such as a beautiful natural landscape are subject to 'rival' consumption. If you try to enjoy the solitary contemplation of a scenic view, my trying to do likewise from the same vantage-point spoils your experience of solitude – just as you spoil it for me! There are many other kinds of positional goods besides those defined by natural scarcity, such as *luxury goods*, which construct scarcity in social terms. Tourism provides many prime examples of such positional goods, of rival consumption and of 'coercive consumption' – where one is compelled to try to outrun the crowd, yet destined never to escape it. It highlights the tyranny of individual decisions. Nonetheless, Urry argues that such debates are frequently oversimplified. To begin with, the form of consumption that tourism represents is primarily about looking and seeing (the words 'visit' and 'vision' share a common Latin root) – as a glance at any tourism marketing or advertising will readily confirm. Yet all too often, the '*tourist gaze*' is simply assumed to benefit from solitude. This 'Romantic' gaze, Urry argues, is all too readily overgeneralized. Much tourism relies not on solitude, but rather on being among others and participating in what he calls the 'collective tourist gaze'. Sometimes an atmospheric experience of place requires solitude; yet on other occasions it is equally likely to benefit from the presence of others, giving the sense of a 'shared experience'. So, although tourist 'honeypots' may be a bad experience for some people, that does not mean that they cannot be a good experience for others. While the *physical* 'carrying capacity' of places may arguably set an absolute limit to the number of tourists a place can handle, the *perceptual* carrying capacity depends on which kind of tourist gaze we are dealing with and the dispositions of the tourists themselves (see **Bourdieu**). Too much of the literature on tourism adopts an élitist attitude and an uncritical view of what tourists are seeking. While 'authenticity' may be a sought-after quality for certain kinds of tourists, it leads almost invariably to a kind of 'staged authenticity' in any case – and perhaps we have already reached the era of the 'post-tourist', who revels in a postmodern form of the consumption of place (see **Zukin**). After all, an *authentic* tourist experience is an authentic *tourist* experience: writing postcards and eating cotton candy, no less than communing with nature and appreciating edifying works of art. Whatever the case may be, an oversimplified assessment of tourism will surely lead to an unsophisticated appreciation of the peculiarities of the consumption of place.

Of course, people do not only consume places and spaces as tourists, even if this is the most obvious and explicit instance of the consumption of geography. We are, one might argue, *constantly* consuming space and place in the course of our everyday lives. One of the most dramatic recent developments that has helped to bring this background, taken-for-granted fact out into the open has been the process of *gentrification* in certain areas of the city. 'Gentrification' refers to the regeneration of rundown inner-city areas caused by the return of a new generation of affluent middle-class residents from which the middle classes once fled. But after years of decentralization and suburbanization, why and how has this reversal of earlier trends taken place? This is the geographical conundrum **Paul Redfern** sets out to address. Redfern claims that previous explanations of gentrification have offered a very limited understanding – particularly because they have tried to focus on *why* gentrification occurs. To get to the nub of the issue, Redfern focuses his attention on *how* gentrification occurs. His highly original explanation places changes in domestic-consumption technologies at the heart of the process. Earlier contributions to the gentrification debate tended to fall into one of two categories. They emphasized changes either in *demand* – such as the desire for city-centre housing by young professionals – or in *supply* – such as the 'rent gap' hypothesis, whereby housing stock, ripe for redevelopment, is available at a lower cost than new housing developments on the outskirts of the city. Both explanations, according to Redfern, are thoroughly *tautologous*: that is, they employ 'circular reasoning'. For example, demand-side explanations suggest that the demand for professional housing in central areas of the city explains gentrification – and the fact that gentrification is occurring in such areas is proof that such demand exists. Similarly, a 'rent gap' allows gentrification to take place – and the fact that gentrification is taking place suggests the existence of such a 'rent gap'. While many commentators have suggested that a combination of demand-side and supply-side factors explains gentrification, Redfern sees a need for an altogether different approach – although both of these elements are also crucial to his own account of gentrification. He focuses on the material object at the centre of the process: the *housing* itself – and the different kinds of *household* that can be supported by particular forms of housing. So, while the type of housing favoured by gentrifiers was in the past only maintainable by middle-class households if they had servants, the arrival of 'labour-saving' domestic technologies – such as central heating, dish washers and washing machines – has allowed a new generation of affluent middle-class professionals to modernize and occupy these houses (see **Miller**). Since gentrification is a fairly recent phenomenon, its *timing* is one factor that is explained by Redfern's account: it is the general fall in the cost of domestic technologies, relative to housing costs, that accounts for when gentrification first began to be a viable alternative to new suburban residence. So, while the 'rent gap' hypothesis and the demand by young professionals for housing in central urban locations both play a part in Redfern's account, it is the declining price of domestic technologies that represents the *mechanism* that allows the *potential* for gentrification to be *actualized*.

The *re-embourgeoisement* of the central city by gentrifiers and other agents of urban redevelopment has led ultimately to a far broader consumer-led urban renaissance, as **Sharon Zukin** aims to show. The contemporary city, she argues, is increasingly being made to the measure of the consumer society. Just as **Schivelbusch** dealt with an early phase of the *re-enchantment* of the disenchanted world within the sphere of consumption, Zukin considers the renewed generalization of this process across the contemporary city. Specifically, she focuses on the way in which strategies of urban redevelopment have been based increasingly on consumption – being centred on 'visual attractions that make people spend money' – highlighting the fact that these processes are riddled with tensions and fraught with difficulties. In particular, corporate investment in the 'cultural' or 'symbolic' economy – driven by entertainment, the arts, sports, gambling and retailing – constantly negotiates a tension between standardization and diversity. Paralleling **Cook** and **Crang**'s discussion, Zukin does not deny the potentially homogenizing forces of consumerism, captured in such notions as 'Americanization', 'Coca-colonization' and 'Disneyfication'. She also raises concerns about the privatization of formerly public urban spaces and the social exclusion of the poor by an aestheticization of the city that caters to the tastes of middle-class families and young urban professionals (**Redfern**'s gentrifiers). In addition, she also points to recent attempts by a whole series of actors and institutions to cater for the increasing variety of consumers and the explicit recognition by urban managers that the cultural diversity of the city is a sign of its vitality. Even the major corporate powers have recognized the significance of low-income, minority-group areas of the city and sought to target them (although this also indicates the expansion of private capital to fill partially a vacuum left by a receding welfare state). Consequently, the diversity of urban lifestyles and the vitality of city life make the most recent encounter between consumerism and the city far more complex than is sometimes assumed. To redeploy **Cook** and **Crang**'s terminology, the city might best be considered in terms of an ever-changing geography of displacement. This makes absolute certainties hard to come by, and necessitates a more nuanced analysis of the tensions it carries forward. It also demonstrates neatly that the geography of consumption provides as rich a source of insight into the nature of consumption as does its history. Consumption changes over space as much as it changes over time, revealing not only that consumption matters to geography, but also that geography matters to consumption.

Night Life

Wolfgang Schivelbusch

> On se promène, on flâne dans toutes ces rues, où le commerce entretient tous les soirs une illumination splendide.
> (Julien Lemer, *Paris au gaz*, 1861)

In the seventeenth century the two frontiers of the night – previously *terra incognita* – were discovered and thrown open at once. The police conquered and controlled the night by installing street lighting. Simultaneously with this *lighting of order*, a *lighting of festivity* developed in the form of baroque festivals of light and firework displays. 'There is no more brilliant spectacle, and none that is more popular at public celebrations', wrote Michel de Pure in his *Idée des Spectacles Anciens et Nouveaux* (1688). 'In almost all nations it serves to express joy at a great victory.'[1]

Using fire to express joy is an ancient custom. In its oldest form, it was a bonfire – usually a pyre that burned itself out in a wild blaze of light. Here fire, destruction and illumination merged to create an engulfing and complex experience that far exceeded that of simple brightness. The archaic bonfire could be described as a saturnalian version of the hearth fire: an outpouring of ecstasy as everyday restraints were suddenly lifted. Festive illuminations and fireworks replaced the bonfire, cultivating, ordering and disciplining its wild power. Thousands of candles and fireworks formed precisely calculated geometric patterns, a glowing transcription of court ceremonial into light, executed by a fireworks' master to display the brilliance of his sovereign's rule. None the less, these displays retained some of the pyromaniacal satisfactions of the original bonfire, so much so that in seventeenth-century France they were still known by the same name: *Feux de Joye*. After all, they consisted of balls of fire, showers of sparks and wonderfully exploding rockets – in short, a continuously changing spectacle of blazing light, just like a bonfire. Even when illuminations and fireworks had become aestheticised, the primeval power of fire would sometimes flare up in an unintended firework display.[2]

Spectacles of light were part of the festive culture of the baroque period. They lit up celebrations held at night – probably the most significant innovation of baroque courtly culture. In the Middle Ages and during the Renaissance, festivities had taken place in broad daylight. Now they began after sunset. 'At 8 or 9 the theatre starts, at midnight there is a supper . . . followed by dancing to daybreak. And when the coaches leave the court to go home at dawn, they meet the burghers in the streets, just going to work.'[3] Richard Alewyn's description hints at the motives behind this shift in the time of festivities. To enjoy oneself while working people slept and to go to bed when artisans and burghers were just starting their working day reversed the normal order of things. It was a social privilege that gave the evening's entertainment extra spice. Added to this were the qualities and states that, since time immemorial, had been associated with night as the antithesis of day: at night, regions that remained closed to people during the day were open to them; night-time brought one into a more direct relationship with the cosmos; it dissolved fixed forms and blurred the distinctions between reality and fantasy. When the night was magically lit up during a festive illumination the removal from reality – almost as if through the effects of a drug – was complete. The 'scene of a second, symbolic life', as Alewyn puts it, was created.

The baroque culture of the night spawned modern night life, which, since its conception in the cities of eighteenth-century Europe, has grown into a characteristic feature of present-day urban life. It began around 1700 in England with the creation of pleasure gardens such as Vauxhall and Ranelagh. They are best described as commercial imitations of courtly festive culture. The entertainments on offer – for the price of an admission ticket – included concerts, illumination and fireworks. Food and drink were available, and sometimes there was dancing. Although these gardens were open during the day, they only really came alive at night – at Vauxhall, between 6 and 8 o'clock. As the years passed, these times were pushed back further and further.[4] 'The present folly is late hours', remarked Horace Walpole in 1777. 'Everybody tries to be particular by being too late, and as everybody tries it, nobody is so. It is the fashion now to go to Ranelagh two hours after it is "over".'[5] In the eighteenth century, this was not merely a passing fashion; it was part of a wider movement.

Court society had underlined the distance that separated it from the bourgeoisie by ostentatiously keeping late hours, day and night. Now the middle classes tried to distance themselves from the petty bourgeoisie and the artisan class in the same way. The later one began the day, the higher one's social rank. Consequently, everything began to happen later and later. Since then, getting up early and going to bed early has become the mark of a simple life. Getting up late, getting to the office late (perhaps staying there late into the evening hours as well), taking late meals and, after an extended *soirée* (instead of simply 'knocking off work', like the lower classes), going to bed late – this has come to characterise the better social circles. What has happened to mealtimes makes this clear, as in the following description of Paris customs in 1801:

> Two hundred years ago, Parisians ate their main meal [dinner] at 12 o'clock midday; today, the artisan eats at 2 p.m., the merchant at 3 and the clerk at 4; the *nouveau riche*, entrepreneurs and bill brokers at 5; Ministers, Deputies and rich bachelors at 6. The latter normally finish their dinner at the hour when our fathers used to sit down to their evening meal. Three-quarters of Parisians no longer eat at night; half of them have adopted this custom for reasons of economy. Those who do eat an evening meal start at 11 p.m. and go to bed when the workers rise.[6]

At the time of Louis XIV, theatres began between 4 and 7 p.m.; in the eighteenth century, the starting time of 5.15 p.m. became established. Performances finished around 9 p.m.[7] The late hours that we know from Balzac's novels did not become customary until after the Revolution. The theatre or the opera were followed by supper, or a visit to a casino, a ball or a brothel. The evening usually ended at about three o'clock in the morning, when the revellers on their way home met the first workers going about their business.[8]

This new order of the day – or rather, of the night – marked not only the social gulf between the leisured classes and the working population, but also the difference between the metropolis and the provinces. In the early nineteenth century, German travellers, princes and artisans alike, frequently expressed their surprise at how late Paris and London remained awake. 'The opera does not finish until after 1 a.m.', Prince Pückler reported from London to his home in Saxony, 'one rarely gets home before 3 or 4 a.m. . . . But then high society does not come alive again before 2 p.m.'[9] But the elevated circles in which high-ranking people such as Pückler moved were not the only ones that stayed awake and lively so long; so did the commercial and amusement centres of the masses: 'Warehouses and shops mostly stay open until midnight, but then they usually do not open until 9 a.m. . . . The day, for business or entertainment, really lasts until midnight; not until then does some peace – in some quarters, total peace – descend.'[10]

What we think of as night life includes this nocturnal round of business, pleasure and illumination. It derives its own, special atmosphere from the light that falls onto the pavements and streets from shops (especially those selling luxury goods), cafés and restaurants, light that is intended to attract passers-by and potential customers. It is advertising light – commercialised festive illumination – in contrast to street light, the lighting of a policed order. Commercial light is to police light what bourgeois society is to the state. As the state, in its appropriately named 'night-watchman' function, guarantees the security that bourgeois society needs to pursue its business interests, so public lighting creates the framework of security within which commercial lighting can unfold. When shop lights go out after business hours, the light of the street lanterns, whose weak glimmer was drowned in the sea of advertising lights, becomes visible again and the lanterns go into action as the guardians of order that they have always been.

Unlike police lighting, which is uniform and homogeneous, commercial light is fed by heterogeneous sources. That is why, even today, it lends the city colour. Commercial lighting began in shops, and it still draws the bulk of its light from them.

SHOP WINDOWS

Until the late seventeenth century, shops for retail trade were little more than anterooms of the warehouses behind them. Indeed, their plain and simple furnishings made them almost indistinguishable from warehouses. But they compensated for their austerity with magnificent signboards, which hung out in the street showing what was on sale. When it was discovered, during the seventeenth and eighteenth centuries, that the signboards obstructed the traffic, these imaginative precursors of modern advertising gradually disappeared from the streets. As Sombart comments, 'the disappearance of these signboards, one after the other, almost as symbols of a dying age, was one of those momentous steps out of the cheerful world of words and colours into the grey world of figures'.[11] But this is only half the story. The colourful and aesthetic display of shop signs disappeared from the streets only to reappear, in a different form, inside the shop. A new combination of aesthetics and business was developing in the capitals of Europe. The luxury trade was in the hands of bourgeois merchants, but their customers were almost exclusively members of the court aristocracy. In the cities, the trade in luxury goods depended almost entirely on the court. To quote Sombart again: 'The elegant luxury shops, in particular, which had multiplied in Paris and London since the seventeenth century, served as popular meeting places for high society, for people who were happy to spend an hour of the day there, chatting, looking at the newest goods available and buying a few things, rather like at fashionable art auctions today.'[12]

The new social role of shops was reflected in their interior design. Catering for the taste of their customers, shop owners made their sales-rooms look like reception rooms in a palace. The most popular materials were rare woods, marble, brass and especially the high-status materials used at Court: glass and mirrors.

The new splendour was foreign to the bourgeois, puritan morality of traders who experienced this transformation in the early eighteenth century. 'It is a modern custom, and wholly unknown to our ancestors, who yet understood trade, in proportion to the Business they carried on, as well as we do, to have tradesmen lay out two-thirds of their fortune in fitting up their shops', writes Defoe in *The Complete Tradesman*.

> By fitting up, I do not mean furnishing their shops with wares and goods to sell; for in that they came up to us in every particular, and perhaps beyond us too; but in painting and gilding, in fine shelves, shutters, boxes, glass doors, sashes and the like, in which they tell us now, 'tis a small matter to lay out two hundred or three hundred pounds, nay five hundred pounds to fit up a Pastry-Cook's, or a Toy-Shop. The first inference to be drawn from this must necessarily be, that this age must have more fools than the last, for certainly fools only are most taken with shews and outsides . . . but that a fine shew of shelves and glass windows should bring customers, that was never made a rule in trade till now.[13]

As an example of a luxurious fitting-out, Defoe refers to a pastry shop in which £300 was spent on furnishings, while the stock was only worth £20. Defoe's description is one of the very few that have survived from the early period of the luxury trade:

1. Sash windows, all of looking-glass plates, 12 inches by 16 inches in measure.
2. All the walks of the shops lin'd up with galley tiles, and the Back-shop with galley-tiles in pannels, finely painted in forest-work and figures.
3. Two large Peir looking-glasses and one chimney glass in the shop, and one very large Peir-glass seven foot high in the Back-shop.
4. Two large branches of Candlesticks, one in the shop and one in the back-room.
5. Three great glass Lanthorns in the shop, and eight small ones.
6. Twenty-five sconces against the wall, with a large pair of silver standing candlesticks in the back room.[14]

The glass, mirrors and lights used in its furnishing must have made this shop a remarkably sparkling, reflecting, brilliant room – a miniature hall of mirrors.[15] In Defoe's time, all this splendour was limited to the interior of the shop. This changed with the social profile of the customers, as increasingly anonymous buyers replaced what had been a largely personal clientele. The more the streets could supply potential customers, the more the shops opened up to them. The display window, that began to develop as an independent part of the shop around the middle of the eighteenth century, was the scene of this interchange. While previously it had been little more than an ordinary window that permitted people to see into and out of the shop, it now became a glassed-in stage on which an advertising show was presented. 'Behind the great glass windows absolutely everything one can think of is neatly and attractively displayed in such abundance of choice as almost to make one greedy', wrote Sophie von La Roche from London in the 1780s. 'There is a cunning device for showing women's materials. They hang down in folds behind the fine, high windows so that the effect of this or that material, as it would be in a woman's dress, can be studied.'[16]

In the eighteenth and nineteenth centuries, shop display windows still looked like ordinary windows. They consisted not of a single sheet of glass, but of several smaller panes, separated by a number of ribs. Around 1850 it became technically possible to produce large sheets of glass and so to have a glass shop-front which presented 'an uninterrupted mass of glass from the ceiling to the ground', as an observer pointed out admiringly in 1851.[17] This had a profound impact on the appearance of the wares on display. The uninterrupted, transparently sparkling surface acted rather like glass on a framed painting. 'Dull colours receive . . . an element of freshness, sparkle and refinement, because glass as a medium alters appearances and irritates the eye' – this is how Hirth explains the phenomenon. He adds in a footnote: 'Putting paintings under glass makes them appear better than they really are. The protective glass confers upon good copies an additional element of deception. The plate glass of shop windows, too, has an "improving" effect on some goods.'[18]

Artificial light also helped to make the wares on display look more attractive. Its importance grew as business hours were extended into the late evening. During a visit to London in 1775, Lichtenberg observed how shops drew attention to their windows by special effects with coloured lights: 'Apothecaries and grocers display glasses . . . filled with coloured spirits and cover large areas with crimson, yellow, verdigris and skyblue light. Confectioners dazzle the eye with their chandeliers and tickle the nose with their wares at no greater effort or cost than turning both in their direction.'[19] Berlin pastry cooks displayed in their windows 'artificially lit scenes, populated by small, three-dimensional figures, often artificially animated, the whole thing resembling a diorama'. In fact, these displays are thought to have inspired the development of the diorama.[20] Mostly, however, shop window lighting followed the path taken by stage lighting. As long as lights were too weak to be used indirectly, that is with the aid of reflectors, they were placed among the goods in the window. When gas and electricity increased the range over which light could be cast, the source of the light itself disappeared from view. Around the middle of the nineteenth century gaslights on London shops were 'fixed outside the shop, with a reflector so placed as to throw a strong light upon the commodities in the window'.[21] The introduction of electric light, which was not a fire hazard and therefore no longer had to be installed outside the display window, finally made it possible to achieve the sort of lighting effects that were used on the stage. A 1926 advertising handbook states that shop windows should not be 'evenly lit up. Individual spots and objects are to be highlighted by means of strong, concealed reflectors.'[22]

The illuminated window as stage, the street as theatre and the passers-by as audience – this is the scene of big-city night life. As the boulevard at night developed in the nineteenth century, it did in fact look like an interior out of doors. 'Always festively illuminated, golden cafés, a stylish and elegant throng, dandies, literati, financiers. The whole thing resembles a drawing-room' – this is Emma von Niendorf's 1854 description of the Parisian Boulevard des Italiens late at night.[23] There is a simple psychological explanation for the fact that the street looks like an 'Intérieur', to borrow Walter Benjamin's expression.[24] Any artificially lit area out of doors is experienced as an interior because it is marked off from the surrounding darkness as if by walls, which run along the edges of the lit up area. The same applies to the 'ceiling'. Common usage shows that we step *out* of the darkness *into* a circle of light – be it the small one of a camp fire or the larger one of a lighted boulevard. The 'side walls' of the boulevard, as a 'room', were defined by the housefronts – shop windows, restaurants and café terraces; its 'ceiling' was at the limit of the commercial lighting, that is, at about first-floor level.

Before shop lighting created an 'interior' space out of doors, however, it went through a transitional phase, developing in an area that was bigger than the individual shop, but not yet as big as the open boulevard: the glass-roofed arcade, gallery or passage. This type of commercial space was most highly developed in Paris. The Galérie Orléans in the Palais Royal, *the* centre of Paris night life between 1790 and 1830, was the first of its kind. 'It makes a splendid sight indeed', reported a German traveller in 1800 (i.e. before the introduction of gas lighting), 'tasteful Argand lamps illuminating the shops in the evening and at night, luxury goods sparkling with a heightened brilliance, and bright *réverbères* lighting up a packed, surging crowd in the arcades'.[25] Here is the same view under gaslight, described thirty years later: 'A thousand lights are reflected in the surface of polished mahogany and in the large mirror walls. . . . The stranger dazzled by all this begins to think of the Palais Royal as a bazaar.'[26] When the Palais Royal fell out of favour after 1830, Paris's numerous arcades took over the role of the Galérie Orléans in a process of decentralisation that was also an expansion of night life. Contemporary descriptions of these new venues are couched in exactly the same terms as descriptions of the Galérie Orléans. People seem to have been fascinated by the interplay between the brilliance of the light and the wares on display, and the lively crowd. 'A labyrinth of iridescent passages, like rainbow bridges in an ocean of light. A totally magical world. Everything, or rather, much more than the imagination could devise.'[27]

There is a final step in this progression of light: the emergence of light from a roofed-in space into the open air. The following descriptions are of boulevards, but they could just as well be about arcades or the Galérie Orléans. 'Glittering shops everywhere, splendid displays, cafés covered in gilt, and permanent lighting. . . . The shops put out so much light that one can read the paper as one strolls.'[28] 'The gas-lamps sparkle and the suspended lamps glow, and in between there are the tobacconists' red lanterns and the chemists' blue-glass globes – transparent signs announce the marvels of the Paris night in large, fiery letters, and the crowd surges back and forth.'[29]

Impressions like these can be found in city guides and travel reports published between 1850 and 1870, with titles such as *Gas-Light and Day-Light*; *Paris au gaz*; *Paris bei Sonnenschein und Lampenlicht*; and *New York by Gas*.[30] These two decades were the heyday of gas lighting. It had become firmly established in the cultural and psychological structure of Western European and American society. Earlier reservations and fears had disappeared, and its modern successor, electric light, had not yet arrived on the scene. Gaslight, like the railway, reigned supreme as a symbol of human and industrial progress.

All the same, gaslight still burned with an open flame. However functional, neutral and rational it seemed in comparison with earlier forms of lighting, it retained the lively, magical quality of an open flame. It was both a modern, expansive source of light that illuminated incomparably larger spaces than any earlier form of lighting, and an 'old-fashioned' light still bound to the flame. This combination was probably the source of its appeal as a medium of night life in the capitals of Europe between 1850 and 1870. Lighting up the night with gas stirred people's feelings because it represented a triumph over the natural order, achieved without the lifeless hardness of electric light. Gaslight offered life, warmth and closeness. This was true also of the relationship between light and the shop goods upon which it fell. They were close to each other, indeed, they permeated each other, and each enhanced the effect of the other, to judge by descriptions of illuminated luxury shops.

Here, too, electric light injected an element of rigidity, coldness and distance. It burst open the 'ceiling' of the boulevard 'salon' by lifting it to roof level. From now on, commercial light shone down from this distant position, detached from the display windows, in its own, independent sphere. In 1928 Ernst May describes how this symphony of advertising lights in Times Square affected him: 'Here the eye does not read any writing, it cannot pick out any shapes, it is simply dazzled by a profusion of scintillating lights, by a plethora of elements of light that cancel out each other's effect.'[31]

NOTES

1 Michel de Pure, *Idée des Spectacles Anciens et Nouveaux* (Paris, 1668; reprint Geneva, 1972), p. 183.

2 'In France, the connection between the bonfire and fireworks remained alive for a long time. As late as the eighteenth century, Midsummer Day was celebrated with a "Feu de joye complet", that is with a bonfire plus fireworks, in every French town with any claim to significance. And in Paris around the middle of the eighteenth century a rough pyramid of wood would be piled up next to the elaborate edifice for the fireworks in front of the Hôtel de Ville on Midsummer Eve and lit, just as a bonfire had been lit on this day in ancient times to celebrate the summer solstice' (ibid., p. 57). In the perfected firework displays of the baroque period, the edifice from which the fireworks were let off took over the function of the bonfire. 'At the end, the beautiful monument that had served its purpose was burnt, together with the boats that had carried it, in a huge bonfire that rounded off the festive day' (A. Lotz, *Das Feuerwerk* (Leipzig, n.d.; reprint 1940), p. 61, describing a floating firework display at Versailles in 1674).

3 Richard Alewyn, *Das Große Welttheater* (Hamburg, 1959), p. 31.

4 Warwick Wroth, *The London Pleasure Gardens of the 18th Century* (London, 1896), p. 305. These times apply to the second half of the eighteenth century. Before 1760, concerts began around 6 or 7 p.m. (ibid., p. 303).

5 Quoted from Walter Sidney Scott, *Green Retreats: The Story of Vauxhall Gardens 1661–1859* (London, 1955), p. 15.

6 Jean Baptiste Pujoulx, *Paris à la fin du 18e siècle* (Paris, 1801), pp. 141–142; see also Wolfgang Nahrstedt, *Die Entstehung der Freizeit, dargestellt am Beispiel Hamburgs* (Göttingen, 1971), pp. 115ff and 186ff.

7 Gösta Bergman, *Lighting in the Theatre* (Stockholm, 1977), pp. 144 and 149; Louis Sébastien Mercier, *Tableau de Paris*, (Amsterdam, 1782), Vol. 3, p. 88.

8 This is how books on night life generally finished. They were published mainly in the 1850s and 1860s – for example, Julien Lemer, *Paris au gaz* (1861), Alfred Delvau, *Les Heures parisiennes* (1866), and Julius Rodenberg, *Paris bei Sonnenschein und Lampenlicht* (1867).

9 *Fürst Pückler reist nach England. Aus den Briefen eines Verstorbenen* (Stuttgart, n.d.), pp. 144–145.

10 August Jäger, *Der Deutsche in London. Ein Beitrag zur Geschichte der politischen Flüchtlinge unserer Zeit* (Leipzig, 1839), 2 vols in one, pp. 188–189.

11 Werner Sombart, *Der moderne Kapitalismus* (1st edn, Munich, 1902), Vol. 2, p. 402.

12 Ibid., p. 463; see also Sombart's study, *Kapitalismus und Luxus*. In it, he argues that the court aristocracy's consumption of luxury goods ruined it financially while at the other end of the same process, the bourgeoisie's role as supplier of luxury goods was the thing that allowed the bourgeois economy to begin to blossom.

13 Daniel Defoe, *The Complete Tradesman* (2nd edn, London, 1727; reprinted New York, 1969), Vol. 1, pp. 257–258.

14 Ibid., p. 259.

15 The use of large mirrors in seventeenth-century palaces has been explained as an attempt to create *an illusion of space*. 'High walls, doors, even ceilings are more and more often covered with mirrors. Their purpose, however, is not to reflect any particular, delimited image, but to give the impression of a scintillating, rather kaleidoscopic and uncertain sum of light and decoration' (Georg Hirth, *Das deutsche Zimmer der Gotik und Renaissance, des Barock- Rokoko- und Zopfstils* (Munich and Leipzig, 1899), p. 154).

16 Quoted from Dorothy Davis, *A History of Shopping* (London and Toronto, 1966), p. 192.

17 Charles Knight (1851), quoted from Alison Adburgham, *Shops and Shopping 1800–1914* (London, 1964), p. 96.

18 Hirth, *Das deutsche Zimmer*, p. 152. Albert Smith gives us a good example of what a glance through such a window revealed: 'How richly falls the drapery of those emblazoned shawls through the fair plate-glass. How the rows of loves of bonnets . . . gladden and sadden at the same moment the bright female eyes. . . . How gorgeously shines the plate' (*Sketches of London Life and Character* (London, 1859, p. 117), quoted from Wilfried B. Whitaker, *Victorian and Edwardian Shopworkers* (Newton Abbot, 1973, p. 31)).

19 Letter to Boie, 10 January 1775, in *Lichtenbergs Werke in einem Band* (Stuttgart, 1924), pp. 356–357.

20 Marianne Mildenberger, *Film und Projektion auf der Bühne* (Emsdetten, Westphalia, 1961), p. 22.

21 Charles Knight, quoted from Adburgham, *Shops and Shopping*, p. 96. 'Reflectors made of nickel silver or mirror glass with a parabolic cross-section are used to illuminate shop windows. They reflect the light perpendicularly down, so that the goods on display in the window are very brightly lit up' (C. Muchall, *Das A-B-C des Gas-Consumenten*, Wiesbaden, 1889, p. 23).

22 Bruno H. Jahn, *Reklame durch das Schaufenster* (Berlin, 1926), p. 130.

23 Emma von Niendorf, *Aus dem heutigen Paris* (Stuttgart, 1854), p. 171.

24 Walter Benjamin, *Charles Baudelaire. Ein Lyriker im Zeitalter des Hochkapitalismus* (Frankfurt, 1969), p. 37.

25 *Reise nach Paris im August und September 1789* (no place of publication, 1800), p. 180.

26 *Le livre des Cent-et-Un* (Paris, 1831), Vol. 1, pp. 19–21.

27 Niendorf, *Aus dem heutigen Paris*, p. 169.

28 Lemer, *Paris au gaz*, p. 15.

29 Julius Rodenberg, *Paris bei Sonnenschein und Lampenlicht* (Leipzig, 1867), p. 45.

30 Detail on p. 88, n. 8, plus George G. Foster, *New York by Gas-Light* (New York, 1850). Lemer gives the following list of words relating to night life that were in vogue in Paris around 1860: *noctivague, noctilogue, noctiphague, noctiurge, physiologie de l'existence de nuit à Paris*.

31 Quoted from W. Lotz (ed.), *Licht und Beleuchtung* (Berlin, 1928), p. 44.

What's Modern About the Modern World-system?

Introducing Ordinary Modernity Through World Hegemony

Peter J. Taylor

UNEVEN SOCIAL CHANGE: WORLD HEGEMONY

Although rapid change is the norm in the modern world-system, the rate of social change that has to be coped with is not a constant. There are many social theories that emphasize the unevenness of change both over space, as core-periphery at different scales, and over time, as cycles of varying lengths. In the modern world-system, world hegemony defines a particularly acute uneven pattern of social change.

World hegemony is seen as a property of the whole system and not just of the hegemon itself (Arrighi, 1990; Hopkins, 1990). Hegemonic states are particular core states that appear at specific conjunctures in the development of the world-system and are implicated in the overall development of the system. In short, the capitalist world-economy has evolved through rather long cycles which we term hegemonic cycles. These are the eras that encompass the rise, achievement and subsequent decline of a hegemonic state and which define the changing nature of the whole system.

World hegemons are ultimately defined by their pre-eminent success in capital accumulation. Wallerstein (1984) emphasizes their initial productive economic edge which feeds into commercial advantage and culminates in becoming the financial centre of the world-economy. This centre has moved from Amsterdam to London to New York over the last 400 years. 'High hegemony', the apogee of the cycle, is defined when the three economic advantages are located in a single state. Given such a dominance of the world market, the state in question becomes a liberal champion in world politics: the Dutch promote *mare liberum*, the British free trade and the Americans free enterprise. In each case the special interests of the hegemon are presented as universal interests of the whole system in classic Gramscian mode.

High hegemony is more than quantitative economic advantage, however. Each case of world hegemony has arisen out of a major world war, what Arrighi (1990) calls systemic chaos, when the hegemon not only becomes the productive arsenal of the eventual winning side but also shows itself to be an astute balancer of power. The end result is that while all main rivals, both allies and enemies, are devastated by war, the hegemon has a 'good war' with its economy greatly boosted by it. Hence, the hegemon emerges from the war qualitatively different from all other states, the great exception that shows the way to postwar reconstruction. High hegemony is achieved when the capitalist class use this opportunity to project their state's unrivalled economic power and create a 'new world' in their own image and supportive of their purposes. This is expressed in Arrighi's (1990) treatment of world hegemony as stages in world-system development: first Dutch commercial capitalism, second British industrial capitalism, and third US consumer capitalism. High hegemony, therefore, is associated with periods of the most intense social change as new political economies are constructed by the world leader.

THREE MODERNITIES

Berman (1988: 16–17) identifies three main phases of modernity, the first extending from the sixteenth to the eighteenth centuries, the second in the nineteenth century, and a contemporary modernity in this century. At the cultural-ideological level these phases can be defined as intellectual attempts to come to terms with a rapidly changing world by ordering knowledge so that people are subjects as well as objects of change. Three periods of major ordering are usually identified: in the seventeenth century a cartesian world was devised with 'man' at the centre; the nineteenth century was the high mark of change interpreted as human progress; and in our century the idea of change has been globalized and repackaged as development in modernization theory. Each of these is a theoretical taming of the perpetually new by application of a rationality privileging science and technology.

The reader will note immediately the temporal correlation with the three hegemonic cycles. The proposition of this article is that the crucial aspects of the social-cultural dimension we call modernity have evolved with and through the hegemons. This follows from our discussion of hegemony above. Since the rate of change will accelerate precisely when hegemons are creating a new political economy through their restructuring of the world-economy, we can expect this to stimulate a basic need for intellectual reassessments: a new rationalization of the newness has to be created. Hence the three new forms of modernity can be interpreted as the social-cultural reaction to hegemonic creation of uncertainties. Each hegemon has been responsible for creating its particular version of what is modern about the modern world-system.

The first modern world is that of the merchants, the everyday life of commerce and the massive coterie of activities that it generated. There had, of course, been many influential networks of merchants in the past but in the seventeenth century a new rationality came to dominate success in the world-system. In it the calculating behaviour of the merchant was the archetypical practical form: navigation became the great enabling applied science. The Dutch more than anyone else created this modernity: they made making money respectable. In the nineteenth century great commercial ports gave way to massive industrial towns as the archetypical modern. In this industrial world, modern society became mass society: the modern became, in Marx's terms, an alienated way of life. As the merchant gave way to the industrialist, mechanical engineering became the great enabling applied science. There is no doubt that Britain is the country most implicated in bringing this second modern condition about. In the twentieth century the social effects of alienation have been countered in selected countries by a spread of affluence to ordinary people. Suburbia and its ubiquitous shopping mall have become the focal modern place in the new consumer society. Here it is management science with its communication and computer technologies that has been the enabling applied science of our era. There can be no doubt that Americans have been the major purveyors of contemporary modernity. Hence we conclude that the hegemons have been directly instrumental in creating new ways of life that have dominated the modern world-system, they have been the leading architects of modernities.

We should expect nothing less from world hegemons, of course. As the prime actors in the history of the modern world-system, it is they who should define what is modern about the system in their eras. Because each hegemon is indisputably a great success story in its path to high hegemony, its rivals try and emulate it if they can. After the success of the Dutch, other states commercialized their activities in a process known as mercantilism. After the success of the British, other states industrialized to prevent 'falling behind'. After the success of the Americans, other states that could created their own 'affluent societies' based upon mass consumption. This is what is meant by hegemonic 'intellectual and moral leadership' at the world-system scale. This emulation implies no coercion, merely a general consensus about what it is to be modern in a given era [. . .]

SOCIAL PRACTICE: HOME COMFORTS FOR ORDINARY PEOPLE

Ordinary modernity as comfort

To say that modernity has a 'dark side' has become commonplace in the twentieth century. In particular, the 'industrialization' of war has created a litany of horror – the gas warfare of the trenches in the First World War, the Holocaust and the atomic destruction of Hiroshima and Nagasaki in the Second World War, the threat of global nuclear annihilation in the Cold War – far worse than anything experienced in any other world-system. No wonder those writing about modernity insist on viewing it as 'a double-edged phenomenon' (Giddens, 1990: 7). But writers are untypical in their reaction to modernity. For most people, although certainly aware of the horrors, to be modern is considered a good thing. The idea of something being modern is that it does what it does better than what it replaces. Although as applied to bombs this has a terrible outcome, for most of the time the modern's most widespread effect has been to create a better life for many ordinary people. Whether they enjoy the good life or cherish hopes for a good life, for so many people across the world in the twentieth century it is this idea of progress to a better future that has made modernity so popular despite its dark side. Of course, advertisers understand this when they present the commodity they are promoting as 'modern'; unlike academic writers, this profession of scribes interprets being modern as a single-edged sword for good(s).

What is this 'ordinary' modernity? In general we can note that modernity is a very complex, multifaceted concept enabling many different interpretations of its core characteristics. Most commonly its 'rationality' is emphasized, especially in historical arguments: being rational lay behind the scientific revolution of the seventeenth century, provided the cutting edge for the Enlightenment in the eighteenth century, and created the technology of the nineteenth century that made universal progress seem so inevitable. Recently Anthony Giddens (1990), with his eye firmly focused on contemporary globalization, has argued that modernity depends upon new relations of trust. All world-systems separate everyday life experiences from important determinants that are system-wide but the modern world-system deals with this 'disembedding' distinctively. Whereas pre-modern societies treated 'extra-local' effects in terms of divine providence, under modernity such risks are dealt with through trust in ordered systems of social relations that operate to integrate local practices into the system. The contemporary explosion of reliance on 'plastic' money, for instance, is only possible because of widespread trust in the system that operates this banking innovation. But both the philosophers' and historians' arguments for rationality and the sociologists' and economists' arguments for trust are 'top–down' interpretations of modernity that seem far from what we will term 'ordinary modernity'. Studying modernity should focus on the fact that our society is self-consciously modern. The popularity of modernity in the twentieth century has been crucial, not only for sustaining modernity, but also in influencing its changing nature. So can we identify a single concept, like rationality or trust, to capture the essence of ordinary modernity? I think we can, and I offer comfort as my candidate for the concept to take on this onerous role.

What can we mean when we elevate the commonplace notion of comfort to be a core concept of modernity? Like rationality and trust, once we begin to look at comfort seriously it becomes a very complex idea. Although basically meaning to make life easy, in practice this can and has been viewed in many different ways. Witold Rybczynski (1986: 231) writes of the 'mystery of comfort' because it is impossible to measure it objectively: 'it may be enough to realize that domestic comfort involves a range of attributes – convenience, efficiency, leisure, ease, pleasure, domesticity, intimacy, and privacy – all of which contribute to the experience; common sense will do the rest.' But, of course, this common sense is only common to modern society. In the invention of medieval society as a foil for modernity, this pre-modern setting may have been romantically enchanting but it was never viewed as comfortable. Every suburbanite appreciates that her or his life has a degree of comfort unattainable in medieval Europe, even for a king and queen in their (draughty) castle. Of course knowledge of sumptuous lifestyles in other civilizations, both classical Greece and Rome and the great oriental empires, is part of the modern historical consciousness but it is understood that these extreme images of comfort were enjoyed by the very few, a very small elite living on the backs of the many. Modern comfort, in contrast, is available to the ordinary man and woman, the descendants of serfs and slaves, not kings and emperors. Modernity, therefore, is for ordinary people.

Home, sweet home

Contemporary houses where ordinary people live usually have a hall, a small room behind the front door where the stairs begin and outside clothes are hung. Medieval houses of merchants and the nobility also had halls but these were very different and dominated the whole house. These two house arrangements represent two completely different worlds. The great halls of the past were multiple-use spaces that were essentially public in nature: a place for transacting business, for cooking and eating, for entertainment and for sleeping at night (Rybczynski, 1986: 25). Today the hall has been reduced to a small transition zone between a private home and a public outside. It is the home's frontier, where we enter and leave, where visitors are dealt with. Though important for providing a home's initial image to the outside world, the hall does not have to be comfortable like other rooms because it is not a place where you are expected to stay for long. If the visitors are friends they may be invited beyond the hall to specialized rooms that are prized for their comfort and convenience – living-room or lounge, dining-room, kitchen, bathroom and bedrooms. It is this private space of the home that is the essential locus of ordinary modernity.

This modernity beyond the hall had to be created and the three hegemons are intimately implicated in the invention of the home and making it comfortable. The modern sense of the word 'comfort' only came into usage in the eighteenth century but 'there was one place', according to Rybczynski (1986: 51), where the 'seventeenth century domestic interior evolved in a way that was arguably unique, and that can be described as having been, at the very least, exemplary'. This was, of course, the United Provinces where economic successes went hand in hand with new cultural behaviours. Dutch merchants were family-oriented and lived in small households with few or no servants. Children were integral to the new family life and, instead of being apprenticed, stayed at home and attended school. In this way the Dutch invented what we understand today as childhood (Rybczynski, 1986: 60). New town houses were built or converted to adapt to these new practices and therefore eliminated the traditional mix of work with residency. The ground floor was still treated as a public space to entertain visitors but the upper floors were a special family place. Friends allowed access to the upstairs were required to remove their street shoes so as not to disrupt the cleanliness of the home.

This boundary between public and private spaces was a new idea which enabled houses to become homes (Rybczynski, 1986: 66). Here the Dutch could express their personal likes and dislikes in the context of individual family tastes such as decorating their walls with maps or paintings. A contemporary critic of the Netherlands, Sir William Temple, after noting the propensity for monetary tightness among the Dutch, gives one exception: expenditure on furniture and other commodities for the home (Rybczynski, 1986: 61). All in all, the Dutch produced the first modern homes in the world, cosy settings for the family life of ordinary people.

In the eighteenth century, interior comfort advanced in two ways. In France, aristocratic and royal tastes were converted into sumptuous fashions for palaces. In more bourgeois England where the court had little cultural influence, a simpler comfort was developed for the country houses of the middle classes. This Georgian style is a fashion that refuses to become outdated because of its suitability for comfortable home life. This was the era of the Chippendale chair and the ubiquitous 'Windsor chair', both popular because they were so comfortable to sit in (Rybczynski, 1986: 115–116). English economic successes provided the enabling wealth for new cultural practices related to leisure, and the home became the locus of card games, dinner parties, various entertainments such as billiards and dances, and, of course, it was where the latest novels were read. Rybczynski (1986: 112) thinks it highly relevant that Jane Austen so frequently used the words 'comfort' and 'comfortable' in what he describes as her 'domestic genre of novel writing', focusing on the home life of England's middle classes. She even coins the phrase 'English comfort', two words that seemed naturally to go together. In fact it was the English who brought carpets down from the walls to make floors quieter and nicer to walk on. They even introduced wall-to-wall carpeting for the first time. But their main contribution to domestic comfort was to bring furniture out from the walls. It is among the eighteenth-century English that the arrangement of sofas and easy chairs around the fireplace, the epitome of warm cosiness, makes its appearance in the home (Rybczynski, 1986: 118). It is this 'English taste' that spread through Europe and America as ordinary homes with their many personal 'knick-knacks' become the hallmark of the 'Victorian age'.

It is very interesting that the cultural practices associated with British hegemony are named after a monarch, Queen Victoria. The 'Victorian age' evokes the set of social values that mark out the bourgeois ordinariness of nineteenth-century British civil society. With Victoria's accession to the throne in 1837, the tenor of the court changed to match a new decorum at variance with the old Regency manners. This more sober court was confirmed by her marriage to Albert whose concern was for new industrial progress rather than traditional aristocratic pursuits. The result was that while remaining the pinnacle of Britain's social class system, the Queen also cultivated a domesticity: Victoria, Albert and their children came to personify the ideal family. The royal family was conceived as both special because it was royal and 'just like us' because it was a family. 'Victorian values' became a synonym for what were considered ordinary bourgeois qualities: the Queen's devotion to duty was equated with middle-class hard work, her simple lifestyle without ostentation was seen as middle-class thrift, and although she was constitutionally head of the Anglican Church her court's probity and good behaviour matched the 'Low Church' high-minded morals of her middle-ranked subjects. In this way a monarch gave her name to some quite unroyal patterns of behaviour. But not only in Britain; this middle-class culture had a much wider scope than the land Victoria actually ruled. In fact, one of the ironies of history is that Americans call their culture in this period 'Victorian America', only two or three generations after freeing themselves from the kingship of Victoria's grandfather. In his book *Victorian America*, D.W. Howe (1976: 3) admits being embarrassed by the term but uses it because it has become firmly part of the historical lexicon. Such is the cultural dominance of the hegemon.

Of course, even as an icon of domesticity, Victoria had residences which were palaces, castles and great houses. Nevertheless, the popular image of the Queen, especially in her later years, was a homely one, with her shawl over her clothes as she drew her chair up to the fire to keep warm. This was a ritual all her subjects engaged in and although the image is a cosy one, it also indicates limitations of the English home in the nineteenth century. In fact, there is a major 'comfort gap' between 'English comfort' and what we have come to expect as comfort today. Lighting, heating and plumbing lagged behind furniture in the development of the home. For most of the eighteenth century artificial lighting still came from wax candles and although new oil lamps appeared by 1800, it was not until the 1840s that gas lighting in homes became popular. But gas lighting was dirty and added to the smoky atmosphere created by poorly fumed fires. There were many inventions to try and make the home more efficient but they all foundered on one key stumbling block: the lack of access to power in a form small enough for domestic needs. Comfort depended upon servants until the introduction of the electric motor.

The introduction of electricity into the home illustrates well the transition from British to American cultural leadership. The first domestic light bulbs were invented independently in Britain and the USA in 1877 by Joseph Swan and Thomas Edison respectively. But there the similarity ended. In 1880 Swan's invention was used to light up Cragside, the Northumbrian home of Lord Armstrong, the first house to have electric lights (Rybczynski, 1986: 150–151). This was exceptional, however. By now social support for such innovation was disappearing in England; Rybczynski writes of 'a curious situation' in England whereby modern conveniences came to be seen as vulgar. In contrast, by 1882 Edison had laid electric cables in part of New York, so that 200 houses were using 5,000 Edison bulbs for their lighting (Rybczynski, 1986: 150). This was the lead the rest of the world followed and by 1900 major cities throughout America and Europe had electric companies supplying current through underground cables. Once such distribution networks were in place, 'home appliances' could be produced with electric motors finally to extend comfort from domestic leisure to domestic work – what Rybczynski (1986: 158) refers to as 'the great American innovation in the home'. At the 1893 Chicago World Fair an 'electric kitchen' was exhibited and very soon a large range of appliances was widely available from such well-known companies as Westinghouse and Hoover. Irons, vacuum cleaners, washing machines, sewing machines, fans, toasters, cookers, hotplates – convenience came to the home, especially the American home. By the middle of the 1920s over 60 per cent of American houses had electricity and these constituted more than half the world market (Rybczynski, 1986: 153). It was this scale of operation that reduced the price of electricity, so stimulating the market for appliances; this meant more electricity was used, creating a spiralling market. Hence domesticity was at the forefront of the USA, developing the first mass-consumption society. But with all this convenience basic comfort was never forgotten: it is America that invented the bathroom as a three-fixture space of bath, water closet and hand basin (Rybczynski, 1986: 164). This ubiquitous place of ordinary luxury illustrates how far modernity has improved the lives of millions of people in the twentieth century.

Ordinary modernity as comfort was therefore a product of hegemonic civil societies where wealth was accumulated and distributed widely enough to create new domestic worlds. To a large extent the process was cumulative as bourgeois influence spread throughout the world-economy. But the three hegemonic episodes in this development were very different in their geographies. In particular, the Dutch and the British created new homes in very different locales: small town houses and Georgian country houses respectively. In some ways America's contribution to ordinary modernity represents a geographical compromise, with suburbia integrating conveniences of urban life with the atmosphere of country life. But suburbia is much more than a blend of past locales. It represents a distinctive new geographical phenomenon which we may call spaces of concentrated comfort. The Dutch and British contributions to ordinary modernity focused at the level of the house. American suburbia is many such houses in a homogeneous environment of comfort writ large.

Suburbia as concentrated comfort

In strictly geographical terms the suburbs are as old as urban growth. Indeed, residential growth on the edge of the city was distinctive in pre-modern urban development but its characteristics were the opposite of those we associate with suburbs today. Suburbs were for urban outcasts, they were dens of iniquity outside respectable urban society. In contrast, the modern suburb oozes respectability. This is a classic example of a modern concept and reality completely subverting their historical antecedent. Suburbia is a special place for family living that takes homes out of a corrupting urban influence and relocates them in new communities in rural settings. The classic modern suburb consists of single family houses in large gardens set back from tree-lined roads. Perhaps the key feature is the front lawn which is a private space that has a public function (Fishman, 1987: 146–147). Maintained by the homeowner, the lawn has a communal effect: by separating houses from their road, the sequence of green front lawns converts an urban street into a country lane, producing the ideal suburban landscape.

Although, significantly, American writers believe the modern suburb to be a US invention, Robert Fishman (1987: 116) has shown that its origins may be traced back to the growth of eighteenth-century London. These early high-class suburbs were converted into a general middle-class residential pattern in mid-nineteenth-century Manchester where people used new wealth to move out of the industrial city, producing the first example of the now familiar urban structure of working-class inner city and middle-class outer city. The British invented the suburb as an exclusionary zone, therefore. The American contribution was to popularize the idea of the suburb as a goal for every family. This notion of 'suburbs for all' can be found as early as 1868 although it is only with the coming of the 'affluent society' in the twentieth century that it achieved widespread credibility (Fishman, 1987: 129). Americans did not eliminate exclusion from suburbia, of course, but they produced gradations of exclusion so that there were different suburbs for different family budgets. This culminated in the 1950s, the decade of greatest suburban growth in US history; the 1960 census recorded 19 million more people living in the country's suburbs than in 1950 (Hall, 1988: 294).

Large-scale suburbanization began in the late nineteenth century in both Britain and the USA as a result of improvements in urban transport. The resulting railway and streetcar suburbs created linear patterns of growth in the outer city combined with congestion in the inner city. It is Los Angeles that broke away from this structure and released suburbia to invade all available space and hence to dominate the modern city. Fishman (1987: Chapter 6) designates Los Angeles as a 'suburban metropolis' which he views as the climax of suburban history. Although its original growth was by streetcar suburbs like other US cities, by the 1920s Los Angeles already had more automobiles per capita than anywhere else in the world. In addition, as the fastest-growing American city in the first decades of the twentieth century, Los Angeles realtors and developers were especially powerful in local politics. The result was the 1926 approval of a massive bond issue to fund a large road network to overcome development constraints (Fishman, 1987: 166). Los Angeles became the city of the automobile, the city that could spread out across the landscape for mile after mile. As a city of suburbs, differentiation was achieved through concentrating more expensive homes on hills looking down on cheaper suburbs: Hollywood stars lived in their mansions in Beverly Hills. But the idea was still suburbia for all.

With the building of urban freeways across all American major cities in the early post-Second World War years, suburbia becomes the dominant settlement form in the USA. At the other end of the social spectrum from Beverly Hills came the 'Levittowns' of the north-east which brought suburban living to almost anybody with a regular job. The Levitt building company specialized in the mass production of suburbs. Peter Hall (1988: 295) describes their methods as based upon 'flow production, division of labour, standardized designs and parts, new materials and tools, maximum use of prefabricated components, easy credit, good marketing'. The latter facilitated mass consumption; people queued for hours to buy their piece of suburbia and in one development alone 17,000 homes were built and bought. In his survey of 'Levittowners', Herbert Gans (1967: 37) finds that their reason for moving to this suburbia was that 'they wanted the good and comfortable life for themselves and their family'. In terms of aspirations, Gans (1967: 39) identifies two that stand out: 'comfort and roominess' and 'privacy and freedom of action'. These are quintessentially concepts of ordinary modernity. It is in Levittowns and developments like them that achievements of American hegemony reach the lives of ordinary Americans. Living in suburbs enabled manual workers to identify themselves as middle class (Halle, 1984): in suburbia the middle class became the universal class.

Initially, the suburban route to middle-class comfort was not itself universal. Fishman (1987: Chapter 4) describes it as an Anglo-American phenomenon in the nineteenth century, with the middle classes in the rest of Europe and the Americas staying much more loyal to traditional urban residential preferences. But with American hegemony the middle-class suburb has spread to cities across the world. The particular form that suburbs take varies between countries but the ideal of suburbia as collective comfort remains: people who can afford it move to suburbs for a better family life. Throughout the world, 'however modest each suburban house might be, suburbia represents a collective assertion of class, wealth and privilege' (Fishman, 1987: 4). It is the true monument of the modern world, our equivalent to the cathedrals of medieval Europe. But it is also the omen of world impasse.

SOCIAL LIMITS: AMERICANIZATION AS WORLD IMPASSE

In his work on the world between the fourteenth and eighteenth centuries, Fernand Braudel (1981: 31) analysed the everyday lives of the population to provide 'an evaluation of the limits of what was possible in the pre-industrial world'. It is in this spirit that we consider ordinary modernity as defining the measure of what is possible in the modern world. The coterie of processes constituting world hegemony and ordinary modernity have imposed a critical influence on the overall trajectory of the modern world-system. For the first time in its history the idea of social limits has seriously entered the collective consciousness of the modern world.

The idea of limits to progressive social change is anathema to the whole notion of being modern. Defining ourselves as subjects rather than objects of social change has required the expectation of a better world in the future. And this social-cultural edifice is supremely consonant with the political economy imperative of ceaseless capital accumulation. The capitalist world-economy requires continual economic expansion to survive; every periodic slowdown in economic growth creates a problem of realization or recession which may develop into a crisis of the system or depression. The new political economies created by the hegemons are the major restructurings that have resolved past systemic crises. But the problem will never go away. In a very real sense capitalism *is* economic growth, it cannot stop. The question arises, therefore, whether the earth is ultimately too small for capitalism.

REFERENCES

Arrighi, G. (1990) 'The three hegemonies of historical capitalism', *Review* 13: 365–408.

Berman, M. (1988) *All that is Solid Melts into Air: the Experience of Modernity*, New York: Penguin.

Braudel, F. (1981) *The Structures of Everyday Life*, London: Collins.

Fishman, R. (1987) *Bourgeois Utopias. The Rise and Fall of Suburbia*, New York: Basic.

Gans, H.J. (1967) *The Levittowners*, London: Penguin.

Giddens, A. (1990) *The Consequences of Modernity*, Cambridge: Polity.

Hall, P. (1988) *Cities of Tomorrow*, Oxford: Blackwell.

Halle, D. (1984) *America's Working Man*, Chicago: Chicago University Press.

Hopkins, T.K. (1990) 'A note on the concept of hegemony', *Review* 13: 409–412.

Howe, D.W. (1976) 'Victorian culture in America', in D.W. Howe (ed.) *Victorian America*, Philadelphia: University of Pennsylvania Press.

Rybczynski, W. (1986) *Home. A Short History of an Idea*, London: Penguin.

Wallerstein, I. (1984) *The Politics of the World-Economy*, Cambridge: Cambridge University Press.

10

The Empire's Old Clothes

Fashioning the Colonial Subject

Jean Comaroff

AFRICAN ADORNMENT

The Western trope of 'nakedness' – which implies a particular idea of bodily being, nature and culture – would have made little sense to Tswana prior to the arrival of the missions. In South Africa, what the nineteenth-century missionaries took to be indecent exposure was clearly neither a state of undress nor impropriety in indigenous eyes (although local notions of unclothedness existed; uncovered genitals and undressed hair were considered uncouth in Tswana adults). African dress and grooming *were* scanty by European standards, but they conveyed – as such things do everywhere – complex distinctions of gender, age, and social identity. In their seeming nakedness, the Africans were fully clothed.

What was most unsettling to the evangelists was the place of apparel in the whole Tswana social order. In the European world, discerning consumption was the major index of social worth. In fact, consumption was increasingly set off from production as a gendered and markedly female sphere of practice. Women's domestic demesne centred on the display of adornments that would signal the status of their male providers, men whose own attire, as befitted their endeavours, was relatively sober and unelaborated (Turner n.d.). Moreover, while men of the bourgeoisie controlled the manufacture and marketing of clothes, the labour which produced textiles and garments was largely that of poor women and (in the early years) children, members of the lower orders who were conspicuously excluded from the stylish self-production that engrossed their more privileged sisters.

OTHER KINDS OF CLOTHES

Above all else, it struck the evangelists as unnatural that, while Tswana women built houses, sowed, and reaped, 'men ma[d]e the dresses for themselves and the females' (LMS 1824). Refashioning this division of labour was integral to reforming 'primitive' production in all its dimensions; and this, in turn, required the creation of a distinct – feminine – domestic world centred on reproduction and consumption. In this regard, the churchmen were disturbed by the fact that, although it was marginally distinguished by rank, female attire was largely undifferentiated. In direct contrast to bourgeois fashion, it was mainly men's clothes that signalled social standing here (cf. Kay 1834, 1: 201). In fact, European observers pronounced male dress to be quite varied – even dandyish (LMS 1824, 1828).

Such distinctions apart, however, Tswana costume seemed to be unremittingly rude and rudimentary. For the most part, those of the same sex and age dressed alike (Schapera 1953: 25). Nonetheless, it soon struck the Europeans that, albeit in a register of their own, indigenous clothes also spoke volubly of status. By contrast to infants (who wore little besides medicated ornaments), adults of both sexes wore long skin cloaks (*dikobó*; singular *kobó*) that were significant 'sign[s] of wealth' (LMS 1824).[1] Cloaks were first donned at the conclusion of male and female initiation, denoting the onset of sexual and jural maturity (J. Comaroff 1985: 105ff.); interestingly, during lapses from full participation in social life – such as after bereavement – people put on their *dikobó* inside out. Royal males wore especially fine karosses, often incorporating the pelts of wild beasts, although that of the leopard was reserved for reigning chiefs (Philip 1828, 2:126). The skin cape was to prove extremely durable in this economy of signs, surviving amidst a riot of market innovations to give a distinctive stamp to Tswana 'folk' style, where it lived on, in the form of the store-bought blanket, as a crucial element of 'tribal' costume.

Early accounts suggest that Tswana were especially creative in fashioning new ornaments which seemed to radiate personal identity. They favoured shining surfaces (recall the glossy cosmetics) and a gleaming visibility that would contrast markedly with the dullness of mission modes, which countered 'flashiness' with a stress on personal restraint and inward reflection. There is plentiful evidence of novel adornments made with the sparkling buttons and glass beads that found their way into the interior, for by the early nineteenth century, the latter had become a widespread currency linking local and monetized economies.

But bright beads were not all equally desirable; Campbell (in LMS 1824) noted that, by the 1820s, Tswana 'greatly prefer[red] the dark blue colour'. This is intriguing for, as we shall see, dark blue was to be the shade favoured for the dress of converts by the mission. If Campbell was correct, the European's chosen hue had a fortuitous precedent, having already been associated with prestige of foreign origin. Blue beads were globules of exchange value, imaginatively congealed into local designs. Clear blue appears to have had no other place in indigenous artistic schemes: patterns on housefronts, pottery, and ritual artefacts tended to play on the three-way contrast of black, red, and white (J. Comaroff 1985: 114). It is tempting to suggest that blue – so clearly the colour of the mission and its materials (as well, in Tswana poetics, of the pale, piercing eyes of whites) – was the pigment of exogenous powers and substances. The Christians would certainly wield the blues in their effort to counter 'heathenism', for when it came to heathens, they saw red.

CIVILITY, CLOTH AND CONSUMPTION

Above all, the evangelists would try to force Tswana bodies into the strait-jacket of Protestant personhood. The Nonconformists acted on the implicit assumption that, in order to reform the heathens, it was necessary to scramble their entire code of body management; thus 'decent' Western dress was demanded from all who would associate with the church. Tswana soon appreciated the role of clothes in this campaign. When Chief Montshiwa of the Tshidi Rolong perceived that the Christian influence in his realm had begun to extend even to his own kin, he ordered his daughter publicly 'to doff her European clothing, . . . to return to heathen attire' (Mackenzie 1871: 231). His royal counterparts elsewhere also fastened on to such discernible signs of allegiance, and many struggles ensued over the right to determine individual dress. From the first, Southern Tswana tended to treat objects of Western adornment as signs of exotic force; those introduced by the mission were soon identified as *sekgoa*, 'white things' (Burchell 1824, 2: 559). But some items of European clothing had preceded the mission into the interior,[2] where people often seem to have regarded them as vehicles of alien power (1824, 2: 432). An early report from Kuruman tells how the Tlhaping chief addressed his warriors prior to battle in a 'white linen garment', his heir wearing an 'officer's coat' (Moffat 1825: 29). In a published account of this incident, Moffat (1842: 348) revealed that the garment was actually a chemise of unknown origin. Such attire seemed to lend potency to indigenous enterprise, in part because its qualities resonated with local signs and values. White, the usual colour of the baptismal gown (itself, to the untrained eye, much like a chemise), was also the colour of the transformative substances placed on the human body during indigenous rites of passage (J. Comaroff 1985: 98). Similarly, the military uniforms carried inland from the Colony by Khoi soldiers might have had cogent connotations associated with this forceful frontier population. But the interest which they evoked seems also to have been fed by what appears to have been a long-standing Tswana concern with the dress of combat (J. Comaroff 1985: 112; Comaroff and Comaroff 1991: 164).

European costume, in short, opened up a host of imaginative possibilities for Southern Tswana. It offered an enhanced language in which to play with new social identities, a language in which the mission itself would become a pole of reference. In the early days, before the Christians presented a palpable threat to chiefly authority, royals monopolized the Western garments that travelled into the interior. These were worn in experimental fashion, often in ceremonial audiences with visiting whites (Philip 1828, 2: 126–127). Already at this point, several aspects of the synthetic style that would be much in evidence later on in the century seem to have taken shape – among them, the combination of European garments with skin cloaks. This was a form of mixing which the evangelists abhorred, yet would never manage to eradicate.

But the missions would expend great effort and cost to ensure that, in Moffat's telling phrase (1842: 505), the Africans would 'be[come] clothed and in their right mind' (cf. Luke 8.35). As Western dress became more closely associated with expanding evangelical control, the early phase of playful experimentation came to an end. By the 1830s, once a regular mission presence had been established, most senior royals had discarded the dress of *sekgoa*, identifying with an ever more assertively marked *setswana* (Tswana ways). Some were said to 'ridicule . . . and even abuse' those kin who 'laid aside' the dress of their 'forbears' (Smith 1939, 1: 337).

I have noted that the campaign to clothe black South Africa was inseparable from other axes of the civilizing mission, especially the effort to reform agricultural production. Thus in order to dress Tswana – or rather, to teach them to dress themselves – women had to be persuaded to trade the hoe for the needle, the outdoor for the indoor life (Gaitskell 1988). In this endeavour, the Nonconformists relied largely on the 'domesticating' genius of the 'gentler sex' – on their own wives and daughters (cf. Hunt 1990), most of whom started sewing schools almost at once (Moffat 1842: 505); these also served as a focus for the exertions of female philanthropists in Britain, who sent pincushions and needles with which to stitch the seams of an expanding imperial fabric. Recall that, in pre-colonial times, clothing was made of leather; an extension of animal husbandry, it was produced by men. It is not surprising, then, that sewing schools had limited appeal at first. In the early years, moreover, there was no regular supply of materials. But by the late 1830s, once merchants had been attracted to the stations, those Tswana women most closely identified with the church had begun to take in sewing for payment (1842: 17). This was one of several areas in which the evangelists encouraged commercial relations well ahead of a formal colonial labour market.

But even if the missionaries had succeeded immediately in persuading Tswana to clothe themselves, local manufacture would have fallen short of the task. Thus the Christians appealed to the generosity of the great British public. The growth of the fashion industry encouraged obsolescence, and by this time had already provided a steady supply of used garments (or recycled commodities) for the poor and unclad at home and abroad. When, in 1843, the Moffats returned to Cape Town from a visit to the United Kingdom, they sailed with fifty tons of 'old clothes' for the Kuruman station (Northcott 1961: 172). The famous David Livingstone, sometime missionary among the Tswana, was scathing about the 'good people' of England who had given their cast-off ballgowns and starched collars to those 'who had no shirts' (1961: 173). But a letter from Mrs Moffat to a woman well-wisher in London shows that she had thought carefully about the adaptation of Western dress to African conditions:

> The materials may be coarse, and strong, the stronger the better. Dark blue Prints, or Ginghams . . . or in fact, any kind of dark Cottons, which will wash well – Nothing light-coloured should be worn outside. . . . All the heathen population besmear themselves with red ochre and grease, and as the Christians must necessarily come in contact, with their friends among the heathen, they soon look miserable enough, if clothed in light-coloured things . . . *I* like them best as Gowns were made 20 or 30 years ago. . . . For little Girls, Frocks made exactly as you would for the Children of the poor of this country, will be the best.
>
> ([1841]1967: 17–18)

Clothing women and children was her priority. And while any European clothes, even diaphanous ballgowns, were better than none, more sombre, serviceable garb was ideal. Dark blue garments, especially, resisted the stains of a red-handed heathenism that threatened to 'rub off' on the convert. Indigo-dyed prints, now being mass-produced with raw materials drawn from other imperial outposts, conformed well with the long-standing European association of dark hues with humility, piety and virtue. Ochre and grease aside, Mrs Moffat suggested, African converts were like the virtuous British poor, whose inability to produce their own wealth was marked by their exclusion from the fashion system, and by the dismal durability of their dress. Great efforts would also be made to stir a desire for self-improvement in these neophyte Christians.

The fact that Mary Moffat wrote about such matters to a woman was itself predictable. Not only the acquisition and maintenance, but also the dispatch of clothing in the form of charity had become a key element of a feminized domestic economy (Davidoff and Hall 1987). But such recycling carried its own dangers: it could inhibit ambition in the poor. Care had to be taken not to evoke indigence. Here the Protestants put their faith in the sheer charm of commodities. Comfortable and attractive garments, they hoped, would awaken the desire for property and self-enhancement, for a life of righteous getting and spending.

And so, through the effort of mission wives and their European sisters, the germ of the fashion system arrived on the African veld. It bore with it the particular features of the culture of industrial capitalism: an enduring impetus towards competitive accumulation, symbolic innovation, and social distinction (Bell 1949). But its export to this frontier also underscored the deep-seated contradictions in the material expression of the Protestant ethic. Ascetic angst focused most acutely on female frailty. For, in as much as the fashion system made women its primary vehicles, it strengthened the association of femininity with things of the flesh. Willoughby was far from alone in grumbling that many Tswana women were soon in thrall to ridiculous hats and expensive garments.[3] Also, while the Nonconformists might have striven to produce an élite driven by virtuous wants, they had also to justify the lot of the less fortunate majority. They had, in other words, to sanctify poverty and the postponement of physical pleasure in the interests of eternal grace. Their most humble adherents remained the deserving recipients of charity, dressed in the strong dark blue cottons whose colour and texture were to become synonymous with the mission rank-and-file. This would be the nucleus of rural 'folk', whose style and predicament would come to typify Tswana peasant-proletarians in modern South Africa.

As the century wore on, the evangelists would devote their energies increasingly to the cultivation of a black petite bourgeoisie. But in the early years, they encouraged 'improvement and self-reliance' as an ideal for all; hence their attempts to bring traders and, with them, the goods needed to make Christians. Revd Archbell began to pursue merchants for his Wesleyan station among the Seleka Rolong in 1833; and by 1835, Moffat had persuaded David Hume, a factor catering to the 'demand for British commodities', to establish himself at Kuruman. The mission played a large role in stimulating that demand, not just for ready-made garments, cotton prints, and sewing goods, but for all the elements of the European sartorial economy. The Nonconformists, for instance, stressed the fact that, unlike 'filthy skins', clothes had to be washed and repaired, binding wives and daughters to an unrelenting regime of 'cleanliness' – epitomized, to this day, by the starched and laundered uniforms of the black women's Prayer Unions. It was a form of discipline that the evangelists monitored closely, ensuring brisk sales of soap and other cleansing agents (cf. Burke 1990).

From 1830 onwards mission reports speak with pleasure of 'decent raiments' worn by their loyal members. They also note that trade was healthy, and that there was a growing desire among Southern Tswana to purchase European apparel (Moffat 1842: 219; Read 1850: 446). Not only was the campaign to clothe the heathen masses under way, but a distinct and sedately styled Anglophile élite was increasingly visible in the interior.

SELF-FASHIONING ON THE FRONTIER: THE MAN IN THE TIGER SUIT

The growing supply of Western apparel in the interior towards the mid-century also had another effect on the Tswana, one less palatable to the mission. It incited what the Christians saw as an absurd, even promiscuous syncretism:

> A man might be seen in a jacket with but one sleeve, because the other was not finished, or he lacked material to complete it. Another in a leathern or duffel jacket, with the sleeves of different colours, or of fine printed cotton. Gowns were seen like Joseph's coat of many colours, and dresses of such fantastic shapes, as were calculated to excite a smile in the gravest of us.
>
> (Moffat 1842: 506)

Such descriptions give a glimpse of the Tswana *bricoleur* tailoring a brilliant patchwork on the cultural frontier. To the evangelist, they offered a disconcerting distortion of the worthy self-fashionings he had tried to set in motion. Such 'eccentric' garb caused the Christians much anxiety. As Douglas (1966) might have predicted, it came to be associated with dirt and contagion. State health authorities by the turn of the century were asserting that 'Natives who partially adopted our style of dress' were most susceptible to serious disease (Packard 1989: 690). If the selective appropriation of Western attire flouted British codes of costume and decency, it also called into question the authoritative norms of Nonconformism. This was particularly evident in the counterpoint between the colourful, home-made creations of most people and the 'uniforms' introduced by the mission to mark the compliance of those in its schools and associations. (The latter attire, being both novel, yet closed to stylish innovation, anticipated subsequent 'folk' dress in several respects.) But the creative couture contrived by so many Southern Tswana suggests a riposte to the symbolic imperialism of the mission at large. It speaks of a desire to harness the power of *sekgoa*, yet evade white authority and discipline. The *bricoleur* contrasted, on the one hand, with those who ostensibly rejected everything European and, on the other, with those who identified faithfully with church aesthetics and values. Style, here, was clearly implicated in the making of radically new distinctions. And as the colonial economy expanded into the interior, the means for such fashioning was increasingly available through channels beyond the control of the mission.

Indeed, as the century progressed, the growing articulation of the Southern Tswana with the regional political economy was tangible in their everyday material culture. For a start, the volume of goods pumped into rural communities rose markedly. A visitor to Mafikeng in 1875 (Holub 1881, 2: 14) reported that, apart from a small élite, the population persisted in its patchwork of indigenous and European styles. But the make-up of the mixture had subtly altered. Mafikeng was by then a Christian Tshidi-Rolong village that had no white mission presence. Still, store-bought commodities comprised a growing proportion of its cultural *mélange*. British aesthetics were being used in ever more complex ways; both in the honour and in the breach they marked widening social and economic differences.

The deployment of Western style was particularly evident in the changing garb of 'traditional' rulers and royals. As noted above, they had responded to the earlier missionary challenge by reverting, assertively, to *setswana* costume – and by insisting that their Christian subjects do likewise. By the late nineteenth century, however, with the colonial state ever more palpably upon them, few but the most far-flung of Tswana sovereigns harboured illusions about the habits of power. Some, in fact, sought to outsmart the evangelists at their own game; Mackenzie (1883: 35) records the fascinating case of Chief Sechele who, in 1860, had a singular suit tailored from 'tiger' (i.e. leopard) skin – all 'in European fashion'. According to the missionary, many of the Kwena ruler's subjects thought he wished 'to make himself a white man'. But the matter was surely more complex. In crafting the skin, itself a symbol of chiefly office, the chief seems to have been making yet another effort to mediate the two exclusive systems of authority at war in his world, striving perhaps to fashion a power greater than the sum of its parts! Other rulers, most notably the Tshidi and Ngwaketse chiefs (Holub 1881, 1: 291), took another tack, now choosing to dress themselves in highly fashionable garb, clothes whose opulence set them off from their more humble Christian subjects – missionaries included! These early examples of royal dandyism involved only male dress; but the nascent local bourgeoisie had already begun, like its European counterpart, to signal status on the bodies of its women, whose clothing became ever more nuanced and elaborate (cf. Willoughby n.d.: 25, 48).

MIGRANTS, MERCHANTS AND THE COSTUME OF THE COUNTRYSIDE

In the closing decades of the century, it was labour migration that had the greatest impact on Southern Tswana dress. Whites in the interior had insisted, from the beginning, that 'natives' with whom they sustained contact should adopt minimal standards of 'decency' – covering at least their 'private' parts. Men who interacted regularly with whites soon took to wearing trousers, and those who, in later years, journeyed to the new industrial centres had little option but to conform to the basic rules of respectability pertaining to public places. By then, however, the Christians had already established a widespread 'need' for European garments, if not necessarily for European styles. Schapera (1947: 122) is not alone in suggesting that the desire for such commodities as clothes was powerful in initially drawing migrants to urban areas.

But desire is seldom, in itself, a sufficient explanation for large-scale social processes. The migration of Southern Tswana to the cities occurred in the wake of regional political and ecological forces which impoverished large sections of local populations. Nor was the consumption of European fashions a specifically urban affair. Willoughby indicates that, by the late nineteenth century, Tswana living near rural mission stations had learnt very well how to craft themselves with commodities; some spent 'as much on clothes in a year as would keep my wife well-clad for Ten Years'.[4] None the less, it is clear that those who did migrate to the industrial centres were immediately confronted by an array of 'Kaffir Stores' that pressed upon them a range of 'native goods' designed especially for the neophyte black proletarian.[5]

Advertisements attest that clothes were by far the most significant commodities sold by urban 'Kaffir storekeepers'.[6] The sheer volume of this trade at the time suggests that migrants were devoting a high proportion of their earnings to self-fashioning. And the standardization and range of goods indicate that some customers, at least, were putting on the dress of industrial capitalism, with its distinctions between labour and leisure, and manual and non-manual toil. Contemporary advertisements also invoked class distinctions: texts aimed at literate Africans, for instance, suggested that discerning taste conferred social distinction. The moral economy of mission and marketplace overlapped ever more neatly.

NOTES

1 See also W.C. Willoughby, 'Clothes' (Willoughby Papers, The Library, Selly Oak Colleges, Birmingham, Unfiled Notes, Box 14).

2 Some linen goods seem to have found their way to Northern Tswana peoples from the east coast in the early nineteenth century, probably via Arab traders.

3 W.C. Willoughby, 'Clothes' (Unfiled Notes, Box 14).

4 Ibid.

5 For a more detailed account of the African clothing market in Kimberley at the time, see Comaroff and Comaroff (n.d.: Chapter 5).

6 Such advertisements were common in papers like *The Diamond Fields Advertiser* in the late 1860s (see Comaroff and Comaroff n.d.: Chapter 5).

REFERENCES

Bell, Q. (1949) *On Human Finery*, New York: A. A. Wyn.

Burchell, W. J. (1822–4) *Travels in the Interior of Southern Africa* (2 vols), London: Longman, Hurst, Rees, Orme, Brown & Green.

Burke, T. (1990) '"Nyamarira that I love": commoditization, consumption, and the social history of soap in Zimbabwe', paper read at the Africa Studies Seminar, Northwestern University.

Comaroff, J. (1985) *Body of Power, Spirit of Resistance: The Culture and History of a South African People*, Chicago: University of Chicago Press.

Comaroff, J. and Comaroff, J. L. (1991) *Of Revelation and Revolution: Christianity, Colonialism, and Consciousness in South Africa* (vol. 1), Chicago: University of Chicago Press.

Comaroff, J. and Comaroff, J. L. (n.d) *Of Revelation and Revolution* (vol. 2), (in preparation).

Davidoff, L. and Hall, C. (1987) *Family Fortunes: Men and Women of the English Middle Class, 1780–1850*, Chicago: University of Chicago Press.

Douglas, M. (1966) *Purity and Danger: An Analysis of the Concepts of Pollution and Taboo*, London: Routledge & Kegan Paul.

Gaitskell, D. (1988) 'Devout domesticity? Continuity and change in a century of African women's Christianity in South Africa', paper read at Meeting of African Studies Association, Chicago.

Holub, E. (1881) *Seven Years in South Africa: Travels, Researches, and Hunting Adventures, between the Diamond-Fields and the Zambesi (1872–79)* (2 vols), trans. E. E. Frewer, Boston: Houghton Mifflin.

Hunt, N. R. (1990) '"Single ladies on the Congo": protestant missionary tensions and voices', *Women's Studies International Forum* 13: 395–403.

Kay, S. (1834) *Travels and Researches in Caffraria* (2 vols), New York: Harper & Brothers.

London Missionary Society (LMS) (1824) 'Kurreechane', *Missionary Sketches*, No. XXV, April, London: London Missionary Society. [South African Public Library: South African Bound Pamphlets, No. 54.]

—— (1828) 'Sketch of the Bechuana Mission', *Missionary Sketches*, No. XLIII, October, London: London Missionary Society. [South African Public Library: South African Bound Pamphlets, No. 54.]

Mackenzie, J. (1871) *Ten Years North of the Orange River: A Story of Everyday Life and Work among the South African Tribes*, Edinburgh: Edmonston & Douglas.

—— (1883) *Day Dawn in Dark Places: A Story of Wanderings and Work in Bechwanaland*, London: Cassell. Reprinted 1969, New York: Negro Universities Press.

Moffat, M. (1967) 'Letter to a well-wisher', *Quarterly Bulletin of the South African Library* 22: 16–19.

Moffat, R. (1825) 'Extracts from the journal of Mr Robert Moffat', *Transactions of the Missionary Society* 33: 27–29.

—— (1842) *Missionary Labours and Scenes in Southern Africa*, London: Snow.

Northcott, W. C. (1961) *Robert Moffat: Pioneer in Africa, 1817–1870*, London: Lutterworth Press.

Packard, R. M. (1989) 'The "healthy reserve" and the "dressed native": discourses on black health and the language of legitimation in South Africa', *American Ethnologist* 16: 686–703.

Philip, J. (1828) *Researches in South Africa; Illustrating the Civil, Moral, and Religious Condition of the Native Tribes* (2 vols), London: James Duncan.

Read, J. (1850) 'Report on the Bechuana Mission', *Evangelical Magazine and Missionary Chronicle* 28: 445–447.

Schapera, I. (1947) *Migrant Labour and Tribal Life: A Study of Conditions in the Bechuanaland Protectorate*, London: Oxford University Press.

—— (1953) *The Tswana*, London: International African Institute. Revised edition, I. Schapera and J. L. Comaroff (1991), London: Kegan Paul International.

Smith, A. (1939) *The Diary of Dr Andrew Smith, 1834–1836* (2 vols), P. R. Kirby (ed.), Cape Town: The Van Riebeeck Society.

Turner, T. S. (n.d.) 'The social skin', unpublished manuscript. Published in abridged form in J. Cherfas and R. Lewin (eds) (1980) *Not Work Alone*, Beverly Hills, CA: Sage.

Willoughby, W. C. (n.d. [*c.* 1899]) *Native Life on the Transvaal Border*, London: Simpkin, Marshall, Hamilton, Kent.

11

Plurality of Taste

Food and Consumption in West Germany During the 1950s

Michael Wildt

Consumption, where food is concerned, involves not only quantifiable purchase, but also 'production'. Food has to be supplied, prepared, cooked, and – last but not least – served. 'The investigation of consumption', as Alf Lüdtke points out, 'is meaningless if cooking, eating and going hungry are not included as forms of social practice.'[1] So the horizons of this inquiry must be widened to take in not only the purchase of food, but how it was cooked and served, and the 'production' of cultural meanings connected with food and consumption.

For example, the quantity of potatoes as an item in the household budget certainly fell during the 1950s (as it had early in the century); but, at the same time, ready-made items using dehydrated potato like dumplings or pancakes took their place, and consumption of potato crisps and chips later increased markedly. Or again, while consumption of rice remained steady, it was increasingly being served as a savoury component of a hot meal, replacing potatoes or cereals, rather than as rice pudding or sweet soup.

Baking provides another example. The Federal Statistics Board household budgets show declining purchases of flour, and increased purchase of biscuits and cakes. The obvious inference is confirmed in the records of the Oetker Company, leading West German producer of baking powder. Their product, *Backin*, experienced declining sales in the 1950s as, according to the sales manager, 'the joy of home baking was trickling away'.[2] Home baking was indeed in decline: what the managers and the marketing board (presumably male) described as 'joy' was hard work, and one can easily imagine why housewives might choose to abandon kneading dough or beating eggs and patronize the bakery instead. But they did not give up home baking altogether. Indeed, the Oetker Company was surprised by the success of *Tortenguss*, a new glaze for fruit tarts which they put on the market in 1950. Fruit tarts were the hit of the 1950s: unlike the old-fashioned cakes, which involved much beating, they could be prepared easily and fast, and their fresh, fruit taste was valued as healthy and light. At the same time they reassured the housewife that she did not feed her family entirely on shop food, but still had the skill to do her own baking.

This tension, between making work easier on the one hand and retaining the housewife's sense of competence on the other, also affected the use of electric applicances. Such new durable goods appeared first in the kitchen. At the beginning of the 1960s more than ten per cent of all four-member households in West Germany had an electric kitchen machine. The most coveted item during the 1950s was the refrigerator. In a 1955 opinion poll only ten per cent of all households owned a refrigerator, but nearly fifty per cent dreamed of buying one. Even in 1958 the refrigerator topped the list of desired goods.[3] Ownership of refrigerators rose from nineteen per cent in 1958 to thirty-nine per cent in 1961, and two years later more than fifty-one per cent of West German private households had a refrigerator.[4]

Most housewives appreciated any technical innovation which eased their daily work. During the 1950s a housewife in a household of four to six people did over seventy-two hours' housework a week.[5] In several opinion polls taken during the 1950s women said that their work was easier with kitchen appliances, and that they would save time and energy. Those who acquired the new labour-saving machines had good reason to be proud of them. And owning these gadgets and machines also stamped them as modern.

Yet there could be a difference between owning such equipment and using it. Jakob Tanner cited the comments of a Swiss housewife on her new machine.

> Crushing food with lightning rapidity seems brutal and shocking to me. I see hard nuts, apples, lemon peel cut to pieces and transformed into an unrecognizable mass. In only a few moments cabbage and carrots, onions and potatoes, bacon and fish are indistinguishable. Something inside me rebels against this bringing of food into line. . . . I disappointed my husband with my reserve. He expected me to be delighted with his present, because usually I approve of all innovations likely to make the work of housewives easier. Still, he didn't have to wait too long: once I had tried it out a few times my hostility changed to honest admiration.[6]

This story suggests that the uses of new kitchen equipment cannot be taken for granted. Survey findings in the early 1960s likewise indicate that women made their own decisions about the use of gadgets and machines, and did not simply follow the manufacturers' instructions.[7] Tiring work, like mixing dough or beating eggs, was gladly given over to machines, but women preferred to peel potatoes or scrub vegetables by hand. 'Potatoes have to be peeled properly', one of the interviewers quoted. 'Peeling potatoes in the kitchen machine is too imprecise. So most of the housewives I spoke to still peeled potatoes by hand.'[8]

Kitchen appliances and ready-made food meant that knowledge and skill gained through experience became less essential. At the same time the appliances formalized understanding of weights, quantities and timing, as well as requiring proper handling. Nevertheless, the complex and differentiated practices involved in daily production of meals could not be reduced to a purely technical process. Despite the industrial law of efficiency, it was still necessary to 'waste' time or energy to improve the taste or to discover new ones.

Tension can again be seen between making use of 'modern' industrial products like tinned food, and maintaining traditional but labour-intensive skills such as preserving food. In 1953, seventy-six per cent of all private households in West Germany bottled or canned their own fruit. Country households were more likely to do this than town ones, and households with several members more than those with only one or two; while younger housewives were slightly less likely than older ones to do their own preserving.[9] Their main reason for home preserving was unequivocal – they wanted to save money, and in the early 1950s these home products were much cheaper than bought equivalents, even when the produce was bought not home-produced. (Well over half the housewives preserved food though they did not have a garden of their own.) The second reason was the superior taste of home preserves. This superiority justified the continued practice of home preserving even when rising incomes allowed consumers to buy the dearer commercial products more frequently.

From the mid-1950s on, the sample working-class households bought more and more tinned food. The opinion polls taken by the Allensbach Institute suggest why.[10] For more than half the housewives asked, tinned food was convenient because they were able to prepare a quick meal every time. A third of them believed tinned food tasted better, and a third thought it looked fresh and delicious. Most of them had bought tins when the vegetables they wanted were out of season, so clearly it was a way to overcome seasonal dependencies. Those who went out to work and also had domestic responsibilities were especially likely to use tins to save time.

With tinned fruit, however, there was a different rationale. Pineapple or tangerine slices were served on Sundays or feast days. Tinned fruit was undoubtedly used to mark meals on special occasions.

One tinned product was always top of the list: evaporated milk. Consumption rose tenfold between 1950 and 1963! The market was dominated by four brands: *Glücksklee* and Libby's, which were ultimately US-owned, and Nestlé and *Bärenmarke*, ultimately Swiss.[11] Analysis of consumer behaviour was commissioned by the West German subsidiary of Libby's in the 1950s.[12] They found that nearly all consumers used evaporated milk for coffee; while a third used it to prepare salad, sauces, puddings or mashed potatoes. When asked what they most liked about it, most answered, first that it tasted creamy, second that it was best with coffee, and third that it gave the coffee a beautiful colour. The original advantage of evaporated milk – that it stayed fresh – was noted less and less often.

This shift from practical to aesthetic justification also figured in the advertisements, here taken from *Die kluge Hausfrau* (see below). In 1950 evaporated milk was praised for its various applications:

> Libby's milk . . . the creamy one, makes the cake light and tasty, makes coffee aromatic and gives cocoa flavour. It improves the taste of sweet dishes, soups and many meals. Because Libby's is concentrated milk it contains all of the nutritional qualities that make fresh milk so valuable.

It is noteworthy that while this advertisement emphasized taste and aesthetic aspects, it also dwelt on nutritional content. Within a year this had changed; the emphasis on varied uses gave way to an exclusive focus on the association between Libby's milk and coffee:

> Libby's milk . . . the creamy one! makes the coffee aromatic, flavoursome and delicious.

Nutritional quality was still indicated, with the sentence 'Libby's milk is concentrated full-cream milk', but the rhetorical shift from nutritional towards aesthetic is clear. By 1952 the aesthetic dominated:

> What do you need for coffee? Libby's milk . . . the creamy one! Even the best coffee can be still more aromatic and delicious with Libby's milk. With just a few drops it goes the most delectable brown.

In this sequence of advertisements we see intrinsic value being split off from nutritional value. Part of the initial sales-pitch, that evaporated milk not only substitutes for full-cream milk but is even healthier, fades into the background, and is replaced by continual subliminal reference to another desired food, cream. The linking of tinned milk and coffee had made it possible to initiate an aesthetic discourse about the milk; now the sales-pitch was all about the 'golden colour of the coffee' and the 'creamy flow'.

This shift in the rhetoric of consumption was one of the most important changes related to consumption during and after the 1950s. It involved the increasing invocation of desire through concealed and coded references, especially (though not only) in advertising. This language of semiotic reference, which consumers were having to learn, can be examined in the recipes of the consumer magazine, *Die kluge Hausfrau* (the Clever Housewife), a free weekly produced by the *Edeka* company.[13] Far from being repetitive and standardized, the rhetoric of the *kluge Hausfrau* was diverse and multi-faceted.

In the first years after the 1948 currency reform, the recipes published in the *kluge Hausfrau* were rather simple and frugal. They included such dishes as a macaroni pudding made with mince and mushrooms, and beef or butter were seldom mentioned. Their limits were geographical as well as pecuniary, lacking any intentional flair or extravagance, and bounded by domestic cuisine within the former limits of the German Reich. Dishes from Hamburg, Silesia or East Prussia appeared along with recipes such as 'cheap pumpkin', recommended for strict economy in the household budget.

In 1951 the *kluge Hausfrau* began to widen its horizons. For a Sunday meal it proposed a pork cutlet prepared *a la Milano*. Subsequent dishes *a la Milano* did not share a uniform way of preparing the food, but were all in various ways connected to the referents cheese and tomato. From the mid-1950s the recipes became increasingly international. In 1954 lamb was offered as 'Caucasian shashliks', 'Italian lamb', lamb cutlets 'provençale' or 'Viking-style'. Cabbage was no longer just cabbage, but 'Swiss cabbage', 'Scottish' or 'Norman' cabbage, even 'cabbage à la Strasbourg'.

In 1958 the *kluge Hausfrau* invited its readers on a 'culinary journey around the world': 'Italy: fish Milanese'; 'Portugal: Portuguese spinach roll'; 'France: omelette parisienne'; 'Netherlands: soup hollandaise' – and even 'Africa: banana salad'! It is clear that Eurocentrism was firmly internalized. Africa was depicted as a single country, and it was moreover represented through bananas – the stereotypical white assumption about what black people in Africa would eat. Through this caricature it becomes clear that all the international connotations were simply cultural constructs. Just as the '*milano* style' dishes had little to do with the authentic cuisine of northern Italy, so too the recipes for the culinary journey round the world represented not authentic local cuisines but German longing for international rapport, for acceptance back into the family of nations. The rhetoric of the recipes was less concerned with practices of preparation than with the stirring of dreams.

This phenomenon of internationalization was not specifically German, but following the failed attempt at world conquest it was especially noticeable in Germany. In the famous West German periodical *Magnum*, in 1960, Klaus Harpprecht described this post-war mentality:

> The Germans long to be part of the 'family of nations'. They are sick of standing apart, being alone, whether in a brilliant or in a miserable state. . . . The yearning to be assimilated to the international standard of taste, desires and needs has engulfed their architecture as well as their menus. (No architect would dare to build an office-block in any but the same style as his colleagues in Louisville, Nagasaki or Lyon. No urban restaurant would relinquish serving 'Steak à la Hawaii' or Nasi Goreng.) The Germans desperately want to strike lucky, and the world wants finally to have better luck with the Germans. So we are resolved to be happy and mediocre.[14]

So through the *kluge Hausfrau* recommendations West Germans could sample the mediterranean atmosphere by eating 'lamb cutlets à la Murillo' long before they made their first trips to southern Europe; and they could show their modernity and 'American life-style' by serving light low-calorie meals. After the early 1950s, and the frugal recipes full of references to shortages and the economical use of money, came timid excursions into specialities within everyday cuisine, then the first attempts to open up the rhetoric to little luxuries and international dishes. The multitude of little snacks, the miniaturization of meals, and finally the 'new conscience' eating, 'healthy and light', were all part of the developing rhetoric and show that these recipes did not stand outside social reality in West Germany, but can be read as a text for West German mentality.

The *kluge Hausfrau* also took up 'healthy nutrition' in an increasingly 'scientific' way, which parallelled the discourse in medical magazines. In the early 1950s the term 'healthy nutrition' was connected with an anti-modern perception and critique of civilization as creating sickness. During these years the *kluge Hausfrau* recommended whole-grain bread and sport, in line with earlier concepts of *Volksgesundheit* [literally folk health]. In the late 1950s the *kluge Hausfrau* had to respond to changing living conditions and became increasingly realistic. So the discourse on 'healthy nutrition' shifted: now the focus was either on fitness at work, especially for husbands and for children at school, or on 'the slim line'. This linking of nutrition to new socially defined standards (fitness at work or a specific, male-oriented imagining of the female body) signalled an important change in the discourse about eating during the 1950s. The 'modern housewife' as portrayed by the *kluge Hausfrau* knew all about calories, vitamins and other essentials of healthy nutrition; she was busy rationalizing her household and made extensive use of appliances; and thus she saved enough time to make herself pretty. The modern housewife was not only housewife and mother but good with machines and an attractive wife as well.

The shift in consumption practices towards concealed and coded references which invoked and expanded desires was reinforced by the new self-service store, which brought fundamental changes in the way people bought food. In 1951 there were still only thirty-nine self-service stores in West Germany, but they soon began to spread. By 1955 there were 203; and by 1960, only five years later, their number had shot up to 17,132. By 1965 West Germany had more than 53,000 self-service shops.[15]

The rupture could hardly have been greater. From the nineteenth century until the 1950s it had been an everyday experience to buy in shops where the shopkeeper or an assistant stood behind a counter, asked what you wanted, fetched the articles from behind the counter, weighed or measured them out, wrapped them, added up the prices and took the cash. Not only did the counter now disappear, the whole shop was reshaped for self-service. Now everything was in reach, ready to grasp at the level of eyes and hands; the arrangement of goods, lighting, decor – everything was organized around the presentation of commodities. In the grocery there had always been a chance to talk with neighbours, to hear local gossip, to 'waste' time. The new self-service stores instead represented modern ideas about the efficient use of time. Time-saving was the justification housewives quoted most often when asked about the advantages of self-service.[16]

When people entered a self-service store for the first time they were overwhelmed by the wealth of goods on offer. Their second important impression, after the initial confusion, was that of freedom of choice in this store staged as a glittering world of merchandise.[17] The attraction of 'self-service' was being able to choose and purchase on your own. Paradoxically, as options widened and the choice of goods became more and more complicated, the customer's desire for advice did not increase. Instead of the personal relationship between the shopkeeper (or assistant) and the customer, the goods now spoke directly to the customers and had to compete with their 'rivals' on the shelves. The communication previously effected by shop staff (often sales girls) had now to be achieved by the actual articles, and their presentation. This transfer of influence from the personal to the semiotic constituted a decisive change in consumption in West Germany from the end of the 1950s.

The qualitative change in consumption at the end of the 1950s was not simply a matter of further economic growth, higher rates of consumption, and the buying of better appliances. It was above all a multiplication of options and a diversification of practices. The new world of goods had not yet been fully disclosed for West German customers. But the diversification of practices associated with its development can already be seen in these early stages. Entering a self-service shop, buying industrially produced food, taking home frozen food for the weekend because the refrigerator would keep it fresh, preparing snacks ('TV dinners') in order not to interrupt the television programme for a family meal, cooking 'light and healthy' so as not to endanger the 'slim line' – all these were becoming familiar practices. Consumers, or more specifically housewives, had to learn all sorts of new ways. They had to find their way in a complex, unstable and confusing new world of commodities, to learn the new language of advertisers' descriptions and to decipher the various semiotic codes underlying the presentation of goods.

The memory of hunger, ever-present in the minds of the older generation, was disappearing. The stores overflowing with goods, the abundant displays in the butcher-shop windows – all proved to be not just a transient dream of prosperity but permanent affluence. Of course many households still had to stick to budgets, still had to watch the pennies; but at the end of the 1950s the need for modest frugality was passing away, along with previous structural restrictions and the cultural limitations of an older way of living. With the extension of the 'universe of goods' and the variety of consumer options, working-class households began to abandon their 'proletarian' life style.

The 'mass consumption society' was by no means the society predicted by the West German sociologist Helmut Schelsky, in which standards of living would increasingly converge around the middle-class level.[18] Social disparity continued, but traditional consumer hierarchies gave way to finer 'subtle differences' (in Bourdieu's title phrase). Heterodoxy and plurality marked the 'mass consumption society' that West Germany was becoming from the late 1950s onwards. Social inequality was no longer defined by occupation and by position in the processes of production. Instead what came to count more were conditions of work and leisure time, social security, and the potential for individual development. Gender, however, continued (and does still) to play a part in social inequality.[19]

It has been argued, for instance by Max Horkheimer, that food in industrialized societies loses its regional and seasonal contracts.

> The process of civilization can be discerned in culinary taste. Artificial methods in agriculture, butchery, and transportation have smoothed out distinctions in food as in other areas. As asparagus nowadays tastes like peas, so the unambiguous specific tastes of ham, sausage, lettuce or potatoes are vanishing because of the same manipulations. Fermentation of wine is to be interrupted and sulphuric acid added in aid of more rapid, rationalized and extensive production. Consequently the sense of taste is flattened; and a carrot of former times would surely confuse today's civilized palate, much like a bourgeois entering a garlic-saturated tenement in Lennox Avenue.[20]

Contrary to this pessimistic view, instead of this uniform homogenization of taste what happened was that taste itself changed fundamentally. True, traditional contrasts, such as the distinction between Sunday and everyday dishes, or across seasons and regions, were disappearing. But this was because the number of options increased, the choice of food broadened and the international agro-market offered fruit and vegetables of all kinds during the whole year. Traditional taste, it can be argued, was in fact being refined and expanded. The British historian Stephen Mennell described this process (though for the nineteenth century) in appropriate terms:

> underneath the many swirling cross-currents, the main trend has been towards diminishing contrasts and increasing variety in food habits and culinary taste. One trend, not two: for in spite of the apparent contradiction between diminishing contrasts and increasing varieties, these are both facets of the same process.[21]

In becoming part of the 'consumer society' everyone had, in the words of Pierre Bourdieu, to pay attention 'not to differ from the ordinary, but to differ differently'.[22] From within the increasing range of options consumers had to learn to choose and to develop their own distinctive style. Where previously and for so long consumption practices had involved making much out of little, now they meant learning to construct individuality out of the mass.

The practice of consumption becoming familiar from the late 1950s was to create new patterns and expectations. Where the process by which food and other commodities are produced and brought to the shelves of the self-service store involves so many people, at local, regional, and national level, the consumer is unlikely to be aware of the physical labour involved in producing the food purchased and consumed. Cash nexus and semiotic presentation in the world of commodities together obliterate the physical effort and skill through which the shelves are filled. The new 'consumer subject', then, is far removed from the production and sets a high value on individual freedom of choice. But because consumption and happiness are still supposed to be equivalent, after more than thirty years of the 'affluent society' and an abundance of goods that no one dreamed of in the early 1950s, this new consumer, skilled in the reading of semiotic references and codes and in the exercise of individual choice, is still consumed by unsatisfied desires.

NOTES

1 Alf Lüdtke, 'Hunger, Essens-"Genuss" und Politik bei Fabrikarbeitern und Arbeiter-frauen. Beispiele aus dem rheinisch-westfälischen Industriegebiet, 1910–1940', pp. 194–209 in his *Eigen-Sinn. Fabrikaltag, Arbeiter fahrungen und Politik vom Kaiserreich bis im olen Faschismus*, Hamburg, 1993, p. 195.

2 Jahresberichte der Verkaufsabteilung, 1955ff: Firmenarchiv Oetker, Bielefeld, P1/431.

3 Institut für Demoskopie, Allensbach, 'Wunsch und Besitz', 1958; typescript: Bundesarchiv Koblenz Zsg 132–707.

4 'Ausstattung der privaten Haushalte mit ausgewählten langebigen Gebrauchsgütern 1962/63', Stuttgart/Mainz 1964: Statistisches Bundesamt, Fachserie M. Reihe 18.

5 Ursula Schroth-Pritzel, 'Der Arbeitszeitaufwand im städtischen Haushalt', pp. 7–22 in *Hauswirtschaft und Wissenschaft* 6, Jg 1958, Heft 1.

6 Jakob Tanner, 'Grassroots-History und Fast Food', pp. 49–54 in *Geschichtswerkstatt* 12, 1987, p. 52.

7 Gesellschaft für Konsumforschung e.V., 'Hausfrauenbefragung über Küchenmaschinen', 1962; typescript: Archiv der GfK, Nürnberg, U 778.

8 Ibid.

9 Institut für Demoskopie, Allensbach, 'Das Einmachen', Umfrage 1953/54: Bundesarchiv Koblenz ZSg 132–284 I/II.

10 Institut für Demoskopie, Allensbach, 'Gemüse- und Obstkonserven', Marktanalyse, 1956: Bundesarchiv Koblenz ZSg 132–544 I.

11 See Max Eli, *Die Nachfragekonzentration im Nahrungsmittelhandel*, Berlin/München 1968, pp. 30–32.

12 Institut für Demoskopie, Allensbach, 'verschiedene Untersuchungen zum Dosenmilchverbrauch, 1950–1958': Bundesarchiv Koblenz ZSg, 132–188, –165, –280, –392, –465, –630.

13 *Die kluge Hausfrau* had already begun to appear before World War Two. It was revived in 1949, and by the end of the 1950s had a circulation of over a million. It was the most widely read consumer magazine of the food trade, and can be compared with famous public magazines like *Stern*, *Quick*, or *Constanze*.

14 Klaus Harpprecht, 'Die Lust an der Normalität', *Magnum* 29, April 1960 (pp. 17–19), p. 18.

15 Disch, *Der Gross- und Einzelhandel in der Bundesrepublik*, p. 60.

16 Gesellschaft für Konsumforschung, 'Einkaufsgewohnheiten in Deutschland', Nürnberg, 1953; typescript: Archiv der GfK, Nürnberg, U 183.

17 Ibid.

18 Helmut Schelsky, *Auf der Suche nach Wirklichkeit*, Düsseldorf/Köln, 1965, pp. 331–336.

19 Cf. Stefan Hradil, *Sozialstrukturanalyse in einer fortgeschrittenen Gesellschaft. Von Klassen und Schichten zu Lagen und Milieus*, Opladen, 1987.

20 Max Horkheimer, 'Bürgerliche Küche', in his *Notizen 1953–1955* (Gesammelte Schriften, vol. 6), ed. Werner Brede, Frankfurt am Main, 1991, pp. 237–238.

21 Stephen Mennell, *All Manners of Food: Eating and Taste in England and France from the Middle Ages to the Present*, Oxford/New York, 1985, p. 322.

22 Pierre Bourdieu, 'Klassenstellung und Klassenlage', pp. 42–74 in his *Zur Soziologie der symbolischen Formen*, Frankfurt am Main, 1975, p. 70.

12

The World on a Plate

Culinary Culture, Displacement and Geographical Knowledges

Ian Cook and Philip Crang

> The world on a plate. From Afghani ashak to Zimbabwean zaza, London offers an unrivalled selection of foreign flavours and cuisines. Give your tongue a holiday and treat yourself to the best meals in the world – all without setting foot outside our fair capital.
>
> (*Time Out*, 16 August 1995)

GEOGRAPHIES OF DISPLACEMENT

At one level the plate *Time Out* lays before us conforms to and re-presents a long-established conceptualization of cultural geographies, one rooted in the figure of the '*cultural mosaic*' (see also Friedman, 1994; Hannerz, 1992; Rosaldo, 1993). Comprised of bounded cultural regions or areas, this figure has long inspired geographical and anthropological imaginations committed to documenting Herder's plurality of cultures and relativizing the more evolutionist and hierarchical senses of a single cultured state (see Hatch, 1983). Here, it is constructed in terms of a range of national and regional cuisines, a range that literally exhibits an A–Z of placed tastes. This draws on and reinforces a long history of constructed associations between foods, places and peoples, associations epitomized in conceptions of national, regional and local cuisines, and in the use of foods as emblems and markers of national, regional and local identities (see Murcott, 1995). But the mosaic is not the only spatial figure that this globalized plate draws on. It also depends upon, and refers to, a variety of '*cultural flows and networks*' (Appadurai, 1990; Chambers, 1990), in particular of migrations and of tourisms. We can 'give our tongues a holiday' because a world of 'foreigners' and 'foreign flavours' has come to cosmopolitan London. One of the key conceptual agendas raised by this fragment of material culture is therefore how these two differing figurations of the geographies of culture are to be articulated.

Perhaps the most usual relationship constructed is one of a mutually supportive opposition of locally meaningful cultural artefacts and practices and homogenizing invasive flows. In pessimistic portraits of culinary culture the spectre of McDonaldization looms large here (Ritzer, 1993), as fast-food standardization provides a contemporary echo of much longer fears of global homogenization, and, at least in Britain, Americanization (Hebdige, 1988). The growth of globally distributed fast-food franchise chains is claimed to 'erase [*sic*] the differences between "this place" and "that"' (MacClancey, 1992: 193) through the establishment of 'a new order and scale of experience . . . a powerful culture which overwhelms local and regional experience' (Peet, 1989: 176). In more optimistic accounts the cultural outcomes are viewed differently, but the logic remains the same, as local cultural creativities are seen as indigenizing the standardized cultural materials of global commodity flows (see, for example, Hannerz, 1992). In culinary culture particular emphasis might be laid on how the arts of cooking and presentation allow food ingredients to be locally and creatively reworked. Evidential support might be drawn from histories of how 'foreign' foods and 'imported' cuisines have been adapted to fit in with the availability of key ingredients and culinary expectations in their new settings (Levenstein, 1985; MacClancey, 1992).

However, it may be more profitable to break out of the mutually supportive opposition of homogenizing and invasive commodity flows and either submitting or resisting, but always distinctive, place-based traditional cultures altogether. Instead, we want to suggest that we might think of foods, and other material cultures, as geographically constituted through processes of '*displacement*' (see also Crang, forthcoming). Never a tightly defined concept, the notion of displacement has nonetheless been used by a number of writers to evoke a sense of a geographical world where cultural lives and economic processes are characterized not only by the points in space where they take and make place, but also by the movements to, from and between those points (Clifford, 1988, 1992; Robertson *et al.*, 1994). To elaborate, in terms of food consumption the figure of displacement might be used to suggest an understanding whereby: processes of food consumption are cast as local, in the sense of contextual; but where those contexts are recognized as being opened up by and constituted through connections into any number of networks, which extend beyond delimiting boundaries of particular places (see also Massey, 1991a, 1991b, 1992, 1993, 1995); furthermore, where imagined and performed representations about 'origins', 'destinations' and forms of 'travel' surround these networks' various flows; and where consumers (and other actors in food commodity systems) find themselves socially and culturally positioned, and socially and culturally position themselves, not so much through placed locations as in terms of their entanglements with these flows and representations.

More specifically, this suggests three interrelated geographies that constitute food-consuming worlds. First, the geographies of the local places of food consumption or usage, operating as 'spaces of identity practice' (Friedman, 1994). Second, the spatial structures of often globally extensive 'systems of provision' (Fine and Leopold, 1993: 20) that stretch beyond places of food consumption but which are vital in their constitution, providing resources not only of foods themselves, but also of knowledges about how to value and use them, domestic technologies for their use, and non-domestic sites for their consumption. And third, there are the geographical knowledges associated with the materials that flow through these systems of provision, which for consumers form part of the discursive complexes within which they are increasingly asked reflexively to manage their food consumption habits and their selves. These three constitutive geographies suggest something other than the opposition of placed culinary cultures and imported/exported global commodity flows. The geographies of these foods and culinary cultures are not to be cast in terms of location in fenced-off spatial arenas, and need to be divorced from what Doreen Massey calls introverted senses of place in which a them and us mentality is sustained through constructions of there and here (Massey, 1993). Instead, emphasis is laid on an extroverted sense of place in which boundaries are seen as contestable and contested social constructions and where any here/us is constituted through its connections into the there/them. Thus, any placed cuisine depends upon those connections, not simply historical accretion or stasis within that place. And, in turn, there is no simple or unconstructed association of foods and places; rather, placings of foods are active social constructions (Mintz, 1985), as borne witness to most obviously by the contemporary fabrication or simulation of many 'ethnic' cuisines (e.g. *chilli con carne* as a Texan construction of a Mexican dish) (see Smart, 1994: 177) but, crucially, equally true of all such placings including those more commonly valued as authentic.

As an aside, this means that the figure of displacement is not synonymous with conceptions of cultural *creolization* or *hybridity* (see Hannerz, 1992), inasmuch that it is not about a cultural mixing due to a 'leaky mosaic' (Friedman, 1995: 85). Instead, it suggests that there are no pure cultures to mix, if purity means bounded exclusivity. And it emphasizes how processes of displacement are not some recent disturbance of past cultural forms. Indeed, it is important to note that these geographies of 'displacement' are not solely a contemporary phenomenon, though they may be increasingly notable and noted in the modern world's cultural economies. For example, Jack Goody has pointed out how

> It is difficult to conceive of Italian food without pasta and tomato paste. But the use of pasta may have arrived from China via Germany only in the 15th century . . . Stouff's study of the 14th and 15th centuries concludes by denying that there was an original Provencal cuisine in the late Middle Ages. . . . The outstanding feature of 'traditional' Provencal cooking of the 19th and 20th centuries, olive oil, was used only for eggs, fish and frying beans. Otherwise it was the fat of salted pork, used particularly to flavour the soup of peas, beans

> and above all cabbage. This was the basic food, he claims, of the ordinary folk of Provence, just as it was in the rest of Europe at that time. . . . 'Traditional Provencal cooking', like many other folk-ways, only emerged in recent times, a salutary thought for those attached either to the holistic or timeless view of culture.
>
> (Goody, 1982: 36, cited Murcott, 1995: 12)

Thus regional cuisines are invented traditions (inventions in which the genre of cookery books often seem to have played a particularly important role; see Appadurai, 1990). Many of the most basic, and 'traditional', ingredients in European culinary cultures such as tomatoes, potatoes, vanilla and chillies were 'discovered' overseas in the early stages of imperialist 'adventure', brought back 'home', and 'domesticated'. And many characteristically European foods – for example the English 'cuppa' of tea – were produced through, and continue to depend upon, networks of imperial connections, connections that comprise 'the outside history that is inside the history of the English' (Hall, 1991: 49) (see also Smith, 1992). In these cultural spaces, designations of cultural hybridity, with their emphasis on previously separate cultures mixing, are not inherently illegitimate, but they must themselves be recognized as constructed geographical knowledges, locally produced as part of situationally specific identity projects, and, like all such knowledges, constructed from within the spaces of material culture and not from some Olympian viewpoint above them. Jonathan Friedman makes a similar point using the example of the dominant non-hybrid understandings of pasta-based Italian cuisine:

> The introduction of pasta into the cuisine of the Italian peninsula is a process of globalization, and the final elaboration of a pasta-based Italian cuisine is, in metaphorical terms, a process of cultural syncretism, or perhaps creolization. But such mixture is only interesting in the practice of local identity. . . . Thus the fact pasta became Italian, and that its Chinese origin became irrelevant is the essential culture-producing process in this case. Whether origins are maintained or obliterated is a question of the practice of identity.
>
> (Friedman, 1995: 74)

So, foods do not simply come from places, organically growing out of them, but also make places as symbolic constructs, being deployed in the discursive construction of various imaginative geographies. The differentiation of foods through their geographies is an active intervention in their cultural geographies rather than the passive recording of absolute cultural geographic differences.

REFERENCES

Appadurai, Arjun (1990) 'Disjuncture and Difference in the Global Cultural Economy', *Theory, Culture and Society* 7: 295–310.

Chambers, Iain (1990) *Border Dialogues*. London: Routledge.

Clifford, James (1988) *The Predicament of Culture. Twentieth-century Ethnography, Literature, and Art.* London: Harvard University Press.

Clifford, James (1992) 'Traveling Cultures', in Lawrence Grossberg, Cary Nelson and Paula Treichler (eds) *Cultural Studies*, pp. 96–116. New York: Routledge.

Crang, Philip (forthcoming) *Displacement: On the Geographies of Modern Cultures.* London: Arnold.

Fine, Ben and Leopold, Ellen (1993) *The World of Consumption*. London: Routledge.

Friedman, Jonathan (1994) *Cultural Identity and Global Process.* London: Sage.

Friedman, Jonathan (1995) 'Global System, Globalization and the Parameters of Modernity', in Mike Featherstone, Scott Lash and Roland Robertson (eds) *Global Modernities*, pp. 69–90. London: Sage.

Goody, Jack (1982) *Cooking, Cuisine and Class.* Cambridge: Cambridge University Press.

Hall, Stuart (1991) 'Old and New Identities, Old and New Ethnicities', in Anthony King (ed.) *Culture, Globalisation and the World System*, pp. 41–68. Basingstoke: Macmillan.

Hannerz, Ulf (1992) *Cultural Complexity*. New York: Columbia University Press.

Hatch, Elvin (1983) *Culture and Morality. The Relativity of Values in Anthropology*. New York: Columbia University Press.

Hebdige, Dick (1988) 'Towards a Cartography of Taste 1935–62', in *Hiding in the Light: On Images and Things*, pp. 45–76. London: Comedia.

Levenstein, Harvey (1985) 'The American Response to Italian Food', *Food & Food-ways* 1: 1–24.

MacClancey, Jeremy (1992) *Consuming Culture* London: Chapmans.

Massey, Doreen (1991a) 'A Global Sense of Place', *Marxism Today* 35(6): 24–9.

Massey, Doreen (1991b) 'The Political Place of Locality Studies', *Environment and Planning A* 23: 267–281.

Massey, Doreen (1992) 'A Place Called Home?', *New Formations* 17: 3–15.

Massey, Doreen (1993) 'Power-Geometry and a Progressive Sense of Place', in Jon Bird, Barry Curtis, Tim Putnam, George Robertson and Lisa Tickner (eds) *Mapping the Futures: Local Cultures, Global Change*, pp. 59–69. London: Routledge.

Massey, Doreen (1995) 'The Conceptualization of Place', in Doreen Massey and Pat Jess (eds) *A Place in the World? Places, Cultures and Globalization*, pp. 45–85. Oxford: Oxford University Press.

Mintz, Sidney (1985) *Sweetness and Power: The Place of Sugar in Modern History*. Harmondsworth: Penguin.

Murcott, Ann (1995) 'Food as an Expression of National Identity', unpublished paper, South Bank University.

Peet, Richard (1989) 'World Capitalism and the Destruction of Regional Cultures', in Ron Johnston and Peter Taylor (eds) *The World in Crisis*, 2nd edn, pp. 175–199. Oxford: Blackwell.

Ritzer, George (1993) *The McDonaldization of Society*. Thousand Oaks, CA: Pine Forge Press.

Robertson, George, Mash, Melinda, Tickner, Lisa, Bird, Jon, Curtis, Barry and Putnam, Tim (eds) (1994) *Travellers' Tales: Narratives of Home and Displacement*, London: Routledge.

Rosaldo, Renato (1993) *Culture and Truth. The Remaking of Social Analysis*. London: Routledge.

Smart, Barry (1994) 'Digesting the Modern Diet: Gastro-Porn, Fast Food and Panic Eating', in Keith Tester (ed.) *The Flâneur*, pp. 158–180. London: Routledge.

Smith, Woodruff (1992) 'Complications of the Commonplace: Tea, Sugar and Imperialism', *Journal of Interdisciplinary History* (Autumn): 259–278.

13

The 'Consumption' of Tourism

John Urry

THE SOCIAL LIMITS TO TOURISM

The economist Mishan presents one of the clearest accounts of the thesis that there are fundamental limits to the scale of contemporary tourism (1969). These limits derive from the immense costs of congestion and overcrowding. He perceptively writes of: 'the conflict of interest . . . between, on the one hand, the tourists, tourist agencies, traffic industries and ancillary services, to say nothing of governments anxious to augment their reserves of foreign currencies, and all those who care about preserving natural beauty on the other' (1969: 140). He quotes the example of Lake Tahoe, whose plant and animal life has been destroyed by sewage generated by the hotels built on its banks. A 1980s example would be the way in which the coral around tourist islands like Barbados is dying, both because of the pumping of raw sewage into the sea from the beachside hotels, and because locals remove both plants and fish from the coral to sell to tourists.

Mishan also notes that here is a conflict of interest between present and future generations which stems from the way in which travel and tourism is priced. The cost of the marginal tourist takes no account of the additional congestion costs imposed by the extra tourist. These congestion costs include the generally undesirable effects of overcrowded beaches, a lack of peace and quiet, and the destruction of the scenery. Moreover, the environmentally sensitive tourist knows that there is nothing to be gained from delaying their visit to the place in question. Indeed if anything the incentive is the other way round. There is a strong pull to go as soon as possible – to enjoy the unspoiled view before the crowds get there! Mishan's perspective as someone appalled by the consequences of mass tourism can be seen from the following: 'the tourist trade, in a competitive scramble to uncover all places of once quiet repose, of wonder, beauty and historic interest to the money-flushed multitude, is in effect literally and irrevocably destroying them' (1969: 141). His middle-class, middle-aged elitism is never far from the surface. For example, he claims that it is the 'young and gullible' who are taken in by the fantasies dreamt up by the tourist industry.

However, Mishan's main criticism is that the spread of mass tourism does not in fact produce a democratisation of travel, it is an illusion which destroys the very places which are being visited. This is because geographical space is a strictly limited resource. Mishan says: 'what a few may enjoy in freedom the crowd necessarily destroys for itself' (1969: 142). Unless international agreement is reached (he suggested the immensely radical banning of all international air travel!), the next generation will inherit a world almost bereft of places of 'undisturbed natural beauty' (1969: 142). So allowing the market to develop without regulation has the effect of destroying the very places which are the objects of the tourist gaze. Increasing numbers of such places come to suffer from the same pattern of destruction.

This pessimistic argument is criticised by Beckerman who makes two important points (1974: 50–52). First, concern for the effects of mass tourism is basically a 'middle-class' anxiety (like much other environmental concern). This is because the really rich 'are quite safe from the masses in the very expensive resorts, or on their private yachts or private islands or secluded estates' (Beckerman 1974: 50–1). Second, most groups affected by mass tourism do in fact benefit from it, including even some of the pioneer visitors who return to find services available that were unobtainable when the number of visitors was small. Hence Beckerman talks of the 'narrow selfishness of the Mishan kind of complaint' (Beckerman 1974: 51).

This disagreement over the effects of mass tourism is given more theoretical weight in Hirsch's thesis on the social limits to growth (1978: see the collection Ellis and Kumar 1983). His starting point is similar to Mishan's when he notes that individual liberation through the exercise of consumer choice does not make those choices liberating for all individuals together (1978: 26). In particular he is concerned with the positional economy. This term refers to all aspects of goods, services, work, positions and other social relationships which are either scarce or subject to congestion or crowding. Competition is therefore zero-sum, as any one person consumes more of the goods in question, so someone else is forced to consume less. Supply cannot be increased, unlike the case of material goods where the processes of economic growth can usually ensure increased production. People's consumption of positional goods is relational. The satisfaction derived by each individual is not infinitely expandable but depends upon the position of one's own consumption to that of others. This can be termed coerced competition. Ellis and Heath define this as competition in which the status quo is not an option (1983: 16–19). It is normally assumed in economics that market exchanges are voluntary so that people freely choose whether or not to enter into the exchange relationship. However, in the case of coerced consumption people do not have such a choice. One has to participate even though at the end of the consumption process no one is necessarily better off. This can be summarised in the phrase: 'one has to run faster in order to stay still'. Hirsch cites the example of suburbanisation. People move to the suburbs to escape from the congestion in the city and to be nearer the quietness of the countryside. But as economic growth continues so the suburbs get more congested, they expand and so the original suburbanites are as far away from the countryside as they were originally. Hence they will seek new suburban housing closer to the countryside and so on. The individually rational actions of others make one worse off and each person cannot avoid participating in the leapfrogging process. No one is better off over time as a result of such coerced consumption.

Hirsch argues that much consumption has similar characteristics to the case of suburbanisation, namely that the satisfaction people derive from it depends upon the consumption choices of others. This can be seen most clearly in the case of certain goods which are scarce in an absolute sense. Examples cited here are 'old masters' or the 'natural landscape' where increased consumption by one leads directly to reduced consumption by another (although see Ellis and Heath 1983: 6–7). Hirsch also considers the cases where there is 'direct social scarcity', which are luxury or snob goods enjoyed because they are rare or expensive and possession of them indicates social status or good taste. Examples include jewellery, a residence in a particular part of London, or designer clothes. A further type Hirsch considers is that of 'incidental social scarcity', that is goods whose consumption yields satisfaction which is influenced by the relative extensiveness of use by others. Examples here include the car purchase but with no increase of satisfaction because of increased congestion as everyone else does the same; and the obtaining of educational qualifications and no improved access to leadership positions because everyone else has been acquiring similar credentials (Ellis and Heath 1983: 10–11).

It is fairly easy to suggest examples of tourism which fit these various forms of scarcity. On the first, access to Windermere in the English Lake District is in a condition of absolute scarcity. One person's consumption is at the expense of someone else's. On the second, there are many holiday destinations which are consumed, not because they are intrinsically superior, but because they convey taste or superior status. For Europeans, the West Indies, West Africa and the Far East would be current examples, although these will change as mass tourism patterns themselves change. And third, there are many tourist sites where people's satisfaction depends upon the degree of congestion, currently such as Greece. Hirsch quotes a middle-class professional who remarked that the development of cheap charter flights to a previously 'exotic' country meant that, 'now that I can afford to come here I know that it will be ruined' (1978: 167).

Although I have set out these different types of positional good identified by Hirsch, the distinctions between them are not consistently sustained and they merge into each other. Furthermore, there are a number of major difficulties in his argument. First, it is ambiguous just what is meant by consumption in the case of much tourism. Is it the ability to gaze at a particular object if necessary in the company of many others? Or is it to be able to gaze, without others being present? Or is it to be able to rent accommodation for a short period with a view of the object close at hand? Or finally, is it the ability to own property with a view of the object nearby? The problem arises, as we have noted, because of the importance of the gaze to touristic activity. A gaze is after all visual, it can literally take a split second, and the other services provided are in a sense peripheral to the fundamental process of consumption, which is the capturing of the gaze. This means that the scarcities involved in tourism are more complex than Hirsch allows for. One strategy pursued by the tourist industry has been to initiate new developments which have permitted greatly increased numbers to gaze upon the same object. Examples include building huge hotel complexes away, say, from the coastline itself; the development of off-peak holidays so that the same view can be gazed upon throughout the year; devising holidays for different segments of the market so that a wider variety of potential visitors can see the same object; and the development of time-share accommodation so that the facilities can be used all of the year.

Moreover, the notion of scarcity is problematic for other reasons. I shall begin here by noting the distinction between the physical carrying capacity of a tourist site, and its perceptual capacity (Walter 1982). In the former sense it is clear when a mountain path literally cannot take any more walkers since it has been eroded and effectively disappeared. Nevertheless, even here there are still thousands of *other* mountain paths that could be walked along and so the scarcity only applies to *this* path leading to *this* particular view, not to all paths along all mountains.

However, the notion of perceptual capacity further complicates the situation. Although the path may still be physically passable, it no longer signifies the pristine wilderness upon which the visitor had expected to gaze (Walter 1982:296). Its perceptual carrying capacity would have been reached, but not its physical capacity. However, perceptual capacity is immensely variable and depends upon particular conceptions of nature and of the circumstances in which people expect to gaze upon it. Walter cites the example of an Alpine mountain. As a material good the mountain can be viewed for its grandeur, beauty and conformity to the idealised Alpine horn. There is almost no limit to this good. No matter how many people are looking at the mountain it still retains these qualities. However, the same mountain can be viewed as a positional good, as a kind of shrine to nature which individuals wish to enjoy in solitude. There is then a 'romantic' form of the tourist gaze, in which the emphasis is upon solitude, privacy and a personal, semi-spiritual relationship with the object of the gaze. Barthes characterises this viewpoint as found in the 'Guide Bleu': he talks of 'this bourgeois promoting of the mountains, this old Alpine myth . . . only mountains, gorges, defiles and torrents . . . seem to encourage a morality of effort and solitude' (1972: 74). For example, Stourhead Park in Wiltshire illustrates

> the romantic notion that the self is found not in society but in solitudinous contemplation of nature. Stourhead's garden is the perfect romantic landscape, with narrow paths winding among the trees and rhododendrons, grottoes, temples, a gothic cottage, all this around a much indented lake. . . . The garden is designed to be walked around in wonderment at Nature and the presence of other people immediately begins to impair this.
>
> (Walter 1982: 298)

When I discussed Mishan it was noted that he emphasised that 'undisturbed natural beauty' constituted the typical object of the tourist gaze. But this is only one kind of gaze, the 'romantic'. I shall now set out the characteristics of an alternative, which I shall call the 'collective' tourist gaze.

I will begin here by considering a different Wiltshire house and garden, Longleat, which is

> a large stately home, set in a Capability Brown park; trees are deliberately thinned . . . so that you can see the park from the house, and the house from the park. Indeed the house is the focal point of the park . . . the brochure lists twenty-eight activities and facilities. . . . All this activity and the resulting crowds fit sympathetically into the tradition of the stately home: essentially the life of the aristocratic was public rather than private.
>
> (Walter 1982: 198)

In other words, such places are designed as public places. They would look strange if they were empty. It is in part other people that make such places. The collective gaze thus necessitates the presence of large numbers of other people, as are found, for example, in English seaside resorts. Other people give atmosphere to a place. They indicate that this is *the* place to be and that one should not be elsewhere. Indeed one of the problems for the contemporary English seaside resort is precisely that there are not enough other people to convey these sorts of messages. 'Brighton or Lyme Regis on a sunny summer's day with the beach to oneself would be an eerie experience' (Walter 1982: 298). It is the presence of other *tourists*, people just like oneself, that is actually necessary for the success of such places which depend upon the collective tourist gaze. This is particularly the case in major cities, whose uniqueness is their cosmopolitan character. The presence of people from all over the world (tourists in other words) gives capital cities their distinct excitement and glamour.

A further point here is that large numbers of other tourists do not simply generate congestion as the positional good argument would suggest. The presence of other tourists provides a market for the sorts of services that most tourists are in fact eager to purchase, such as accommodation, meals, drink, travel and entertainment. New Zealand is an interesting case here. Once one leaves the four major cities there are almost no such facilities because of the few visitors compared to the size of the country. The contrast with the Lake District in north west England is most striking, given the scenic similarity.

Thus Hirsch's arguments about scarcity and positional competition mainly apply to those types of tourism characterised by the romantic gaze. Where the collective gaze is to be found then there is no problem about crowding and congestion. And indeed Hirsch's argument rests on the notion that there are only a limited number of objects which can be viewed by the tourist. Yet in recent years there has been an enormous increase in the objects of the tourist gaze, far beyond those providing 'undisturbed natural beauty'. It was reported in a study conducted by the Cabinet Office in the UK that of all the tourist attractions open in 1983, half had been opened in the previous fifteen years (Cabinet Office 1985). And part of the reason for such an increase results from the fact that contemporary tourists are collectors of gazes. They are less interested in visiting the same place year after year. The initial gaze is what counts and people appear to have less and less interest in repeat visits (Blackpool being almost the exception that proves the rule).

There are two concluding points to note here. First, those who value solitude and a romantic tourist gaze do not see this as merely *one* way of regarding nature. They consider it as 'authentic', as real. And they attempt to make everyone else sacralise nature in the same sort of way. Romanticism has become widespread and generalised, spreading out from the upper and middle classes, although the notion of romantic nature is a fundamentally invented pleasure. And yet the more that its adherents attempt to proselytise its virtues to others, the more the conditions of the romantic gaze are undermined: 'the romantic tourist is digging his [*sic*] own grave if he seeks to evangelise others to his own religion' (Walter 1982: 301). The romantic gaze is part of the mechanism by which tourism is spreading on a global scale and drawing almost every country into its ambit, thereby providing uniformity, minimising diversity, and encouraging the 'romantic' to seek ever new objects of the romantic gaze (see Turner and Ash (1975) on this extension of the 'pleasure periphery').

Second, the tourist gaze is increasingly signposted. There are markers which identify what things and places are worthy of our gaze. Such signposting identifies a relatively small number of tourist nodes. The result is that most tourists are concentrated within a very limited area. As Walter says, 'the sacred node provides a positional good that is destroyed by democratisation' (1982: 302). He in turn favours the view that there are 'gems to be found everywhere and in everything . . . there is no limit to what you will find' (Walter 1982: 302). We should get away from the tendency to construct the tourist gaze at a few selected sacred sites, and be much more catholic in the objects at which we may gaze. This has begun to occur in recent years, particularly with the development of industrial and heritage tourism. However, in part the signposts are designed to help people congregate and are in a sense an important element of the collective tourist gaze. Visitors come to learn that they can congregate in certain places and that that is where the collective gaze will take place.

I will conclude this section on the economic theory of tourism by noting the pervasiveness of the romantic as opposed to the collective gaze and the consequential problem of the positional good of many tourist sites,

> professional opinion-formers (brochure writers, teachers, Countryside Commission staff, etc.) are largely middle class and it is within the middle class that the romantic desire for positional goods is largely based. Romantic solitude thus has influential sponsors and gets good advertising. By contrast, the largely working class enjoyment of conviviality, sociability and being part of a crowd is often looked down upon by those concerned to conserve the environment. This is unfortunate, because it . . . exalts an activity that is available only to the privileged.
>
> (Walter 1982: 303)

CONCLUSION

I have tried to demonstrate here that the consumption of 'tourist services' is important yet by no means easy to understand and explain. The importance derives from the centrality of tourist activities in modern societies. Indeed elsewhere it will be argued that the way in which 'tourism' has been historically separated from other activities, such as shopping, sport, culture, architecture and so on, is dissolving. The result of such a process is a 'universalising of the tourist gaze' (Urry 1990).

The difficulty of understanding tourist activities derives from the unclear character of just what is being consumed. I have suggested that it is crucial to recognise the visual character of tourism, that we gaze upon certain objects which in some ways stand out or speak to us. I have also shown that there are two characteristic forms of such a gaze, the romantic and the collective, and that problems of congestion and positionality are very different in these two cases. More work though needs to be undertaken on the impact of these different gazes on particular places, and how the providers of different services structure them in relationship to such different gazes. A particular issue is that of authenticity. It is argued especially by MacCannell that what tourists seek is the 'authentic', but that this is necessarily unsuccessful since those being gazed upon come to construct artificial sights which keep the inquisitive tourist away (MacCannell 1976). Tourist spaces are thus organised around what he calls 'staged authenticity'. Two points should be noted here. First, the lack of authenticity is much more of a problem for the 'romantic gaze' of the service class for whom naturalness and authenticity are essential components. It is less of a problem for those engaged in the collective tourist gaze where congregation is paramount. Second, it has recently been suggested that some tourists might best be described as 'post-tourists', people who almost delight in inauthenticity. The post-tourist finds pleasure in the multitude of games that can be played and knows that there is no authentic tourist experience. They know that the apparently authentic fishing village could not exist without the income from tourism or that the glossy brochure is a piece of pop culture. For the post-tourist there is no particular problem about the inauthentic. It is merely another game to be played at, another pastiched surface feature of post-modern experience.

REFERENCES

Barthes, R. (1972) *Mythologies*. London: Jonathan Cape.

Beckerman, W. (1974) *In Defence of Economic Growth*. London: Jonathan Cape.

Cabinet Office (Enterprise unit) (1985) *Pleasure, Leisure and Jobs. The Business of Tourism*. London: HMSO.

Ellis, A. and Heath, A. (1983) 'Positional Competition, or an Offer You Can't Refuse', in Ellis and Kumar, pp. 1–22.

Ellis, A. and Kumar, R. (eds) (1983) *Dilemmas of Liberal Democracies*. London: Tavistock.Hirsch, F. (1978) *Social Limits to Growth*. London: Routledge & Kegan Paul.

MacCannell, D. (1976) *The Tourist. A New Theory of the Leisure Class*. London: Macmillan.

Mishan, E. (1969) *The Costs of Economic Growth*. Harmondsworth: Penguin.

Turner, L. and Ash, J. (1975) *The Golden Hordes*. London: Constable.

Urry, J. (1990) *The Tourist Gaze*. London: Sage.

Walter, J. (1982) 'Social Limits to Tourism'. *Leisure Studies* 1: 295–304.

TWO

14

A New Look at Gentrification

Gentrification and Domestic Technologies

Paul A. Redfern

Builders working on converting an old building into something new is such a familiar sight that the very fact that such activity can take place at all is taken for granted. Smith (1992, page 113) criticises Hamnett (1991) for limiting the definition of gentrifiers to middle-class individuals and for neglecting to take into account other kinds of individuals 'responsible for the actual physical transformation of urban landscapes'. But only one group out of all the kinds of individuals cited by Smith, 'builders, property owners, estate agents, local governments, banks and building societies', is actually responsible for the physical transformation of the urban landscape, namely the builders (and even then I would suspect the builders are thought of by Smith only in their roles as capitalist employers or developers, not as workers). All the rest provide key services such as finance or plans, but it is the building workers alone who actually engage physically in the transformation of the urban landscape. The rest of the actors may supply the financial capital, but that capital has to be converted into physical capital – capital goods. Builders need the *material*, not simply financial, wherewithal to undertake these transformations. So far as gentrification is concerned, this means the availability of domestic technologies; in such a form as enables the housing services offered by an old property to be brought into line with the services offered by a new property. Yet the question of domestic technologies is almost completely absent as a topic for analysis in gentrification studies.

I follow Du Vall (1988) in defining domestic technologies as covering food production, preservation, cooking facilities and utensils, clothing, cleaning, water and waste disposal, heating, and lighting. Du Vall traces developments in these technologies from Neolithic times onwards. As Cowan (1983) and others have shown, modern developments in these technologies have led to the devolution from the home of clothing, food production, and, to a certain extent, food preparation. Although these are extremely important from the point of view of gentrification studies, the key developments only occur after the introduction of piped water and sewage, and especially external energy sources, gas and electricity, into the home. The term 'domestic technologies' is therefore taken here to apply particularly to cooking, cleaning, water and waste disposal, heating, and lighting in so far as the operation of these technologies depends on piped water, sewage, gas, and/or electricity (Hardyment, 1988; Nye, 1991).

Once pointed out, the relevance of domestic technologies to gentrification is obvious: you cannot have gentrification without being able to do up a house. Doing up a house means putting in all 'mod cons' and you cannot put in all mod cons if they have not been invented yet. Yet the lack of attention paid to this issue is amazing. Ley (1986), for example, searching for the causes of inner-city gentrification in Canadian cities, makes no reference to domestic technologies out of a total of thirty-five variables in his correlation exercise; neither does Bourne (1993). It is possible that the participants in the gentrification debate genuinely do not consider domestic technologies as worthy of attention. But without electricity or household appliances for cleaning, cooking, and heating, what good does it do to spend money on repairing the structure of a house which can only be run with the aid of these technologies, or with the aid of servants? If they do hold the view that domestic technologies are not worthy of attention, one purpose of this paper is to show that this would be a mistake.

As late as 1950 in England, the cost of a vacuum cleaner, fifteen guineas – excluding purchase tax of 25 per cent – was approximately that of the average weekly wage, and the price of a washing machine, around £125. Bendix, the first automatic washing machine, was so expensive that no prices were quoted in *Ideal Home* magazine, the source of these figures, during the first three years following its introduction in 1951. Instead, generous, but unspecified, hire-purchase terms were advertised. One can only speculate that it must have been so expensive that quoting a price would have frightened off even the wealthy. The nearest competitor, the Servis twin tub, introduced in 1953, cost £95, including tax. Hunkin (1989) compared the price of that machine with approximately that of a small car. The price of a house in Canonbury, in the heart of gentrified Islington, at this time was £2,650 (Humphries and Taylor, 1986, page 151), that is, a washing machine alone cost between 5 per cent and 10 per cent of the cost of a gentrifiable house, and about ten times the average weekly wage. If washing machine prices had kept pace with prices of Canonbury properties, they would cost in the order of £12,000 to £15,000 today. Figures such as these render redundant most case studies on gentrifiers' motivations.

The reason why domestic technologies have an impact on gentrification is because of the dramatic fall in prices of these technologies since the early 1950s. Taking the United Kingdom, London and Islington, as the example, the trends in house prices with those of an automatic washing machine over the period 1951 to 1981 were compared. A washing machine is obviously not the be-all and end-all of domestic technologies though, as Cowan (1983) and Hardyment (1988) both make clear, it has been possibly the most significant of all domestic technologies in terms of its impact on the management of domestic work and its class relations. Attempts at creating a mechanical alternative to scrubbing clothes by hand form among the earliest of applications of technological principles to domestic labour. Despite all the attention given to their development, the cost of these machines in 1951 was still extremely high. As the flagship domestic technology, the comparison is therefore instructive.

The price of a fully automatic washing machine (Bendix, Indesit, Hoover Keymatic), incorporating all the latest improvements, fell from around 4 per cent of the purchase cost of housing to less than 1 per cent over this period (less than 0.5 per cent in Islington).

Despite the absence of data for 1961 (as a result of the reorganisation of the LCC into the GLC), the trend is clear. Domestic technology prices have fallen sharply in relation to house prices. It is the existence of domestic technologies at all that permits gentrification to occur. It is, however, the fall in the comparative cost of these technologies that has permitted its diffusion. It is true that these technologies were available in the 1920s and 1930s, but they were so expensive then, that they were only a feasible option in suburban developments, where essentially the house was built around the technology, which was therefore 'embodied' in the property. It is not until the late 1950s and early 1960s that the costs of these technologies falls to the point where it becomes feasible to consider investing them in an existing building, as a permitting 'disembodied' technical progress. And as they have continued to fall in cost relative to the cost of purchasing a building, so also has gentrification spread far and wide beyond the boundaries of Islington.

To argue that the explanation of gentrification lies in the ability to invest domestic technologies to an already existing structure does push the burden of explanation on to the qualities of these technologies themselves, just as the postindustrialist argument for concentrating on the production of gentrifiers pushes the burden of explanation on to the qualities of the gentrifiers, with, as I have argued, disastrous consequences for their claims to account for gentrification. To account for the particular qualities of domestic technologies which permit them to act as the basis of gentrification, the contingencies in the development of domestic technologies must also be considered. There are a number of complementary possible defences to any charge of technological determinism in this account.

Accounting for the qualities of domestic technologies which make gentrification possible means investigating why their provision took the form they did, together with developments in domestic labour associated with their introduction. The available forms of domestic labour, performed by (mainly female) servants in the nineteenth century and housewives in the twentieth, greatly affected forms of housing provision and the development of these technologies (Cowan, 1983; Hardyment, 1988; Nye, 1991). Space constraints prevent further discussion here but see Redfern (1992, 1997). Any such investigation will therefore directly and immediately confront issues of class and gender in gentrification, rather than trying to force these issues into discussions of gentrifiers' so-called characteristics. Bondi (1991), for example, criticises the current state of the gentrification debate for its lack of attention to gender issues. Her proposal for remedying this, and this is indicative of the hegemony of postindustrialism in gentrification studies, is, however, for the production of a gendered gentrifier or member of the new middle class by students of gentrification.

An account which places the explanation of the causes of gentrification in the introduction of domestic technologies into existing houses must also include in its account the forms under which the original housing was provided. In London, it is the possibility of investment of domestic technologies in a housing stock originally built to be operated by servants that permits the 'recolonisation' of that stock by the middle classes and which fuels the rent gap. It is this hypothesis that forms the basis of the model presented in Redfern (1997).

THE NEGLECT OF DOMESTIC TECHNOLOGIES AS A SUPPLY FACTOR IN GENTRIFICATION STUDIES

I have already remarked on the absence of domestic technologies from the multi-variate analyses carried out by Ley (1986) and Bourne (1993). In this neglect of domestic technologies they, however, follow a general trend. Saunders (1989, 1990) contains no reference to this aspect of the history of the home (although there is no end of feminist literature on the impact of domestic technologies on the organisation of the home: e.g. Cowan, 1983; Hardyment, 1988; Nye, 1991). The closest any of the gentrification literature comes to acknowledging the importance of domestic technologies is Hamnett's (1973) examination of the use of 'Improvement grants as an indicator of gentrification in inner London'. Hamnett, however, does not problematise the creation of the possibility of improvements. Smith (1979, page 170) has similar passing comments on improvement grants. These comments are, however, made in the context of a discussion of the role of the state in the gentrification process. Smith (1987, pages 167–169) refers to consumer durables in passing, but only as part of a discussion of suburbanisation, not of gentrification, as does Myers (1992).

The issue of domestic technologies in gentrification is presumably simply dismissed by Smith because of his opposition to consumption-side explanations, and domestic technologies are most typically encountered as *consumer* durables. On the postindustrialist side, however, the existence of the possibilities of home improvements serves merely as a peg on which to hang arguments about class distinctions formed on the bases of conspicuous forms of consumption. Smith summarises the postindustrialist approach here rather well:

> gentrification and the *mode of consumption it engenders* are an integral part of class constitution . . . they are part of the means employed by new middle class individuals to distinguish themselves from the stuffed shirt bourgeoisie above and the working class below.
>
> (1987, page 168, emphasis added)

His mistake is in imagining that the only way in which the contribution of domestic technologies can be analysed is as a part of a mode of consumption.

That gentrification is possible at all is not problematised in the postindustrialist approach but taken for granted, so that the discussion can move on to the meaty business of class. So, for example, P. Williams comments that 'style and the income which makes it possible can in turn be traced to developments around the mode of production, changes in the class structure, and residential differentiation; in other words, it is not an autonomous response but one that *mirrors* continuing social tensions and conflicts' (1984, page 219, emphasis added).

The metaphorical use of the verb 'mirrors' here demonstrates the close connection between the neglect of domestic technologies and the implicit use in gentrification analysis of 'base-superstructure' metaphors. In such metaphors, characteristic of, ironically, Marxist thought, activities which take place in the superstructure (for example, local cultures, including gentrification) and which therefore appear to have the character of agency are theorised in terms of subsidiary metaphors, as being some form of reflections, typifications, or mediations of relations in the economic base. All developments in the superstructure are therefore determined, in the last instance, by developments in the base (R. Williams, 1977, page 81). Thus P. Williams, who otherwise would find himself in the opposite camp to Smith, shares with him the same basic attitude towards gentrification, namely that it *reflects* something else that is going on that is more important than simply gentrification, namely class, or class constitution.

Use of the base-superstructure metaphor shuts off enquiry and replaces it with the demonstration of already and otherwise known truths ['What is already *and otherwise* known as the basic reality of the material social process is reflected, of course, in its own ways, [in the superstructure]' (R. Williams, 1977, page 97) – just like in the story of the blind men and the elephant. In other words, 'There is a persistent presupposition of a knowable (often wholly knowable) reality' (R. Williams, 1977, page 102). Superstructural elements are of interest only as they can be fitted into (and so illustrate the nature of) this reality, not because they have any intrinsic interest in themselves. These metaphors are pervasive in gentrification studies and bear strong parallels with 'Orientalist' studies of the Orient, which according to Said (1991) consisted of an elaboration of ignorance rather than positive knowledge (1991, page 62). The tendency to discuss gentrification in terms of *results*, of achieved housing situation, of elephants, and not in terms of *means*, that is, in functionalist terms (compare Runciman, 1969, pages 40–41, 113), is closely associated with base-superstructure theorising.

Whether gentrification explanations are oriented towards consumption-side accounts such as postindustrialism, or production-side accounts such as the rent gap, they tend to share a common perspective, namely, functionalism and the use of base-superstructure metaphors. Smith is justified therefore in defending his position against the postindustrialists' criticisms (Hamnett, 1984, 1991) but only to the extent that they cannot offer a fundamental critique of his model, as they share so many of the same presuppositions with Smith's own.

REFERENCES

Bondi, L. (1991) 'Gender divisions and gentrification: a critique'. *Transactions of the Institute of British Geographers, New Series* **16**: 190–198.

Bourne, L. S. (1993) 'The demise of gentrification? A commentary and prospective view'. *Urban Geography* **14**: 95–107.

Cowan, R. S. (1983) *More Work for Mother: The Ironies of Household Technology from the Open Hearth to the Microwave* (Basic Books, New York).

Du Vall, N. (1988) *Domestic Technologies: A Chronology of Developments* (G. K. Hall, Boston, MA).

Hamnett, C. (1973) 'Improvement grants as an indicator of gentrification in Inner London'. *Area* **5**: 252–261.

Hamnett, C. (1984) 'Gentrification and residential location theory, a review and assessment', in *Geography and the Urban Environment: Progress in Research and Applications,* Volume 6. Eds D. Herbert and R. Johnston (John Wiley, Chichester, Sussex), pp. 283–320.

Hamnett, C. (1991) 'The blind men and the elephant: the explanation of gentrification'. *Transactions of the Institute of British Geographers, New Series* **16**: 173–189.

Hardyment, C. (1988) *From Mangle to Microwave: The Mechanization of Household Work* (Polity Press, Cambridge).

Humphries, S. and Taylor, J. (1986) *The Making of Modern London 1945–1985* (Sidgwick & Jackson, London).

Hunkin, T. (1989) *The Secret Life of Machines* (Channel 4, London).

Ley, D. (1986) 'Alternative explanations for inner-city gentrification: a Canadian assessment'. *Annals, Association of American Geographers* **70**: 238–258.

Myers, D. (1992) 'Filtering in time: rethinking the longitudinal behavior of neighborhood housing markets', in *Housing Demography: Linking Demographic Structure and Housing Markets.* Ed. D. Myers (University of Wisconsin Press, Madison, WI), pp. 274–296.

Nye, D. (1991) *Electrifying America: Social Meanings of a New Technology, 1880–1940* (MIT Press, Cambridge, MA).

Redfern, P. (1992) *There Goes the Neighbourhood: Gentrification and Marginality in Modern Life.* Unpublished Ph.D. thesis, University of London.

Redfern, P. (1997) 'A new look at gentrification: 2. A model of gentrification'. *Environment and Planning A* 29: 1275–1296.

Runciman, W. G. (1969) *Social Science and Political Theory,* 2nd edn (Cambridge University Press, Cambridge).

Said, E. (1991) *Orientalism: Western Conceptions of the Orient* (Penguin Books, Harmondsworth, Middx).

Saunders, P. (1989) 'The meaning of home in contemporary English culture'. *Housing Studies* **4**: 177–192.

Saunders, P. (1990) *A Nation of Homeowners* (Unwin Hyman, London).

Smith, N. (1979) 'Gentrification and capital: theory, practice and ideology in society hill'. *Antipode* 11: 24–35; reprinted in *Antipode* (1985), 17: 163–173.

Smith, N. (1987) 'Of yuppies and housing: gentrification, social restructuring and the urban dream'. *Environment and Planning D: Society and Space* 5: 151–172.

Smith, N. (1992) 'Blind man's buff, or Hamnett's philosophical individualism in search of gentrification'. *Transactions of the Institute of British Geographers, New Series* 17: 110–115.

Williams, P. (1984) 'Gentrification in Britain and Europe', in *Gentrification, Displacement and Neighborhood Revitalization.* Eds J. Palen and B. London (SUNY Press, Albany, NY), pp. 205–234.

Williams, R. (1977) *Marxism and Literature* (Oxford University Press, Oxford).

15

Urban Lifestyles

Diversity and Standardisation in Spaces of Consumption

Sharon Zukin

STRATEGIES OF URBAN REDEVELOPMENT

Strategies of urban redevelopment based on consumption focus on visual attractions that make people spend money. They include an array of consumption spaces from restaurants and tourist zones to museums of art and other cultural fields, gambling casinos, sports stadia and specialised stores. In older cities, such strategies emerge in the absence of specialised alternative business developments. In cities whose economies are still expanding, such as Orlando and Las Vegas, consumption spaces grow along with new offices and homes.

There is some disagreement about the ethical and social value of this new dependence on urban consumption. Gambling casinos, in particular, are associated with serious social problems and place local governments under the influence of the gambling industry (Goodman, 1995). Neither are these consumption spaces entirely profitable operations – at least, not for the local governments that subsidise construction. Sports stadia are especially questionable as public investments. While owners of teams from the New York Yankees to Manchester United reap profits from sales of box seats and products outside the playing field – for example, refreshment franchises, television broadcasting rights and sales of team paraphernalia – all evidence shows that cities derive mixed economic benefits, at best, from subsidising construction, operating and maintenance costs (Shropshire, 1995). But team owners are tough bargainers. They have been able to persuade local governments that, without new facilities, they will relocate their teams. (In the case of US cities, this often means relocation to a new stadium within the metropolitan area, but at a suburban site.) Rarely do mayors or voters reject their ultimata. Voters in San Francisco, who turned down the chance to build a new football stadium several years ago, are a noteworthy exception.

Neither are there conclusive data about the economic value of expanding resources of art museums and commercial culture, such as theatres. Studies conducted by the Port Authority of New York and New Jersey (1983, 1993) strongly indicated that many tourists come to New York to see art works and performances, and spend many times the cost of theatre or admission tickets in hotels, restaurants and shops. These studies showed, further, that the wages and operating costs of museums, art galleries, theatres and television and film production add up to a considerable sum. The city government's conclusion – to capitalise on cultural resources in order to maintain New York as an international culture capital and tourist centre – implied a firmly rational point of view. Yet not just New York, but almost every city has decided to promote its art museums, and convert old railroad terminals and power stations to cultural complexes. Although it may make sense in New York or London to develop the synergies of an already-strong symbolic economy, other cities face higher risks of failure. But what alternatives do they have? These days, as office buildings proliferate in the suburbs and overseas, cities face a difficult choice between casinos, museums and Hard Rock Cafés: truly, a 'fantasy city' (Hannigan, 1998).

Economic factors, nonetheless, still motivate investors to create new spaces for urban consumption. Since the 1980s, they have been pushed in two directions: by decreases in domestic shoppers' willingness to buy – ascribed by retailers, in the US, to shoppers' boredom with existing stores and by increases in consumer markets overseas – notably, in Japan, China and the city-states of South-east Asia. Under these conditions, developers have built elaborate, new shopping centres in both Asia and the US – from Canal City in Fukuoka, Japan, to Las Vegas, Orlando and New York City. These consumption spaces attempt to revitalise shopping by dramatising the retail 'experience'. They try to capture shoppers' imagination by inviting them to participate in simulated forms of non-shopping entertainment, such as sports (Nike Town), interactive video installations (Viacom) or even 'wilderness' (REI trekking gear stores) and 'nature' (The Nature Experience). Although these spaces are described by the rubric 'entertainment retail', they really sell an easily recognisable 'brand name' – Disney, Nike, Sony, Viacom – in many different product variations.

So far, most of the prototype entertainment retail stores have opened in the largest cities – New York, Los Angeles, Chicago, Boston – where they have become new landmarks on the urban scene. They have replaced the landmarks of the great department stores, many of which went bankrupt or merged during a wave of corporate buyouts in the 1980s. Like the old department stores, entertainment retail stores enjoy favourable coverage in local newspapers for their 'enchantment' of the urban landscape (see, for example, *New York Times Magazine*, Special Issue on 'The Store as Theater, Taste Machine, Billboard', 6 April 1997). They exert a magnetic appeal to tourists, especially more affluent foreigners. But their potential to spur economic development seems limited by the usual market factors: higher prices than outlets and chain stores, eventual overexposure and inevitable reproduction of the same shops in other cities. Entertainment retail complexes in Asia pose a special threat. Using US architects, installing some of the same US store names and financing elaborate, clean and secure facilities (see, for example, 'Japanese Mall Mogul Dreams of American Stores', *Wall Street Journal*, 30 July 1997), these super-shopping centres may eventually keep Asian tourists at home as contented consumers, leaving American and European cities empty.

The future development of urban consumption spaces is predicated on a continuously mobile lifestyle. Neither 'niche' shopping nor 'entertainment retail' fully expresses both the standardisation and diffusion of consumption spaces, and the incorporation of diverse groups of consumers into them. The common denominator of all the new consumption spaces is a sociability dependent on visual coherence and security guards, a collective memory of commercial culture rather than either tolerance or moral solidarity. The Disney Company pointed in this direction many years ago. Perhaps that is, at bottom, what makes them such a formidable presence in contemporary urban redevelopment. As recently as 1990, New York City might have seemed immune to 'the Disney touch'; now, however, the redevelopment of Times Square depends mainly on three Disney projects: a Disney Store (one of several in the city), a legitimate theatre for Disney stage productions and Disney's participation in a portion of a time-share hotel. Disney's agreement to establish a presence in Times Square was sufficient to mobilise financing for other projects, and to encourage the city government's support for the entire theatre district. As a critical New York newspaper observes about 42nd Street, this area

> has undergone a miraculous change from a no-man's land of pornography shops and assorted criminal activities into a neon-drenched mecca of theme restaurants, theaters and other family entertainment fare. Indeed, the transformation is so complete that political figures and real estate brokers have taken to touting the rejuvenated block as the premier symbol of New York City's unquenchable vitality.
>
> (*New York Observer*, 17 March 1997, p. 1)

In return, when the Disney Company wanted to rent a large part of Central Park for the première of the cartoon movie *Pocahontas*, the Parks Department agreed, and when the company wanted to hold a torchlight parade down Fifth Avenue to celebrate the opening of another animated feature, *Hercules*, the Police Department closed the street and provided security. Many New Yorkers protest against the 'Disneyfication' of Times Square, but the greater danger is that a single corporate vision could dominate Manhattan.

This trend is deepened by increased corporate investment in consumption spaces in low-income, minority-group areas such as Harlem. Long ignored by major department stores, big chain stores and retailers selling high-quality goods, urban ghettos have only recently attracted the interest of corporate planners. They now realise that residents of these districts represent large markets for standard, high-price brands – to the extent that, in the late 1980s and early 1990s, certain brands of athletic shoes (Nike) and trekking gear (Timberland shoes) became identified with 'urban' – i.e. 'ghetto' – cultural styles. An investment partnership with the professional basketball-player 'Magic' Johnson has brought Sony Movie Theaters into low-income urban areas. In the past few years, with reductions in social welfare programmes, local governments and community groups have reoriented themselves towards attracting mainstream retailers, including supermarkets, in addition to demanding jobs. Although these urban areas have always been underserved by purveyors of basic consumer goods, encouraging retail stores fits the general social and political context of reducing government's role and enlarging that of the private sector (see, for example, Porter, 1995). The long-term political and cultural effects of bringing new stores and multiplexes into low-income neighbourhoods remain to be seen.

URBAN POLITICS AND CULTURES

Ten or fifteen years ago, urban lifestyles might have been analysed in terms of gentrification and its effects on social class polarisation and displacement of the urban poor (Smith and Williams, 1986). Consumption was viewed as a means of driving a wedge between urban social classes and an indicator – although never a cause – of economic and political realignments. By the end of the 1990s, consumption is understood to be both a means and a motor of urban social change. The reorganisation of world markets has expanded the consumption functions of mature urban economies, creating new jobs and new spaces of consumption. Many of these jobs are low-paying jobs in stores, restaurants, hotels and domestic and personal services. While many of the new consumption spaces rely on a high level of skill and knowledge, and provide cultural products of beauty, originality and complexity, others are standardised, trivial and oriented towards predictability and profit (see Ritzer, 1996).

At the same time, individual men and women express their complex social identities by combining markers of gender, ethnicity, social class and – for want of a better word – cultural style. Many of these markers are created in, and diffused from, cities: on the streets, in advertising offices and photography studios, on MTV. Many of the people who create these markers live in cities, too. They are artists, new media designers, feminists, gays, single parents and immigrants – some of the most visible protagonists of 'urban lifestyles'.

Most women and men live in the spaces between the images manipulated so prominently in the past 30 years by identity politics and 'lifestyle magazines' and the desire to live as good a life as possible in their own neighbourhoods. Yet the diversity of their lives is often submerged by the increasing standardisation of consumption spaces, even at their most spectacular, exemplified by superstores and multiplex movie theatres.

An analytical framework of urban consumption has to be posed in the broad terms of social theory. Like critical interpretations of modernity, this analytical framework should make connections between the production of physical spaces and symbols and between the built environment, sociability and urban lifestyles. Beginning with the various analytical frameworks of gentrification (Zukin, 1987), attempts to think through these connections have generally focused on the urban middle class, especially the educated middle class' tastes or preferences in cultural consumption. As autonomous social actors, this group thinks through, or is self-conscious of, their lifestyle choices; their 'reflexivity' (Lash and Urry, 1994) is assumed to indicate a new mode of collective consciousness. Certainly, there is a fit between demands for more 'aesthetic' consumer goods and the reorganisation of some consumer industries. To some degree, consumer industries have strengthened the role of design in the manufacturing process; they provide a large variety of goods and switch production lines quickly; and they advertise their products in a tone of postmodern self-mockery. But these are not their only strategies. Standardisation and mass production have not been relegated to the ash-heap of industrial history. The enormous popularity of fast food, among all social strata, relies on standardised products made in an assembly-line production system. Despite the choices, around the world, between beefburgers, chicken fillets and vegetable kebabs, fast food belies the aesthetic awareness of reflexive consumption. Yet 'reflexive' consumers, such as they are, do risk political disengagement and even polarisation. The aestheticisation of their tastes implies stylisation and detachment as well as pleasure (Featherstone, 1991; Sennett, 1990). These attitudes may discourage sympathy with other urban groups, including fast-food workers.

In the current retreat from the welfare state, aestheticisation of the urban landscape is associated with a collective abandonment of the homeless and exasperation with public stewardship over public space (Zukin, 1995; Smith, 1996; Mitchell, 1997). Streets, parks and even entire districts have been derogated to control by private associations of property owners and patrons. In New York City, for example, the largest parks – Central and Prospect Parks – are partly financed and wholly administered not by the New York City Parks Department, but by private conservancies comprised of individual and corporate patrons. Commercial districts all around the city, beginning with the most expensive, midtown business areas, are managed by Business Improvement Districts. Although these remain public spaces in the sense that they are open to all, the private associations set rules by which entry can be denied. Abandonment of collective responsibility for others also motivates the construction of gated residential communities – graphically connecting privatisation with aestheticisation of an anti-urban lifestyle (Davis, 1990; Ellin, 1997; Judd, 1996).

Alternatively, the shopping streets frequented by immigrants and native-born minorities are avatars of new urban and ethnic identities. On streets in New York City, Los Angeles, Atlanta, or Toronto, shoppers, peddlers, store owners, managers and clerks are likely to be Africans, 'Caribbeans', Koreans and African Americans. These shopping streets create a new African-American identity by interaction among, and fusion between, various traditions of the African diaspora. Although Asians tend to live separately from other minorities, and increasingly in the suburbs, they are active in these shopping streets as merchants – often with both bad and good results (Min, 1996). Storefront telephone and delivery services feature signs in many languages, with prices of services to many lands. Newspaper stands owned by members of one immigrant group sell newspapers written in other languages. Store owners stock distinctive ethnic goods that will appeal to several different ethnic groups, and some goods, such as clothing and cosmetics, are re-exported to the same or even different countries of origin. 'Aestheticised' commodity worlds are not rejected, but are irrelevant in these streets. Here, 'transnational' consumers interact and develop their own urban lifestyles. They are neither 'detached' nor particularly 'reflexive'. The interaction and juxtaposition among urban lifestyles – especially in spaces of consumption – indicate a 'hybrid' urban culture (Bhabha, 1994) rather than domination by corporations or the middle class. On these streets, diversity thrives.

Questions of lifestyles, public space and sociability return to the theme posed more than 30 years ago by Jane Jacobs: How can cities encourage trust among strangers? For the private-sector managers of public space, the answer lies in aesthetic design and private security guards; for the private-sector managers of entertainment retail, the answer lies in Disneyfication, or selling the experience of pleasure in shopping spaces that are both visually coherent (by branding and themed entertainment) and physically controlled (by cleaning staff, service representatives and private security guards). But on the shopping streets in immigrant and ethnic neighbourhoods, trust among strangers is a result of social interdependence and neighbourhood solidarity. As in the classic visions of modernity defined by both Georg Simmel and Jane Jacobs, this is what urban lifestyle is all about.

CONCLUSION

Cities hit hard by a long-term decline in middle-class residents and the erosion of commitment by business élites have gradually begun to view the diversity of 'urban lifestyles' as a source of cultural vitality and economic renewal. Elected officials who, in the 1960s, might have criticised immigrants and non-traditional living arrangements, now consciously market the city's diverse opportunities for cultural consumption (Lang *et al.*, 1997). They also welcome the employment offered by new culture industries and expanding cultural institutions – as part of the cities' new comparative advantage in the 'symbolic economy'. Yet the diffusion of new 'urban' lifestyles may pose problems for city governments' traditional concerns. These lifestyles bring more pressure on public space, including parks and art museums; less desire to finance such public institutions as schools; and continued instability of employment in service jobs that depend on consumers' disposable income. New York City's high-price restaurants wax and wane in response to the stock market, for example, and the city's growth as a 'culture capital of the world' has not brought new financial resources to the beleaguered public schools.

Cities' receptivity to 'destination retail' sites and entertainment facilities have lured them, moreover, into dependence on property developers and multinational corporations that share the same, endlessly repeated vision. There is a Hard Rock Café – or at least its retail store – in every major city of the world, new suburban-style shopping centres throughout eastern Europe and a Disney Store even in the duty-free zone of Heathrow Airport. Competition among corporations and cities has led to a multiplicity of standardised attractions that reduce the uniqueness of urban identities even while claims of uniqueness grow more intense. The diffusion of 'urban' lifestyles and the expansion of production sites, throughout suburbs and exurbs, further erode historical spatial differences.

Nevertheless, urban cultural diversity holds a curious and yet wonderously creative mirror to the paradox of polarisation: while cities become more like other places, they continue to attract the extremes of poor, migrant and footloose urban populations and the very rich. Their ability to forge 'urban' lifestyles continues to be the city's most important product.

REFERENCES

Bhabha, H. K. (1994) *The Location of Culture*. London and New York: Routledge.

Davis, M. (1990) *City of Quartz*. New York: Verso.

Ellin, N. (ed.) (1997) *Architecture of Fear*. New York: Princeton Architectural Press.

Featherstone, M. (1991) *Consumer Culture and Postmodernism*. London and Newbury Park, CA: Sage.

Goodman, R. (1995) *The Luck Business*. New York: Free Press.

Hannigan, J. (1998) *Fantasy City: Pleasure and Profit in the Postmodern Metropolis*. London and New York: Routledge.

Judd, D. (1996) Enclosure, community and public life, *Research in Community Sociology*, 6, pp. 217–236.

Lang, R. E., Hughes, J. W. and Danielsen, K. A. (1997) Targeting the suburban urbanites: marketing central-city housing, *Housing Policy Debate*, 8, pp. 437–470.

Lash, S. and Urry, J. (1994) *Economies of Signs and Space*. London and Newbury Park, CA: Sage.

Min, P. G. (1996) *Caught in the Middle: Korean Communities in New York and Los Angeles*. Berkeley and Los Angeles: University of California Press.

Mitchell, D. (1997) The annihilation of space by law: the roots and implications of anti-homeless laws in the United States, *Antipode*, 29, 3, pp. 303ff.

Port Authority of New York and New Jersey (1983) *The Arts as an Industry: Their Economic Importance to The New York–New Jersey Metropolitan Region*. New York: Port Authority of New York and New Jersey and Cultural Assistance Center.

Port Authority of New York and New Jersey (1993) *The Arts as an Industry: Their Economic Importance to The New York–New Jersey Metropolitan Region*. New York: Port Authority of New York and New Jersey, Alliance for the Arts, New York City Partnership and Partnership for New Jersey.

Porter, M. (1995) The competitive advantage of the inner city, *Harvard Business Review*, May–June, pp. 55–71.

Ritzer, G. (1996) *The McDonaldization of Society*. Thousand Oaks, CA: Pine Forge Press.

Sennett, R. (1990) *The Conscience of the Eye*. New York: Norton.

Shropshire, K. L. (1995) *The Sports Franchise Game: Cities in Pursuit of Sports Franchises, Events, Stadiums*. Philadelphia: University of Pennsylvania Press.

Smith, N. (1996) *The New Urban Frontier: Gentrification and the Revanchist City*. London and New York: Routledge.

Smith, N. and Williams, P. (eds) (1986) *Gentrification of the City*. Boston: Allen & Unwin.

Zukin, S. (1987) Gentrification: culture and capital in the urban core, *Annual Review of Sociology*, 13, pp. 129–147.

Zukin, S. (1995) *The Cultures of Cities*. Oxford and Cambridge, MA: Blackwell.

PART THREE

Subjects and Identity

INTRODUCTION TO PART THREE

All too often consumers have been characterized in highly *polarized* terms: as *either* passive and entirely at the mercy of big business to manipulate their desires, *or* entirely active and free to determine their spending patterns rationally. Many commentators, for instance, have regarded consumers as little more than gullible dupes – especially those who believe that the forces of consumerism (and particularly the persuasive powers of advertising) can manipulate consumers into buying things they don't really need or even want (see **Falk**). But are people really so easily duped? Can business interests really instil 'false needs' into hapless consumers (and 'satisfy' them with 'ersatz use-values')? Can consumer demand simply be 'created' at will? While a string of business successes – from Rubik's cube to oxygen bars – suggests that it can, a string of notorious business flops – from the Ford Edsel to the Sinclair C5 – suggests that matters are not quite so straightforward. Probably most people feel that they have *occasionally* been duped into buying things that they neither need nor want – but it would be difficult to dupe all of the people, all of the time. At the other extreme, the consumer has often been regarded as *sovereign*. Apologists of the market, from Adam Smith onwards, have suggested that the 'consumer is king' – though this gender-blind expression ignores one of the most important aspects of 'the consumer'. It certainly does not square with the (admittedly outmoded, yet nonetheless telling) business practice of referring to 'the consumer' as 'the housewife', whose aggregate demand has been likened to the power of a global dictator! Though the term 'consumer sovereignty' was not coined until the 1930s, the view it encapsulates is of much older vintage. It is a view holding that business simply *responds* to pre-existing consumer demand; that advertising merely provides *neutral information* that permits rational consumers to match their existing needs and wants with what the market has to offer; and that the fate of business is, in the final instance, in the hands of myriad individual consumers, whose *freedom of choice* ultimately determines what is produced and sold.

It is not difficult to see that the first view discussed above, that business manipulates consumers and creates demand, arose initially to combat the self-serving ideology of the market that portrays the consumer as being 'in command' – particularly when it became evident that business was increasing in both scale and power, and spending ever greater amounts on marketing and advertising. These two views are, in effect, mirror images of one another. It should come as no surprise, therefore, that both are irredeemably partial, neither is particularly accurate, and that – although most people, depending on their political beliefs, will feel attracted to one side or the other – a more sophisticated view of 'the consumer' would be highly desirable. In terms of this polarized debate, a third way between the two extremes exists: consumers are neither instilled with 'false needs' nor entirely rational and 'free to choose'. Consumer capitalism seems eminently capable of generating 'new needs' that are quite real. Life is made that much more difficult if one cannot procure the means of satisfying them. Or, to put it another way, new objects seem to become increasingly indispensable to changing patterns of modern living. Paradoxically, then, consumers may be *forced* to choose rather than *free* to choose: 'freedom of choice' is a 'freedom' that bears the hallmarks of a 'necessity' for fully paid-up members of the consumer society. The view of consumers as zombified dupes relates more to a discourse about business power than about actual consumers. The market ideology of consumer sovereignty is equally incapable of saying much about consumers themselves. For the most part, it remains fixated on the idea that consumption is about the relationship between the needs of individual subjects and specific objects capable of satisfying those needs. This fails to recognize that consumption is far more often about the social relations between people *as mediated* by consumer objects (Part 4). Many classic concepts in consumption studies, such as 'conspicuous consumption' (see **Veblen**), are about how people use objects in order to say something about themselves – about 'who they are' and 'what they (are) like' – for the benefit of *others* (see **Douglas**). It would be fair to say that consumption is much more about *social* relations than it is about the *dyadic* relationship between subjective need and its satisfaction by this or that particular object. If the extracts in this part of the Reader have a common concern, it is to make this situation clearer. They are committed to developing a more sophisticated and nuanced characterization of 'the consumer' than the polarized views allow. In doing so, however, they develop a variety of positions, adhering to or departing from the polarized characterizations to different degrees.

Simon Mohun arguably belongs far more in the 'manipulationist' than the 'consumer sovereignty' camp. Nonetheless, his piece is not a direct attempt to convince the reader of the persuasive and manipulative powers of big business – after the fashion of authors such as J. K. Galbraith and Paul Sweezy. Rather, it aims to provide a *critique* of the notion of consumer sovereignty, seeking to undermine its tenability from the perspective of Marxian political economy (see **Marx**). Mohun summarizes the orthodox (neoclassical) economic position before offering a Marxist account of capitalist production which questions the received wisdom that production occurs primarily to satisfy consumers. Instead, production is seen as a 'capitalist imperative' that is justified ideologically by a disingenuous and dissuasive appeal to the supposed sovereignty of the consumer. This provides a useful exposition of Marx's position. However, its emphasis – its *reliance*, one might say – on the notion of 'use-value' should be contrasted with the view that critiques of this kind do not really get to grips with the nature of consumption, or provide us with a very sophisticated view of the consumer (see **Baudrillard**).

Mary Douglas begins from a very different starting point. While she hardly accepts the ideological view enshrined in the notion of 'consumer sovereignty', she is equally under no illusions that consumers are gullible dupes. She approaches the 'either'/'or' characterization of consumers obliquely, staunchly defending the consumer against both the reductive claims of economic theory and the romantic views of those who lament the decline of a close-knit community. The sanctions that were once available to communities made it very difficult for people to transgress what was demanded of them. Consumption, in contrast, offers a far greater degree of volition. Douglas argues not only that consumption enables people to perform as active, *creative* individuals, but also that their consumption choices allow them to express *solidarity* with other people and alternative visions of society. For Douglas, then, consumption is not merely about the satisfaction of physical and material needs. It is the very language of personal and social identity, both of which tend to be constructed as much through negative differences as through positive identifications. Commonplace phrases about tastes in fashion – such as 'I wouldn't be seen dead in it!' – reveal that consumers make active choices about what they consume, expressing something about 'who they are' and 'how they see themselves'. That expression is, though, as much *reactive* as *active*. What consumers would like to wear is very much defined by what they *would not* like to wear; by how they *would not* like to be seen by others. It permits individuals to express their affiliation with others who hold similar views to themselves, and to distinguish themselves from those who do not. This is a far cry from the stultifying situation of the close-knit community, where failure to toe the line brought all kinds of subtle or not-so-subtle sanctions into play. In her defence of the consumer, one can also detect in Douglas a concern to defend *women*. She clearly senses that the common characterization of consumers as 'passive' and 'fickle' has more often than not been directed towards women, particularly in the guise of 'the housewife', and usually – whether wittingly or unwittingly – by men (see **Bowlby** and **Simmel**).

The view that consumption is an expressive practice that constructs identity through a double process of social alignment and social antagonism is, perhaps, most famously explored in **Dick Hebdige**'s study of *subcultures*. While Douglas provides a general analysis, Hebdige considers one very notable instance of this general process. A 'subculture' is a cultural group that considers itself to be at variance with the majority and its 'mainstream' culture, typically because it is marginalized and disadvantaged within society. As an act of opposition, subcultures typically draw on the *detritus* of the dominant culture in order to set themselves apart, affirm their own counter-culture, and use the 'discarded' objects, symbols and values they recoup in *rituals of resistance* enacted against the expected social norms. In their own eyes, those adhering to particular subcultures manage to translate social and economic disadvantage into cultural and political prestige. Subcultural styles are often associated with particular *generational* groups (one thinks of classic 'youth' subcultures, from mods and rockers to punks), often specifically with *working-class* youth; or defined in *racialized* terms (such as subcultures based around hip-hop and rap music). Or again, as Hebdige suggests, they may well relate to *sexuality*. The key insight to be derived from the study of subcultures is that consumption and taste do not necessarily adhere to the kind of single social pecking order implied by the notion of 'social emulation' (cf. **McKendrick** and **Veblen**). There is no reason to assume that everyone, regardless of their place in society, will accord to the same tastes, values and consumption norms. It seems far more likely that different social locations will imply different tastes, values and norms – often defined *against* the tastes, values and norms of those occupying other social locations (see **Bourdieu**).

Subcultures, as **Andrew Bennett** argues, have proved to be unforgiving concepts. Their very definition presupposes a uniform dominant culture, with powerful values, against which those who are excluded are almost obliged to find a kind of solidarity through resistance. Each side of this great divide consumes people: body and soul. It is governed by a logic of 'all or nothing', which strikes many as unbelievable in a society that appears to be increasingly fluid and ephemeral. For example, although classic subcultures fit their particular mould extremely well, do such 'alternative' social groupings necessarily have to be *reactive* (based on the recycling of detritus)? Might they not be *self-affirming* (based on the pure creative force of new ideas) instead? The notion of *neo-tribes*, first proposed by Michel Maffesoli, begins where subcultures effectively seal themselves off. It is meant to convey the struggle for identity in an age without strong norms and dominant values; or, perhaps, in an age where there is an *excess* of norms, values and lifestyles. Given this proliferation of possible social worlds, and the withdrawal of a clear adjudication on the 'right' course of action, it has become possible – even necessary – for people to find their own way amidst a plethora of competing ways of life. One way to cope with this difficult task is to 'buy into' a neo-tribe that provides a ready-made lifestyle – or, rather, to buy into it for as long as it serves one's purposes. The proliferation of 'lifestyle choices' suggests that it is unwise to invest too heavily in any particular lifestyle, since tomorrow's requirements may be very different from today's. Indeed, the optimum strategy might be to tap into a number of lifestyles, adopting whichever one best fits the situation to hand. So, whereas subcultures worked through the dialectic of power and resistance, and invariably demanded total commitment, neo-tribes are more likely to follow the dictates of fashion – demanding little, if anything, from their affiliates. Little wonder, then, that not only is the membership of neo-tribes fluid, so are the neo-tribes themselves. The fact that, in today's society, middle-aged, middle-class whites might enjoy a musical form such as rap suggests that the concept of the 'neo-tribe' is more appropriate to our current situation than earlier notions of 'subcultures'.

The kind of view of society and social identity emerging from the notion of the neo-tribe implies the increasing importance of consumption in structuring society as a whole. It marks a significant contrast with the dominant role once played by production in social stratification. While acknowledging the importance of consumerism, individualism and lifestyle, **Rosemary Crompton** nevertheless maintains the continuing importance of class. She concerns herself with whether social identity defined in terms of *class* has really been displaced by social identity defined in terms of *consumption*. Crompton finds that matters are not quite so straightforward. For a start, the extent of any such displacement varies across different social classes (this reasoning may appear somewhat circular, but Crompton makes a convincing case that it is not!). Moreover, changes in the world of work have not tended to neutralize class *in itself*, but rather class *for itself*: that is, changes in work practices have eroded *class consciousness*, and the likelihood of people responding to their class position in terms of *collective* action. Consumer capitalism has promoted the idea that *individual* solutions are the most apt response to collective situations – particularly the individualistic 'market solution' offered by *consumption*. Furthermore, the workplace, particularly for those engaged in service-sector occupations, increasingly values exactly the same individualistic skills and habits that are promoted by consumerism. Those who are the most successful consumers tend automatically to find themselves best equipped for success in the workplace. In short, the consumer society has actively reconfigured employment and changed the nature of social class in highly complex ways. Crompton therefore insists that one cannot say that class identity has simply been supplanted by consumption-oriented lifestyles. People have always been defined by both their production and their consumption activities. The most recent changes in Western consumer societies amount not just to a change in the relative importance of the latter over the former, but to a complex set of interaction effects between the world of work and the world of consumption.

Mike Featherstone is convinced that the consumer society marks a far more significant departure from earlier social arrangements than Crompton allows. He reviews a range of literature that points to the emergence of a new 'personality type', associated primarily with the growth of consumerism and a 'culture of narcissism'. Featherstone figures this 'other-directed' personality type in terms of the '*performing self*'. The new centrality of the performing self, he maintains, increasingly focuses the attention of the individual on his or her *body*. He marshals evidence of the way in which an increasingly *narcissistic* attitude has been cultivated and nurtured by a variety of business interests. In his estimation, however, this is not a clear-cut case of the indoctrination of consumers with 'false' needs. Rather, business interests tend to *channel* pre-existing needs and desires in highly specific ways – tapping into genuine needs and desires, but ultimately distorting them, by bringing them into line with their own self-serving ends. The extraction of profit depends on the production of consumers, and cultural change has become geared increasingly towards the kind of narcissistic consumer culture that serves precisely this aim.

Rachel Bowlby gives the kind of issues raised by Featherstone both a longer provenance and a more specific theoretical grounding. Like Featherstone, she focuses on the way in which capitalism has gradually placed increased emphasis on the production of *consumers*, in addition to the production of *commodities*. She shows that it was typically *women* whose capacity to consume was nurtured most strongly by business interests – though Bowlby is equally as sensitive to issues of class as she is to gender. While the gendering of the consumer remained largely implicit in the work of **Mary Douglas**, Bowlby draws on psychoanalytical theory to explain how the *visual* pleasures of narcissism have tended to be cultivated more actively in women than in men. Commenting on the aestheticization of commodity display, particularly in shop windows (see **Schivelbusch**), Bowlby provides a convincing assessment of the relationship between commerce and femininity, offering numerous insights into the deconstruction of 'appearance' and 'reality' in the consumer society (see **Baudrillard**).

Continuing the themes developed by Bowlby, **Hillel Schwartz** focuses on the late nineteenth-century birth of the department store as an 'enchanted' space of consumption. Schwartz addresses the gendering of such spaces and considers the manner in which they gave rise to a number of 'shopping disorders' (see **Abelson**). Such commercial spaces were, above all, devoted to middle-class women. But the captivating display of seductive objects opened up the possibility for: (1) a disturbing encounter with a world of disorderly things and mixed-up values; and (2) theft. The puzzle posed by instances of theft by those middle-class women who generally had the means to buy what they stole led to the *medicalization* rather than the *criminalization* of their actions. 'Kleptomania' was the term coined in an attempt to explain and understand such pathological actions. The discursive construction of such a 'condition' raised all manner of questions concerning the intersection of sex and class. Perhaps most importantly, however, the notion of 'possession' and 'uncontrollable desire' reveals a dark side to the 'pleasures' and 'freedoms' of consumption.

16

Consumer Sovereignty

Simon Mohun

> The basic idea of consumer sovereignty is really very simple: arrange for everybody to have what he prefers whenever this does not involve any sacrifice for anybody else. . . . As a social critic, I may try to change some desires to others of which I approve more, but as an economist I must be concerned with the mechanism for getting people what *they* want, no matter how these wants were acquired.[1]

INTRODUCTION

Consumer sovereignty, first described as such in 1936, and loosely defined by Lerner in the above quotation, is a concept fundamental to modern economic theory and yet discussed little. As a fundamental concept, it is both simple and complex: simple, because it seems self-evidently reasonable; complex, because it is part of both 'positive economics' and 'normative economics' – complex also because it straddles both economic theory and political theory, consumer sovereignty describes for the bourgeois economist both the motivation for production and the axiomatic starting point for its analysis, both the purpose of production and a justification for that production.

[. . .]

As part of positive economics, consumer sovereignty is the postulate that all economic activity is directed towards consumption. People produce and exchange in order to consume. The specialisation of economic activity, the division of labour, factory production, the rapid development of technical progress and the growth of markets – all of this enables man to consume more. Indeed, a standard assumption in economic theory is that economic agents have insatiable appetites for consumer goods and services, and of course economics is conventionally defined as how these unlimited wants shape the economic activity of societies which have only limited resources at their disposal and alternative uses for those resources.

[. . .]

THE ORTHODOX CONCLUSION

This . . . is how bourgeois economic theory analyses the principle of consumer sovereignty: . . . crudely speaking, individual dollar votes in the aggregate determine production priorities. . . . There are problems concerning wants (which are the ends of economic activity) and the ways in which these wants are reflected in actual choices; there are problems concerning how well-informed consumers are, how coherent the structures of their wants are, and how important the spillover effects on other consumers are, and there are problems concerning the depth of integration of wants within the individual personality, and hence concerning the stability of wants, since the principle is largely vitiated if consumers are very suggestible and their wants very volatile. Nevertheless, the principle stands: *the satisfaction of consumers' wants defines the ultimate purpose of all economic activity, simultaneously descriptive and prescriptive, in a market economy.*

And as such it is ideology pure and simple: a complete distortion of the reality of capitalist production. The remainder of the chapter will attempt to substantiate this propositions and the first step must be a proper characterisation of the purpose of capitalist production.

THE PURPOSE OF PRODUCTION: CONSUMPTION?

As has been seen already, the process of exchange in a monetary economy is $C \to M \to C$. Hence it is easy to conceive of production as a process whereby commodities are produced. What enables these commodities to be produced and exchanged? The answer is obvious: they must be useful to people, that is, they are 'use-values'. Only use-values can be exchanged, and the process of exchange is a process whereby something that is not useful to (has no use-value for) its owner is exchanged for something that is (has). The relationship of exchange is obviously symmetrical in this regard. Production for exchange is therefore the production of use-values, of commodities for consumption. These commodities will be consumed precisely because they are use-values, and the purpose of production is to consume some use-values (inputs) in order to get in their place commodities with different use-values, or commodities which can be exchanged for such. The purpose of production is thus consumption. It is easy to conceive of production in this manner (thus bourgeois economics) but it is not very illuminating. And the reason for this is the very symmetry of the relationship of exchange.

This symmetry involves the idea that an exchange takes place between two people each of whom sells in order to buy. These individuals are conceived of solely as exchangers, and it is the relationship of exchange which situates them within the domain of economic analysis. Moreover, as exchangers, their relationship is that of equality: the commodities they exchange are numerically equal in money terms, that is exchanged commodities have the same 'exchange-values'. Economic analysis can therefore only distinguish three fundamental concepts: first, the subjects of the exchange relations – the exchangers; second, the objects of their exchange – equivalents; and third, the act of exchange itself – the process whereby the subjects as exchangers are in an equal relation and the objects as equivalents are likewise equal. Marx powerfully remarks:

> This sphere . . . is in fact a very Eden of the innate rights of man. There alone rule Freedom, Equality, Property and Bentham. Freedom, because both buyer and seller of a commodity . . . are constrained only by their own free will. They contract as free agents, and the agreement they come to, is but the form in which they give legal expression to their common will. Equality, because each enters into relation with the other, as with a simple owner of commodities, and they exchange equivalent for equivalent. Property, because each disposes only of what is his own. And Bentham, because each looks only to himself. The only force that brings them together and puts them in relation with each other, is the selfishness, the gain and the private interests of each. Each looks to himself only, and no one troubles himself about the rest, and just because they do so, do they all, in accordance with the pre-established harmony of things, or under the auspices of an all-shrewd providence, work together to their mutual advantage, for the common weal and in the interest of all.[2]

Clearly the theory is applicable to *any* exchange economy. Now consider one such economy, a capitalist one, an economy in which the capacity for work is itself produced for exchange. But if all economic relationships can fundamentally be subsumed by the exchange relation, a relation of equals exchanging equivalents, there is no explanation of how in a capitalist economy profits exist at all.

THE PURPOSE OF PRODUCTION: PRIVATE PROFIT

Capitalists are people who do not sell in order to buy, but buy at one price in order to sell at a higher one. Rather than $C \to M \to C$, their relation of exchange is $M \to C \to M'$, where M' is greater than M. On the face of it, this is not possible. Abstracting from the use of physical coercion, and from cheating, exchange as the exchange of equivalents seems to preclude the possibility of profits.

The answer to the problem of why and how profits exist is historical and social. What is the commodity that most people sell in the relation $C \to M \to C$? What most people sell is their capacity to produce commodities, or their labour-power, for a given time period, receiving in return a wage. (It is important to note that they do not sell their labour; the only way to do this is physically to sell themselves as slaves.) Why do they do this? They have no choice. They *have* to sell their labour-power in order to obtain the wage needed to buy commodities to keep them alive. So for most people in capitalist society, the relevant relation is labour-power (C) → wage (M) → consumption goods (C). Who then purchases the labour-power sold? Obviously the capitalist does; moreover he thereby acquires the *use* of it for the time period for which he has purchased it. Who supplies the consumption goods? Obviously again the capitalist; but he is not interested in them *qua* consumer goods. It is enough for him that they have use-value for someone else – his purpose is to make a profit. Therefore, for the capitalist, the relevant relation is

initial money (M) → means of production, labour-power (C) → production process (P) → goods produced (C') → money obtained from their sale (M').

The goods *produced* are C' which must have greater value in some sense than C by the rule of equivalent exchange, since they are sold for M'. The riddle therefore lies within the production process itself.

Once the riddle is so posed, the problem is solved: labour-power is itself a commodity, and a peculiarly unique commodity in that its use-value has the property of creating *more* than is necessary to reproduce itself. The capitalist consumes the labour-power he has purchased, but this very consumption (the combining of labour-power with the means of production – machines, and so on) creates *more* than he has to pay the labourer. Moreover, this latter transaction gives to the capitalist as a juridical right both the power to use labour-power and means of production in whatever way maximises profit, and the ownership of the resulting products of the production process. Two further analytical problems are thereby raised.

First, how is it that so many people *have* to sell their labour-power in order to live, and conversely how is it that some few people acquire ownership and control over *all* the means of production? This is a complex question relating to the historical breakdown of feudalism and its gradual supersession by capitalism; generally the process whereby it occurred was one of physical force. In Britain, the enclosures movement from the sixteenth to the eighteenth centuries and the clearance of the Highlands after 1745 are clear examples of this process of expropriation. For capitalism, the labourer must be free: free of both means of subsistence, and juridically a free man. It is this process itself which defines the genesis of capitalism, the specific differentiation between people on the basis of their ownership (or not) of the means of production, and the growth and solidification thereby of the two great antagonistic classes of modern society: the capitalist class and the working class.

Second, what is it that makes C' more than C? What is this 'value' such that the value of C' is greater than the value of C? To answer this is simultaneously to answer the question of why the classes of capitalist society are inherently antagonistic. This answer starts from the proposition that it is impossible to separate what man is from what man does, and what man does transforms what man is. (This is just a different way of saying that one cannot analyse what society is without simultaneously analysing what society is becoming.) What man does is to engage in purposive productive activity, and moreover to co-operate systematically with others in such activity, thereby transforming the environment within which activity occurs.

It follows that it is human labour that creates value; that is, it is only the property they have of being products of human labour that imbues commodities with value. Moreover, since this is the *only* property that every single commodity has in common with every other commodity, it follows that it is *only* human labour that creates value.

The production of commodities is therefore the production of value. As the embodiments of particular expenditures of labour-power in particular forms commodities have use-value, but as embodiments of identical abstract human labour, commodities have value. Since value is only perceptible via the process of exchange, it is this latter which reveals exchange-value as the phenomenal form of value. By definition then, value is 'socially necessary' labour time – it is the labour embodied in commodities measured in units of time, assuming normal production conditions and an average degree of skill and intensity of labour prevalent at that historical moment. How the capitalist appropriates 'surplus' value is now not a difficult problem.

Like any commodity, the value of labour-power is the socially necessary labour time required to produce it; in this case, this means the value of commodities required to reproduce labour-power, or the worker's ability to produce. The worker is not cheated; on the contrary he *is* paid the full value of his labour-power. But the capitalist then consumes labour-power by combining it with means of production in such a way that more value is produced than is consumed in the production process. In this way the capitalist 'exploits' the worker through his ownership of the means of production; part of the working day, the 'necessary' part, the worker works for himself, as it were, reproducing the value of his own labour-power (in the form of commodities), but for the remainder of the working day, the 'surplus' part, he works for the capitalist for nothing. Thus capitalist production is marked by a continual struggle over the relative magnitude of these different components of the working day, the ratio between the two – of surplus labour to necessary labour – comprising the rate of surplus value, or rate of exploitation.

The purpose of capitalist production is thus revealed – it is to expand value. Whereas the simple circulation of commodities characterising the worker's economic behaviour, $C \rightarrow M \rightarrow C$, begins and ends with commodities, and *different* commodities at that, the capitalist starts and finishes at the same point: $M \rightarrow C \rightarrow M'$, M' forming the starting point of yet another circuit. Marx neatly sums up the difference:

> The simple circulation of commodities [$C \rightarrow M \rightarrow C$] – selling in order to buy – is a means of carrying out a purpose unconnected with circulation, namely, the appropriation of use-values, the satisfaction of wants. The circulation of money as capital [$M \rightarrow C \rightarrow M'$] is, on the contrary, an end in itself, for the expansion of value takes place only within this constantly renewed movement. The circulation of capital has therefore no limits.
>
> As the conscious representative of this movement, the possessor of money becomes a capitalist. His person, or rather his pocket, is the point from which the money starts and to which it returns. The expansion of value, which is the objective basis, or main-spring of the circulation $M \rightarrow C \rightarrow M$, becomes his subjective aim, and it is only in so far as the appropriation of ever more and more wealth in the abstract becomes the sole motive of his operations, that he functions as a capitalist, that is, as capital personified and endowed with consciousness and a will. Use-values must therefore never be looked upon as the real aim of the capitalist, neither must the profit on any single transaction. The restless never-ending process of profit-making alone is what he aims at.[3]

Capitalists therefore consume solely in order to produce more, that is to accumulate. Since their control of the production process and their ownership of the products of that process constitute capitalists as the ruling class, it follows that capitalist society consumes solely in order to accumulate. Obviously accumulation does depend on consumption, since surplus value can only be realised in money form through the sale of commodities (here lies the true importance of advertising, salesmanship and the endogeneity of consumer preferences), but *the purpose of production is accumulation*, and accumulation for the sake of further accumulation, all of course under capitalist relations. In fact, and quite paradoxically for bourgeois theory, accumulation for the purpose of *consumption* is a characteristic of *communist* society. As Marx and Engels remark in the Manifesto: 'In bourgeois society, living labour is but a means to increase accumulated labour. In communist society, accumulated labour is but a means to widen, to enrich, to promote the existence of the labourer.'[4]

THE IDEOLOGICAL NATURE OF CONSUMER SOVEREIGNTY

How then is consumer sovereignty a part of bourgeois ideology? The answer to this question primarily centres around the rather peculiar theory of man's essence that emerged side by side with capitalist market society in seventeenth-century England, a theory based on the concepts not only that man was a consumer of utilities but also that his desires were insatiable and therefore infinite. This second concept certainly does not follow from the first, yet the association was and is necessary for capitalist production.

First, an economy comprising competitive markets requires that each economic agent maximises his utility. Each agent must juridically be free to enter into market transactions with his own labour-power and/or property, to dispose of as he thinks best. Each agent must be confronted with prices such that his labour-power, capital or land is continually thrown into the productive process; this involves a large class of agents being forced into the market through the historical process of 'freeing' them from their own means of production. Second, in a society with no traditional, patriarchal or feudal obligation to work, but in which virtually unlimited wealth is potentially attainable through production and exchange, there must be established some incentive to work. This incentive was the right of unlimited appropriation of wealth by individuals, a right to be ideologically justified by its derivation from the essence of man, from supposed eternal and immutable human nature.

Therefore man's essence was seen to be the

insatiable desirer of use-values, or utilities, whether his wants were innate or acquired, and economics became the analysis of how these unlimited wants shaped the allocation of scarce resources with alternative uses through the mechanism of *exchange*. With man's desires infinite, scarcity must always exist, for relative to infinite desire satisfactions must always be scarce. Therefore, through exchange rational economic man engages in the continual and continuous struggle to overcome scarcity, a struggle doomed to failure from the start.

The concept of man as infinite consumer was and is essential for capitalist production. For with private ownership of the means of production, man as the infinite consumer in the exchange process becomes man as the *infinite appropriator* – the capitalist – *in the production process*. And it is the justification of this conclusion, premissed on a theory of being that wants are insatiable, which is inherently ideological and supportive of the *status quo* in capitalist society.

NOTES

1 Abba P. Lerner, 'The Economics and Politics of Consumer Sovereignty', *American Economic Review*, Papers and Proceedings (1972), p. 258.
2 Karl Marx, *Capital* (London: Lawrence & Wishart, 1938), vol. 1, p. 155.
3 Ibid., pp. 129–130.
4 Karl Marx, *The Revolutions of 1848. Political Writings, Vol. 1*, ed. David Fernback (Harmondsworth: Penguin, 1973), p. 81.

17

The Consumer's Revolt

Mary Douglas

Consumerism starts as a liberation. If nothing else, the liberation is from drudgery. In Raymond Williams's great novel about a Welsh village in the 1920s, *Border Country* (1960/1964), there are several conversations about the relative merits of consumer comforts versus the old way. For example, this description of a new bungalow:

> no old stone floors, no muck in the yard, no miles to go to the shops. We got the electric, see, and the gas for cooking, and the car and the good water, and the you know, the proper W.C. I often say to Janie I was born in Glynmawr but I wouldn't go back there to live for anybody. Not if they paid me. Don't I, Janie? . . . Out there, in Glynmawr you know, talking. Why, I ask them, do we put up with dirty old water, and oil lamps, and the buckets, you know the buckets?
>
> (1960/1964, pp. 153–154)

This theme of personal comfort and freedom compared with living at close quarters and under each other's watchful eye runs through the book. Formerly they used to be content to live four or five to a room, now they all want a house for their own family; in the end the valley is going to be destroyed by the people moving out. No one could wish them to stay with the drudgery. It is the same for us and our liking for better kitchens and lighting, easier housework, more freedom to meet each other in distant places. As we extend the lines of our consumption rituals, we too demolish our local communities. The problem is still with us.

But the consumers' revolt is not just against the drudgery. It is a revolt against the despotism of neighbours whose business it is to know and judge everything that is done, what food is eaten, what time the children are put to bed, who is seducing whom and who is wearing clothes that look too seductive by local standards. A consumerist is one who defends the right of a person to be free from a neighbour's tyranny over his or her consumption habits. I insist on consumerism as a form of revolt, not just to be fair to the consumerists, but to put modern consumerism into context. Anyone who feels passionately that consumerism is wrong should be consistent. Are they ready to defend the constraints which hold consumerism in check? The goods that people can buy respond to some extent to the demands that these same people make on each other and these respond to the kind of society.

The basic choice is not between kinds of goods, but between kinds of society, and, for the interim, between the kinds of position in society that are available to us as we line up in the debate about transforming society. When we have made up our mind where we want to be aligned, do we have much free choice about the judgements we are going to make about goods? According to Pierre Bourdieu (1979), no, our preference for kinds of food and drink, housing and clothing is part of the bundle that we initially choose as we align ourselves in the political debate, and even for that alignment, he would say that we do not have much choice.

The scheme works remarkably well for French society because of the social stability provided by the Grandes Ecoles and the control that they ensure for the establishment. It gives a rational explanation of choices among goods: preferences are enrolled in the contest for supremacy. In so far as the French workers are not interested in aesthetic judgements that will never give them access to power, they are off the map. That in itself is a weakness in the theory. Another weakness is that it does not explain how goods are chosen in a community in which everyone is more or less equally well endowed. That is to say, it leaves out of account the kinds of consumption patterns found in many of the egalitarian societies anthropologists study. The scheme is good for making us aware of how politicized our taste is, and so good for attacking the economists' theory of the consumer as an individual sovereignly exercising private preferences. It does explain stable pockets of resistance to the wiles of the media, but it does *not* explain apparently mindless succumbing to the media's suggestions. It is consumer mindlessness that disturbs all the theories and creates the basic reproach to the capitalist system.

THE CONSTITUTIONAL MONITOR

To show consumerism as a kind of cultural revolt we need to take a step back from the northern European scene. Cultural theory is expected to apply to Africa, the Mediterranean, anywhere. It starts with the individual making his or her choices about what company to keep. Other people are the prime problem and on this theory every item of the human and material environment is drawn upon as a resource for dealing with other people. This starting point differs from market research which assumes that each individual is encircled by personal needs of greater or less urgency, physical needs first, then social, then spiritual satisfactions. It is a kind of mad nightmare, as if the average shopper were hungry, naked and roofless, and needs first to assure his next meal, then looks to his clothes from the point of view of warmth and protection from the cold and rain, then he is ready to see to his family and their physical needs, and only when all this is done, turns to the rest of the world in a more benignly philanthropic mood.

Instead of starting from the individual confronting his own basic needs, cultural theory starts from a stable system in which a consumer knows that he is expected to play some part or he will not get any income. In this theory the consumer has what can be called a cultural project. Everything that he chooses to do or to buy is part of a project to choose other people to be with who will help him to make the kind of society he thinks he will like the best. It is as if there were inside his head a little monitor for the constitutional effects that would flow from generalizing the present state of affairs. If he wants not to be dominated, he steers clear of some kinds of persons whose domineering habits he can instantly recognize. If he wants stability, he steers clear of revolutionary others who seem to be inveterate upsetters of apple carts. Steering clear of some, he steers towards others, reading their signals and emitting his own. The signals are, of course, the so-called consumption goods. The constitutional monitor is the elementary background of consumption patterns and consumer behaviour, and even of consumerism.

Another way of presenting this might be to say that the consumer is always engaged in making a collective good. The forms of consumption which he prefers are those that maintain the kind of collectivity he likes to be in. The theory takes account of four kinds of cultural preference.[1]

1. One would be the ideal liberal preference for a society which allows the members uninhibited opportunity to negotiate and transact: the collective good would be a minimum of fair play rules such as ensure the working of a free market.
2. Another would be a preference for the society in which a chap has the right to be left alone, without any wish to negotiate for power or authority. It would be a community of isolates, or drop-outs, or hermits.
3. The third is a preference for collaboration in a rationally integrated society.[2] This pattern restricts opportunism for the sake of protecting the categories and compartments it is prepared to defend.
4. The above can be very restrictive social environments. Rather than be subordinated in the first type by fellow members more competitive than he, or in the second type rather than put up with the poverty and neglect that are generally the lot of drop-outs, and rather than accept the control of a strongly ordered system such as hierarchy, another cultural choice would be to band together with a few like-minded souls to make a protected enclave, protesting against those who want to domineer.

The first two leave the shopper exposed and vulnerable. The last two choices are for forms of corporate grouping, one structured, one unstructured. Both of them have inherent powers of resistance to the temptations of consumerism. In what follows I shall argue that anyone who seriously disapproves of consumerism must logically be bound to support the choice for a communitarian way of life. That would be an uncomfortable choice for many of the most articulate anticonsumerists.

CREAM BUNS AND PRIVATE TRANSPORT

The thesis is that a community controls mindless spending like a person controls mindless consuming of cigarettes or cream buns . . . by making alternative desires effective. The parallel between the person's and the community's objectives has been proposed by Tom Schelling (1978) and worked out by Jon Elster (1985) in an essay that conflates the prisoner's dilemma problem of a community with that of the person who entertains conflicting goals one of which, if reached, will make the others impossible. An example of 'weakness of will', or *akrasia*, is the individual person who wants a svelte, athletic figure but who cannot resist a high-calorie diet. One of his objectives lies further in the future than the other; the cream bun is here and the desire for it is now, the pleasing answer on the weighing machine or in the mirror is in the future, and is postponed. Or the desired cream bun seems a minor interruption of the main plan to reduce weight; it is conceded because it is so inconsequential to the success of the main project. In the same way, free-riders on the collective effort persuade themselves that now is more important than a remote future, or that one little private bonus siphoned off from the community fund for a private purpose will be too inconsequential to matter. So though all Londoners deplore the havoc created by unlimited private transport, those who have cars will use them rather than the buses and underground and there is weakness of will as to how to solve the resultant traffic problem. Likewise, for another London example, most Londoners are ready to deplore the destruction of the small corner shops by competition from the supermarkets. But they show weakness of will whenever they take their car down to the supermarket car park and stock up for the month. Their own contribution to the commons despoiled is too inconsequential to count, and economic rationality seems to win the day, for the corner shop is much more expensive: why should my family be taxed to keep them in business?

On this view consumerism is no more difficult to explain than the failure to control impulse eating or smoking. The curious thing would be that control was ever successfully exerted and sustained. I will argue that if there is going to be a community at all it will impose disciplines, to use Foucault's expression, on the body and on the mind. The disciplines will be painful, and therefore will instigate revolt. If the revolt is successful there are two likely outcomes. One is the revolt of the disciplined, calculated withdrawal, such as refusing to wear hats, an amazingly difficult thing to do in isolation, a sign of great strength of will. The other is to reject both the discipline and the community that exerts the control.

COMMUNITY DESPOTISM

In spite of a large literary documentation of its tyranny, the sentimental idea of community prevails. I have already protested against the distortion this has produced in the theory of consumption (Douglas, 1980, 1986), but it may still be worth making more stark the case that community is repressive and costs a lot. To start with, go to the Musée des Arts Traditionnels et Populaires in Paris, and admire the presentation of a Breton homestead; note the small scale, the paucity of utensils, the economy of storage space, the high degree of order needful for four or five persons to live in the confined quarters; above all, note the standardization of costume, artefacts and decoration. On another wall, arrays of knives, hooks, lace testify to strong local definition. Why is everything so standardized that an amateur can quickly recognize a regional product?

The community, any community, exists because it constitutes a separate claim on the purses and time and energy of its members. There is no community unless its members concede its right to fund itself by levies upon themselves. For their part, the dues they pay are an investment, and they exact corresponding rights. This is a transactional approach to consumption, it requires that the community is one of the agents transacting, or at least that other agents transact on its behalf. The approach requires a distinction between a house and a home, or between a village and a community. 'What a lovely home you have!' exclaimed an American visitor on entering a house for the first time. In England it would be a nonsense sentence, since the house may be lovely without there being any home in it at all. The same for some Cotswold villages where netted thatch and manicured gardens are no guarantee of any community at all.

In *Border Country*, the novel quoted above, the young wife likes to go shopping, but she does not spend on herself: she buys things 'for the house' (Williams, 1960/1964, p. 57). Curtains, cushions, china represent a levy on the householder's purse for the commons. It is taken for granted that the house has a claim on the family budget, a kind of tax on members for their common life together. In the same way a village community has a claim on the purses and energies of its members. It imposes its informal taxes. No one can live in the community and escape paying out for condolences, christenings, name-day celebrations, the police force dance, the lifeboat fund. Some of the dues make a circulating fund, like the Cargas of the Latin American fiestas. Some are centrally collected and disbursed.[3] A Breton colleague told me of how, as a boy at his grandmother's funeral, he was posted at the door of the church to make a list of all the families which had given the priest a contribution for a Mass to be said for her soul: it was *de rigueur* for his family to reciprocate in kind on the death of any member of the families listed. Though the reciprocity went between families, the donations went to the church and would be eventually disbursed among its other charitable *oeuvres*. Paying is not enough: there has also to be a physical presence. In an English farming community everyone has to go to the Hunt Ball, as well as contribute to it. Turning up at the event is one of the ways by which the community knows who its members are, a kind of informal census-taking. Furthermore, when a big crowd turns out, to see the May dances, or the Saint's Day fireworks or the cricket match, community solidarity is made visible. More significantly, the requirement to be present and to pay the equivalent of a community tax channels the earnings of members towards the community itself. The heavier the communal chores and dues, the stronger the sign of commitment.

Some communities recognize that gross disparity in incomes will be disruptive, and so they use the levies for public events for redistributive purposes. Individual surpluses are drawn out of private hands and neutralized or destroyed. In parts of Africa when an important person dies, masses of valuables are buried with him, with the result that debts are cancelled, accumulations run down. The same intention to reduce excessive accumulation may lie behind the Iron Age burials, and certainly they would have had that effect. For example, some Mediterranean villagers display finery in their churches that contrasts with the poverty of their homes: public affluence along with private squalor, to turn Kenneth Galbraith's phrase around. The money that has gone to pay for chandeliers or marble pillars might well have reduced the labour of buckets and scrubbing dirty old stone floors in the kitchens. What Max Weber mistakenly saw as other-worldliness in medieval Christianity was usually a sign of strong this-worldliness, a consumption decision in favour of the community at the expense of the individual.

At the same time, the public demand that absorbs private wealth is the consumers' parallel to the workforce that condemns rate busting. Marshall Sahlins has suggested that many hunting and gathering tribes in Australia and Africa should be called the first affluent societies, because of their high preference for leisure and the shortness of their working day (Sahlins, 1972). In many cases, they are not preferring leisure to the rewards of work, but acceding to the demands of the community that they be present for mourning, rejoicing, eating, praying or dancing together. They have adjusted to a low level of private wants partly because of the high taxes that their community exacts.[4]

I will now speak about the community for convenience as if it were a person with intentions and ideas. The community recognizes that the money earned cannot be stretched to all possible desires. Orlove and Rutz point out that certain expenditures indicate present and future intentions to withdraw: in Indonesia,

> poor peasants will take the wearing of imported cloth by cosmopolitans to suggest a partial withdrawal from sponsorship of ceremonies. . . . The motor-cycles are in some sense an investment, in that money spent on a motor-cycle reduces the expenditures on transport and allows easier access to outside markets; they might well also be a source of individual pleasure, and they indicate to others an orientation beyond the village. They commit the owners to the future acquisition of monetary income beyond subsistence needs, for purchase of gasoline and spare parts.
>
> (1989, pp. 8–9)

Two worries concern the stakeholders who want the community to survive. One is that members may go away. They do go, and emigrants who return to their village are mercilessly milked by the relatives who have stayed behind. They should not have gone; people have died, and they were not there; now that they have come back, they have to be generous. No wonder the migrant worker tries to make huge savings before he can face his family. The other worry is shouldering liability for its old, infirm and indigent. The community applauds those who look after their old. In a way, earning that applause is like a pension fund. Those who have invested in the community all their lives expect to be cared for by someone or other near to them, and they all watch keenly for defectors. The community will be hard on anyone on whom the old or helpless have an unfulfilled claim, and everyone will be engaged in rebutting claims. This one was always drunk, no wonder he has no savings, that one played the horses, it is his own fault if he is stranded now, the other one was quarrelsome, too proud, irreligious. The community finds itself applying standards of good husbandry, health and hygiene, as excuses. If anyone falls into debt through their own fault, the community is not going to be so generous with credit. The acceptance of liability justifies the sumptuary controls, the criticism of extravagance, drinking and gambling. The community withdraws its protection from the deviants. It has to be assured that its members not only pay their dues now, but are likely to be able to continue to do so. Hence a pressure against spending, a time preference for the future, and a general tyranny over private consumption.

Anyone who has invested in such a community has expectations from it. Paying his dues, he does not expect other men to seduce his wife or daughters. Paying his dues, he does not expect other men's wives or daughters to seduce him, or his sons, or otherwise to upset his plans for marriage alliances. Hence some of the disciplines of the body, and the standardization of clothing, decoration and everything to do with sex. To belong to such a community means accepting its standards. Eccentricity is rejected, flamboyancy reproved as much as carelessness. Hence the standardization of objects. Excellence of artisanship has to be achieved without conspicuousness. Economy, wit, proportion, scale, accommodation to function and storage, these are the limits on craft work for the everyday utensils that we so much admire. Artistic display, originality, extravagance may be deployed for objects of public ownership. Everyday objects are minutely graded to their uses. Special objects are endowed with semiotic richness and their consumption is hedged with rules, so that they can mark the occasions when the community celebrates itself. The objects are coded, and to know the coding is a claim to membership. This is the basis of the tyranny which embeds every consumption choice in a communication system. Here there is no problem of how a person finds the strength to resist the lures of commodities; the strength is in the surveillance of neighbours, backed by the security of the local community. Anyone who has been a member of such a community knows that it takes more courage to move out than to remain embedded.

RATIONAL CONSUMPTION

Though this picture draws upon a rather commonplace knowledge of consumption patterns, it has merit in setting up the community chest and the private purse agonistically as rivals. It explains why the community scrutinizes its members' consumption so critically, and judges so harshly any deviation. The more fragile the community bond, the more anxious and severe the scrutiny and the more fuel to the fires of revolt. There is nothing irrational about deciding to invest socially (Orlove and Rutz's expression) or in deciding to invest and spend on a personal basis. Consumerist, or anti-consumerist, both choices are reasonable. However, symmetry is lacking. It is always feasible to have a programme of consumption for oneself. A programme of community investment may be unreasonable, since it depends on ongoing support from other people.

Returning to the four cultural projects which we sketched at the beginning, each of them presents the rational person with a goal, the kind of society he would like to live in. Given the goal, it is rational that consumption should be engaged to serve it, and that conscience should be invoked to uphold the pattern. In the case of the liberal preference for a society upheld by rules of free play, the person's buying pattern may reflect a conscious revolt against community despotism. In the case of the drop-out, who wants a society where he can be left in peace, the prior dignity of a life of contemplation may justify his erratic but low expenditure. In the enclave the conscience of members has been stirred in protest against the liberal pursuit of personal ease and comfort. In the rationally integrated community the conscience is towards collective survival, so that incipient defectors are admonished. So where is there any irrationality?

Most consumer revolts are symbolic, gestures of independence. . . . However, moving out of the range of the community's censorship, there are traps to ensnare the deserter. Choosing to be free of the community censure involves choosing to be a society where each finds his allies by his own efforts, a competitive society in which consumption is inevitably competitive. This is the field well described by Thorstein Veblen and many others. But it would be a great mistake to suppose that competitive consumption is mindless. In a competitive world signals of success attract allies and business; consumption can easily be harnessed to making those signals. Then it is that luxuries and necessities become confused, and total disembedding gives an air of disordered consumerism. Only relatively stable communities can make and keep a distinction between luxuries and necessities. What other people call luxuries appear higgledy-piggledy on the consumerist's plate or in his bedroom and bathroom. The truth is that consumerism is part of a highly competitive way of living, in which everything may be dragged in to the purpose of pleasing a client or ally. Competitive consumerism needs luxuries all the time for a rational deployment of resources. Competition needs to tear down community boundaries, to expand the range of its dealings. There is no surprise that it scoffs at the restraints on spending as well as the disciplines of the body which keep consumption within bounds.

Consumerism is not in itself irrational; what would be irrational would be for the very persons to voice a worry about the environment who demand private transport in the metropolis. The very ones who worry about the absence of community solidarity should not drive the small urban corner shops out of business by shopping in the supermarkets. Rational behaviour puts its money where its mouth is and recognizes community levies and taxes for what they are.

NOTES

1 Cultural theory focuses on four possible kinds of social environment, argues that these are the stable, viable kinds of society and that others are on a transitional course between one or other of them; also that in any community there are pressures to transform the present constitution into one or another type, and that these pressures are manifest in a regular normative debate. Furthermore, each kind of culture defines itself by contrast to and against the attraction of others. There is an extensive literature developing the theory and illustrating and challenging it. For a bibliography, see Thompson *et al.* (1990).

2 The same four cultural types can be defined according to the type of integration, and the amount of integration preferred: the liberal ideal will produce integration based on economic efficiency, and on the power that accumulates where wealth is held; the isolates prefer as little integration as possible, hoping thus to be left alone; the enclavists, because they are integrated on a principle of voluntary protest, are organized with reference to their outer boundary; those who prefer an integration that will organize a wide sweep of positive communitarian goals, are trying to build a collectivity which will in practice focus on several mutually balancing institutional centres.

3 Compare Karl Polyani's analysis of gift exchange patterns in ancient civilizations which he explicitly compared with household distribution. Here I am turning the analogy the other way round.

4 For an interesting discussion of these debates, see Ahrne (1988).

REFERENCES

Ahrne, G. (1988) 'A labour theory of consumption', in Per Otnes (ed.), *The Sociology of Consumption*. Oslo: Solum Forlag, pp. 50–52.

Bourdieu, P. (1979) *La Distinction*. Paris: Editions Minuit.

Douglas, M. (1980) *The World of Goods*. New York: Basic Books.

Douglas, M. (1986) *How Institutions Think*. New York: Syracuse University Press, ch. 3.

Elster, J. (1985) 'Weakness of will and the free-rider problem', *Economics and Philosophy*, 1: 231–265.

Orlove, B. and Rutz, H.J. (1989) 'Thinking about consumption', in *The Social Economy of Consumption: Monographs in Economic Anthropology*, No. 6. Washington, DC: University Press of America.

Sahlins, M. (1972) *Stone Age Economics*. Chicago: Aldine-Atherton.

Schelling, T. (1978) 'Egonomics, or the art of self-management', *American Economic Review: Papers and Proceedings*, 68: 290–294.

Thompson, M., Wildavsky, A. and Ellis, R. (1990) *Cultural Theory*. Boulder, CO: Westview Press.

Williams, R. (1960/1964) *Border Country*. London: Chatto & Windus. Quotations from Penguin edn, 1964.

THREE

18

Subculture and Style

Dick Hebdige

> I managed to get about twenty photographs, and with bits of chewed bread I pasted them on the back of the cardboard sheet of regulations that hangs on the wall. Some are pinned up with bits of brass wire which the foreman brings me and on which I have to string coloured glass beads. Using the same beads with which the prisoners next door make funeral wreaths, I have made star-shaped frames for the most purely criminal. In the evening, as you open your window to the street, I turn the back of the regulation sheet towards me. Smiles and sneers, alike inexorable, enter me by all the holes I offer. . . . They watch over my little routines.
>
> (Genet, 1966)

In the opening pages of *The Thief's Journal*, Jean Genet describes how a tube of vaseline, found in his possession, is confiscated by the Spanish police during a raid. This 'dirty, wretched object', proclaiming his homosexuality to the world, becomes for Genet a kind of guarantee – 'the sign of a secret grace which was soon to save me from contempt'. The discovery of the vaseline is greeted with laughter in the record-office of the station, and the police 'smelling of garlic, sweat and oil, but . . . strong in their moral assurance' subject Genet to a tirade of hostile innuendo. The author joins in the laughter too ('though painfully') but later, in his cell, 'the image of the tube of vaseline never left me'.

> I was sure that this puny and most humble object would hold its own against them; by its mere presence it would be able to exasperate all the police in the world; it would draw down upon itself contempt, hatred, white and dumb rages.
>
> (Genet, 1967)

I have chosen to begin with these extracts from Genet because he more than most has explored in both his life and his art the subversive implications of style. I shall be returning again and again to Genet's major themes: the status and meaning of revolt, the idea of style as a form of Refusal, the elevation of crime into art (even though, in our case, the 'crimes' are only broken codes). Like Genet, we are interested in subculture – in the expressive forms and rituals of those subordinate groups – the teddy boys and mods and rockers, the skinheads and the punks – who are alternately dismissed, denounced and canonized; treated at different times as threats to public order and as harmless buffoons. Like Genet also, we are intrigued by the most mundane objects – a safety pin, a pointed shoe, a motor cycle – which, none the less, like the tube of vaseline, take on a symbolic dimension, becoming a form of stigmata, tokens of a self-imposed exile. Finally, like Genet, we must seek to recreate the dialectic between action and reaction which renders these objects meaningful. For, just as the conflict between Genet's 'unnatural' sexuality and the policemen's 'legitimate' outrage can be encapsulated in a single object, so the tensions between dominant and subordinate groups can be found reflected in the surfaces of subculture – in the styles made up of mundane objects which have a double meaning. On the one hand, they warn the 'straight' world in advance of a sinister presence – the presence of difference – and draw down upon themselves vague suspicions, uneasy laughter, 'white and dumb rages'. On the other hand, for those who erect them into icons, who use them as words or as curses, these objects become signs of forbidden identity, sources of value. Recalling his humiliation at the hands of the police, Genet finds consolation in the tube of vaseline. It becomes a symbol of his 'triumph' – 'I would indeeed rather have shed blood than repudiate that silly object' (Genet, 1967).

The meaning of subculture is, then, always in dispute, and style is the area in which the opposing definitions clash with most dramatic force. . . . As in Genet's novels, this process begins with a crime against the natural order, though in this case the deviation may seem slight indeed – the cultivation of a quiff, the acquisition of a scooter or a record or a certain type of suit. But it ends in the construction of a style, in a gesture of defiance or contempt, in a smile or a sneer. It signals a Refusal. I would like to think that this Refusal is worth making, that these gestures have a meaning, that the smiles and the sneers have some subversive value, even if, in the final analysis, they are, like Genet's gangster pin-ups, just the darker side of sets of regulations, just so much graffiti on a prison wall.

Even so, graffiti can make fascinating reading. They draw attention to themselves. They are an expression both of impotence and a kind of power – the power to disfigure (Norman Mailer calls graffiti – 'Your presence on their Presence . . . hanging your alias on their scene' (Mailer, 1974)). Here I shall attempt to decipher the graffiti, to tease out the meanings embedded in the various post-war youth styles. But before we can proceed to individual subcultures, we must first define the basic terms. The word 'subculture' is loaded down with mystery. It suggests secrecy, masonic oaths, an Underworld. It also invokes the larger and no less difficult concept 'culture'. So it is with the idea of culture that we should begin.

REFERENCES

Genet, J. (1966) *Our Lady of the Flowers*, Panther.
Genet, J. (1967) *The Thief's Journal*, Penguin.
Mailer, N. (1974) 'The Faith of Graffiti', *Esquire*, May.

19

Subcultures or Neo-tribes?

Rethinking the Relationship Between Youth, Style and Musical Taste

Andrew Bennett

THE CONCEPT OF 'SUBCULTURE'

> 'Authentic' subcultures were produced by subcultural theorists, not the other way around. In fact, popular music and 'deviant' youth styles never fitted together as harmoniously as some subcultural theory proclaimed.
>
> (Redhead 1990: 25)

While the essential tenets of the CCCS [Centre for Contemporary Cultural Studies] subcultural theory have been variously criticised and largely abandoned, the concept of 'subculture' survives as a centrally defining discursive trope in much sociological work on the relationship between youth, music and style. In my view, however, the term 'subculture' is also deeply problematic in that it imposes rigid lines of division over forms of sociation which may, in effect, be rather more fleeting, and in many cases arbitrary, than the concept of subculture, with its connotations of coherency and solidarity, allows for. Pondering a similar point, Fine and Kleinman argue that the attempt to reify a construct such as subculture 'as a corpus of knowledge may be heuristically valuable, until one begins to give this corpus physical properties' (1979: 6). Likewise, Jenkins suggests that 'the concept of subculture tends to exclude from consideration the large area of commonality between subcultures, however defined, and implies a determinate and often deviant relationship to a national dominant culture' (1983: 41). In this respect, Jenkins's argument has much in common with McRobbie's (1994) observation concerning the absence of any discussion of the shifting behaviour patterns of members of 'subcultural' groups as they move between subcultural setting and family home. However, while McRobbie suggests that such omissions from subcultural studies were the product of male sociologists' lack of interest in the home and family environments of 'subculture' members, it is equally possible to argue, in line with Jenkins's (1983) observation, that these omissions also conveniently paper over the cracks in the CCCS's attempts to depict 'subcultures' as tight, coherent social groups. Indeed, at one point in *Resistance Through Rituals*, John Clarke *et al.* come very close to admitting such a point when they suggest that although 'sub-cultures are important . . . they may be less significant than what young people do most of the time' (1976: 16).

As previously noted, despite the problems which can be associated with 'subculture' the term continues to be widely used. At the same time, however, 'subculture's' continuing currency as a grounding theoretical base deepens the questioning of the term's sociological validity as it is applied in increasingly contradictory ways. In *Reconstructing Pop/Subculture*, Cagle (1995) takes issue with the Marxist interpretation of youth subcultures employed by the CCCS in view of its conceptualisation of subculture as existing outside the mainstream. According to Cagle, youth groups discounted by the CCCS, for example, glitter rock fans, could also be counted as 'subcultures' despite their mainstream tastes in music and style. In certain respects Cagle has a very good point in that the CCCS did indeed discard a great deal of music and style-centred youth activity, which, in addition to glitter rock, also included 'Rollermania' and heavy metal, presumably on the grounds that its mainstream centredness somehow removed its potential for counter-hegemonic action which the Centre so readily associated with mods, skinheads and punks, etc. However, while Cagle is right to criticise the CCCS on these grounds, at the same time his approach has considerable implications for the term 'subculture' in that it is left meaning everything and nothing. Thus, if we are to accept that there are both *mainstream* and *non-mainstream* subcultures, what are the differences between them, and how do we go about determining such differences?

Thornton's (1995) solution to the mainstream/non-mainstream debate and its bearing upon notions of subcultural authenticity is to introduce the issue of media representation. According to Thornton, 'authentic' subcultures are largely constructed by the media, members of subcultures acquiring a sense of themselves and their relation to the rest of society from the way they are represented in the media. Again, Thornton identifies a very important shortcoming in the CCCS conceptualisation of 'subculture'. Given the centrality of the media in all institutions of late modern social life, there can be little questioning of Thornton's contention that: "subcultures" do not germinate from a seed and grow by force of their own energy into mysterious "movements" only to be belatedly digested by the media' (1995: 117). It seems to me, however, that Thornton's work raises a much more fundamental point in relation to subculture and its validity as an objective sociological concept. Thus, if subculture has acquired a plurality of meaning in sociological discourse, the media's borrowing of the term has increased the problem of definition. In introducing the term 'subculture' into the wider public sphere, the media have completed the process begun in sociological work of reducing subculture to a convenient 'catch-all' term used to describe a range of disparate collective practices whose only obvious relation is that they all involve young people.

NEO-TRIBES: AN ALTERNATIVE THEORETICAL MODEL FOR THE STUDY OF YOUTH

In critically evaluating 'subculture' as a valid framework for the sociological study of youth, music and style, I have identified two main issues. First, there is a problem of objectivity as subculture is used in increasingly contradictory ways by sociological theorists. Secondly, given that in studies which use 'subculture' in relation to youth, music and style there is a grounding belief that subcultures are subsets of society, or cultures within cultures, such a concept imposes lines of division and social categories which are very difficult to verify in empirical terms. Indeed, at the most fundamental level, there is very little evidence to suggest that even the most committed groups of youth stylists are in any way as 'coherent' or 'fixed' as the term 'subculture' implies. On the contrary, it seems to me that so-called youth 'subcultures' are prime examples of the unstable and shifting cultural affiliations which characterise late modern consumer-based societies.

Shields writes of a 'postmodern "persona"' which moves between a succession of 'site-specific' gatherings and whose 'multiple identifications form a *dramatic personae* – a self which can no longer be simplistically theorized as unified' (1992a: 16). From this point of view the group is no longer a central focus for the individual but rather one of a series of foci or 'sites' within which the individual can live out a selected, temporal role or identity before relocating to an alternative site and assuming a different identity. It follows, then, that the term group can also no longer be regarded as having a necessarily permanent or tangible quality, the characteristics, visibility and lifespan of a group being wholly dependent upon the particular forms of interaction which it is used to stage. Clearly, there is a considerable amount of difference between this definition of a group and that which prefigures subcultural theory. Indeed, the term 'group' as it is referred to here is much closer to Maffesoli's concept of *tribus* or 'tribes'. According to Maffesoli the tribe is 'without the rigidity of the forms of organization with which we are familiar, it refers more to a certain ambience, a state of mind, and is preferably to be expressed through lifestyles that favour appearance and form' (1996: 98).

Underpinning Maffesoli's concept of tribes is a concern to illustrate the shifting nature of collective associations between individuals as societies become increasingly consumer orientated (1996: 97–98). Thus as Hetherington, in discussing Maffesoli's work, points out, tribalisation involves 'the deregulation through modernization and individualization of the modern forms of solidarity and identity based on class occupation, locality and gender . . . and the recomposition into "tribal" identities and forms of sociation' (1992: 93). Shields, in a further evaluation of Maffesoli's work, suggests that tribal identities serve to illustrate the temporal nature of collective identities in modern consumer society as individuals continually move between different sites of collective expression and 'reconstruct' themselves accordingly. Thus, argues Shields: 'Personas are "unfurled" and mutually adjusted. The performative orientation toward the Other in these sites of social centrality and sociality draws people together one by one. Tribe-like but temporary groups and circles condense out of the homogeneity of the mass' (1992b: 108). There is some disagreement between Shields and Hetherington as to how the concept of tribalism can most effectively be used. In his foreword to the English translation of Maffesoli's study *The Time of the Tribes*, Shields argues that *tribus* are 'best understood as "postmodern tribes" or even pseudo-tribes' (Maffesoli 1996: x). Hetherington (1992), however, prefers the term 'neo-tribes'. For the purpose of this article, I too refer to *tribus* as neo-tribes, as this seems to me to most accurately describe the social processes with which Maffesoli was concerned.

Interestingly, in Maffesoli's view neo-tribes are a very recent social phenomenon. Indeed, there is a distinctly postmodernist edge to Maffesoli's 'then' and 'now' comparisons between the 1970s and the 'tribalised' 1990s. Thus in describing the nature of neo-tribal society, Maffesoli (1996: 76) observes that

> This 'affectual' nebula leads us to understand the precise forms which sociality takes today: the wandering mass-tribes. Indeed, in contrast to the 1970s – with its strengths such as the Californian counterculture and the European student communes – it is less a question of belonging to a gang, a family or a community than of switching from one group to another.

It seems curious that Maffesoli should place the development of neo-tribalism beyond the late 1960s and early 1970s, a period when conspicuous consumption became synonymous with everyday life in the West. Moreover, Maffesoli's reference to the counter-culture as a stable, coherent cultural entity seems oddly out of place given the loose affiliation of political, aesthetic and stylistic interests which found a fragile and temporal unity under the counter-cultural banner. Indeed, as Clecak suggests, 'counter-culture' was, in effect, an umbrella term which enabled a wide range of different groups, including college students, musicians, mystics, environmentalists, the human-potential movement, peace and anti-war movements 'to find symbolic shapes for their social and spiritual discontents and hopes' (Clecak 1983: 18). Moreover, much of the counter-culture's oppositional stance hinged on forms of expression articulated through commercially available products, such as music and style, themselves a result of the youth market which had been steadily growing in prominence since the 1950s. In my view, then, the process of tribalism identified by Maffesoli is tied inherently to the origins of mass consumerism during the immediate post-Second World War period and has been gathering momentum ever since. That it should become acutely manifest in the closing years of the twentieth century has rather more to do with the sheer range of consumer choices which now exist than with the onset of a postmodernist age and attendant postmodern sensibilities.

LIFESTYLES

In reconsidering issues of social identity and forms of collective expression within the framework of neo-tribes, the related concept of 'lifestyle' provides a useful basis for a revised understanding of how individual identities are constructed and lived out. 'Lifestyle' describes the sensibilities employed by the individual in choosing certain commodities and patterns of consumption and in articulating these cultural resources as modes of personal expression (Chaney 1994, 1996). In this way, a lifestyle is 'a freely chosen game' and should not be confused with a 'way of life', the latter being 'typically associated with a more-or-less stable community' (Kellner 1992: 158; Chaney 1994: 92). Certainly, there are numerous instances of lifestyles which are intended to reflect more 'traditional' ways of life, notably in relation to class background. For example, British pop group Oasis and their fans promote an image, consisting of training shoes, football shirts and duffel coats, which is designed to illustrate their collective sense of working classness. Therein, however, lies the essential difference between the concept of lifestyle and structuralist interpretations of social life in that the former regards individuals as active consumers whose choice reflects a self-constructed notion of identity while the latter supposes individuals to be locked into particular 'ways of being' which are determined by the conditions of class. Moreover, in positing experimentation as a central characteristic of late modern identities, the concept of lifestyle allows for the fact that individuals will also often select lifestyles which are in no way indicative of a specific class background. A fitting example of this is the chosen lifestyle of the New Age Traveller which brings together young people from a range of social backgrounds who share 'an identification with nomadism that is seen to be more authentic than the sociality of modern industrial societies' (Hetherington 1998: 335).

All of this is not to suggest that 'lifestyle' abandons any consideration of structural issues. Rather, 'lifestyle' allows for the fact that consumerism offers the individual new ways of negotiating such issues. Thus, as Chaney observes, 'the indiscriminate egalitarianism of mass culture does not necessarily reproduce the structured oppressions of previous social order. Or rather . . . these oppressions can more easily be subverted by the very diversity of lifestyle' made possible via the appropriation of selected commodities and participation in chosen patterns of consumption (1994: 81). A similar point is made by Willis who suggests (1990: 18) that

> If it ever existed at all, the old 'mass' has been culturally emancipated into popularly differentiated cultural citizens through exposure to a widened circle of commodity relations. These things have supplied a much widened range of symbolic resources for the development and emancipation of everyday culture.

There is a clear correspondence between Willis's observation and Maffesoli's contention that neo-tribalism involves 'a rationalized "social" [being] replaced by an emphatic "sociality", which is expressed by a succession of ambiences, feelings and emotions' (1996: 11). Once again, the central implication here is that a fully developed mass society liberates rather than oppresses individuals by offering avenues for individual expression through a range of commodities and resources which can be worked into particular lifestyle *sites* and *strategies* (Chaney 1996). At the same time, however, Maffesoli's notion of an *emphatic sociality* allows for the fact that such sites and strategies are in no way fixed but may change, both over time or in correspondence with the different groups and activities with which individuals engage in the course of their everyday lives.

REFERENCES

Cagle, V. M. (1995) *Reconstructing Pop/Subculture: Art, Rock and Andy Warhol.* London: Sage.

Chaney, D. (1994) *The Cultural Turn: Scene-Setting Essays on Contemporary Cultural History.* London: Routledge.

Chaney, D. (1996) *Lifestyles.* London: Routledge.

Clarke, J., Hall, S., Jefferson, T. and Roberts, B. (1976) 'Subcultures, Cultures and Class'. In Hall and Jefferson (eds), *Resistance Through Rituals.*

Clecak, P. (1983) *America's Quest for the Ideal Self: Dissent and Fulfilment in the 60s and 70s.* Oxford: Oxford University Press.

Fine, G. A. and Kleinman, S. (1979) 'Rethinking Subculture: An Interactionist Analysis'. *American Journal of Sociology* 85:1–20.

Hall, S. and Jefferson, T. (eds) (1976) *Resistance Through Rituals: Youth Subcultures in Post-War Britain.* London: Hutchinson.

Hetherington, K. (1992) 'Stonehenge and its Festival: Spaces of Consumption'. In R. Shields (ed.), *Lifestyle Shopping: The Subject of Consumption.* London: Routledge.

Hetherington, K. (1998) 'Vanloads of Uproarious Humanity: New Age Travellers and the Utopics of the Countryside'. In T. Skelton and G. Valentine (eds), *Cool Places: Geographies of Youth Culture.* London: Routledge.

Jenkins, R. (1983) *Lads, Citizens and Ordinary Kids: Working Class Youth Lifestyles in Belfast.* London: Routledge & Kegan Paul.

Kellner, D. (1992) 'Popular Culture and the Construction of Postmodern Identities'. In S. Lash and J. Friedman (eds), *Modernity and Identity.* Oxford: Blackwell.

McRobbie, A. (1984) 'Dance and Social Fantasy'. In A. McRobbie and M. Nava (eds), *Gender and Generation.* London: Macmillan.

Maffesoli, M. (1996) *The Time of the Tribes: The Decline of Individualism in Mass Society.* London: Sage.

Redhead, S. (1990) *The End-of-the-Century Party: Youth and Pop Towards 2000.* Manchester: Manchester University Press.

Shields, R. (1992a) 'Spaces for the Subject of Consumption'. In R. Shields (ed.), *Lifestyle Shopping: The Subject of Consumption.* London: Routledge.

Shields, R. (1992b) 'The Individual, Consumption Cultures and the Fate of Community'. In Shields (ed.), *Lifestyle Shopping.*

Thornton, S. (1995) *Club Cultures: Music, Media and Subcultural Capital*. Cambridge: Polity.
Willis, P. (1990) *Common Culture: Symbolic Work at Play in the Everyday Cultures of the Young*. Milton Keynes: Open University Press.

20

Consumption and Class Analysis

Rosemary Crompton

The suggestion that social classes are in part constituted through cultural practices, including patterns of consumption, is not a new one (Veblen, 1934).[1] Bourdieu's work (1984), although by now somewhat out-of-date, remains the most comprehensive sociological treatment of the topic, although there have been recent empirical studies of the 'middle classes' which have also explored in some depth the way in which consumption practices or 'lifestyles' have served to differentiate between social groupings (Savage *et al.*, 1992). Given the importance of the links between social class position and consumption (purchasing) practices, the connections between economic/occupational ('class') position and patterns of consumption have been constantly investigated and reinvestigated by market researchers.

Thus we may agree that in the broadest sense, consumption practices serve to reinforce and reproduce social hierarchies. Whether or not such practices have become *more* important than they once were is difficult to judge, and in any case, the answer would be likely to vary considerably depending on the social group in question. In addition, the simple point that the maintenance of consumption practices is heavily dependent upon economic class position should not be forgotten. However, our subsequent discussion will focus largely on the second area identified – the experience of employment.

THE EXPERIENCE OF EMPLOYMENT, CLASS CONSCIOUSNESS AND IDENTITY

Two further divisions characterize consumption-related changes in employment: (1) developments in the labour market and (2) changes in work practices. In order to explore these factors, it is first necessary to demonstrate the linkages between the sociology of work and employment and the study of class processes. Here the work of Lockwood and Braverman has been crucial.

In British sociology, the 'work' and 'market' situations of employees were taken to be the most significant elements contributing to class situation in employment. Lockwood (1958) argued that (1) although clerical workers were, like manual workers, propertyless employees, nevertheless such 'white-collar' workers could not be said to share the same 'class situation' as manual workers because of their different employment circumstances – including more employment security, better opportunities for upward mobility, and less oppressive working conditions. Furthermore, (2) similar variations existed *within* the clerical category, which served to explain variations in levels of trade unionism among clerical workers. Thus the work and employment circumstances of clerical workers were used as indicators in order to locate these occupations within the 'class structure' as represented in employment. These insights were carried over into the 'Affluent Worker' (Goldthorpe *et al.*, 1969) study, which similarly examined the 'work situation' of 'affluent workers' in researching their 'class situation'.

Braverman's arguments relating to the links between the labour process and processes of class formation has many parallels with Lockwood's work. Like Lockwood, Braverman (1974) argued that the nature of the labour (work) process directly affected the class location of employees – although unlike Lockwood, he argued that clerical workers were undergoing a process of 'de-skilling' and should more properly be allocated to the 'proletarian' class category. A further difference is that Braverman's analytical framework was decisively influenced by a Marxist approach, whereas Lockwood (1988) has been highly critical of aspects of Marx's work.

Another very influential contribution made by Lockwood was his (1966) article on 'Sources of variation in working-class images of society'. On the basis of case study evidence, he suggested that three different types of working-class imagery could be identified; traditional proletarian, traditional deferential, and privatized. These different world views were linked to variations in work (i.e. employment) and locality situations. Here Lockwood may be seen as attempting to describe the empirical link between location in the social (class) structure and propensity for action – or at least, consciousness – of a class kind.

Despite the theoretical differences between Lockwood and Braverman, their work may be seen to have some important similarities of approach. Both held that the labour market was a major source of social structuring, which generated particular kinds of interest groupings and particular forms of social consciousness – and both regarded class concepts (in the most general sense) as central to the analysis of these processes.[2] However, over the past fifteen years or so, it has increasingly been argued that, at the end of the twentieth century, changes in the labour market, and in the nature of employment, have been so profound that the previous links that have been made between employment, interests, and social consciousness no longer apply. As a consequence, it is argued that we have reached a point beyond which 'class' is a useful concept for the analysis of contemporary societies.

It has been argued that the labour market has been progressively fragmented, as a result of which there has been a growth of flexible and non standard employment such as part-time work, short-term contracts, and self-employment (see Beatson, 1995; Gregg and Wadsworth, 1995). These developments have been associated with increasing insecurity of employment and the decline of the long-term career. In Britain, Conservative Government policy has encouraged corporate and organizational restructuring associated with 'downsizing' and job loss, and popular management magazines (such as *Business Age*[3]) have chronicled the subsequent travails of middle management. According to Beck (1992), these kinds of changes have resulted in the development of an 'individualized society of employees' in a 'risk-fraught system of flexible, pluralized, decentralized underemployment' (p. 143). In these 'new forms' of employment (which include homeworking, casual work, and so on), the boundaries between work and non-work are becoming increasingly fluid. 'Work' in the sense of an occupation once provided a focus for the development of class-based identities in industrial societies, but as a consequence of increasing insecurity and 'flexibility' in the labour market, Beck argues, both 'class' and 'status' are losing their significance.

A further set of arguments relates to changes in the nature of work itself, particularly those associated with the expansion of service work and the growth and application of Information Technology. Authors such as Offe (1985), and Lash and Urry (1994), have argued that the growth and development of consumption-related employment, particularly in services, has radically transformed the *meaning* of work for employees, and that as a consequence employment no longer has the capacity to generate 'class'-related consciousness and action. Thus Offe has suggested that 'work' is no longer a 'key sociological concept': 'work', he argues, no longer has 'a relatively privileged power to determine social consciousness and action' (1985: 133). Furthermore, Lash and Urry (1994) have argued that with the development of Information Technology, production systems have themselves become expert systems, and that 'reflexive modernities' are increasingly becoming 'economies of signs and space'. Indeed, they suggest an inversion of the Marxist thesis in relation to consciousness, rather than structure (i.e. classes) shaping consciousness, 'in informationalized and reflexive modernity it is consciousness or reflexivity which is determinant of class structure' (p. 319). These kinds of arguments will be explored in the discussion that follows.

HAS THE EXPANSION OF CONSUMPTION-RELATED EMPLOYMENT IN SERVICES, AND SERVICE PRACTICES IN ORGANIZATIONS AND PRODUCTION, DISSOLVED THE LINK BETWEEN EXPERIENCES IN EMPLOYMENT AND CLASS CONSCIOUSNESS AND ACTION?

That the increase in service employment is a marked feature of recent labour market developments can hardly be contested. In the UK, for example, service employment increased from 53 per cent to 73 per cent between 1973 and 1993, and a similar trend is to be found in other countries (Frenkel *et al.*, 1995). Service employment encompasses a heterogeneous range of occupations among which it is not difficult to find examples of work as routine and 'de-skilled' as is any to be found in manufacturing industry (Gabriel, 1988). Routine and de-skilled work in services is still widespread. However, it is important to move on from the 'de-skilling' debate to focus on changes in the organization and management of paid employment which have been developed over the past one-and-a-half decades. Many of these are associated with the growth of *interactive* service work (retail and other sales including telephone sales and servicing, the leisure industry, and so on), which is the fastest-growing sector of service employment. Besides this growth, 'marketization' and changes in legislation such as the deregulation of the finance sector in the 1980s has also resulted in the transformation of work in other sectors such as finance into *de facto* sales employment.

Two particular consumption-related features of contemporary service work have been identified in making the case that the links between employment and class identity are disappearing: (1) that the boundary between 'work' and 'non-work' is more problematic in contemporary service employment than in other types of work, and (2) that recent developments in the organization and control of service employment have transformed the work relationship, which is in consequence less oppositional and less likely to generate collective 'class' identities. These kinds of managerial techniques, it is argued, are also being applied to non-service employment, with a corresponding decline in the salience of 'class' (i.e. employment) identities.

THE 'WORK'–'NON-WORK' BOUNDARY

Here, it has been argued that the development of service work has led to a blurring of the boundary between producers and consumers as individuals. In respect of employment in retail, Du Gay, for example, has argued that

> In contemporary British retailing there is no longer any room for the base/superstructure dichotomy. As the 'economic' folds seamlessly into the 'cultural', distinctions between 'production', 'consumption' and 'everyday' life become less clear cut.
> (1993: 582–583)

It is being suggested here that the dissolving of the production/consumption boundary means that it is not possible anymore to identify specifically 'class' practices and/or identities in 'reflective modernity' or 'post-modernism', given that the 'producer' (i.e. class) cannot be independently specified (the parallels with Lash and Urry (1994) are evident here).

The observation that particular kinds of employment actively incorporate the personal commitment – indeed, the personality – of the employee is not a new one. This is particularly likely to be the case in respect of interactive service employment, which frequently draws upon roles developed in a non-market context – for example, the work of caring is widely regarded as 'women's work'. In such cases, the employer is not purchasing an undifferentiated capacity to 'labour', but is able to draw upon skills and capacities not usually associated with 'work' as employment. It is often a feature of such work situations that despite the fact that they may be materially 'exploited', the employees in question gain considerable satisfaction from their work (Crompton and Sanderson, 1990).

It does not follow, however, that this blurring of work and non-work identities means that the producer/consumer boundary has thereby been erased. It is well established in economic anthropology and sociology (Davis, 1985) that *social* categories are intrinsically bound up with the operation of (supposedly) rational (or asocial) markets. Indeed, it might be suggested that this fundamental argument constitutes the major point of differentiation between sociology and economics (Ingham, 1996). That is, 'the producer' is socially constructed by and in both the context of employment as well as by employment itself. In short, it is being argued here that the fuzzy and problematic boundary identified by Du Gay has long been recognized in respect of an extensive range of 'work' – indeed, of 'work' in general – and the case remains to be established whether contemporary developments have led to significant changes in this respect.

WORK ORGANIZATION AND CONTROL

The second feature of service work that has been argued to result in the erosion of the links between employment and class identity and action relates to strategies of work organization and control. Here we will focus on recent changes in the nature of employment organization following from the development of sophisticated technological and behavioural techniques. Recent commentaries have suggested that the 'carrying over' of the self into employment discussed briefly above has been consciously ratcheted upwards, as increasingly, employees are trained or persuaded to transform themselves in order to carry out their jobs. With the expansion of mass service provision and the increasing importance of 'emotional labour' (Hochschild, 1983), complex scripts governing interpersonal interactions have to be acquired. This kind of training extends along the full range of the occupational hierarchy, as even the lowest level of employee in the McDonald's hamburger chain is instructed in how to 'treat every customer as an individual in sixty seconds or less' (Leidner, 1993). Such employment is increasingly incorporating a level of *self*-involvement on the part of the employee which would seem to cut directly across the kinds of conflictual and repressive work relationships implied by Braverman and much of the 'labour process' school.

As in the case of the transferable nature of employment and non-employment roles which we have just discussed, it may also be suggested that the observation that contemporary management techniques directly incorporate the identity of the employee is of long standing. In 1951, Mills identified the 'personality market' in respect of sales, and at the same time Riesman (1961: 264) wrote of 'false personalization' – the 'spurious and effortful glad hand' – as a major barrier to autonomy in the sphere of work. Both of these writers were critical of the manipulation of identities within employment, which they saw as an essentially bogus exercise. These kinds of analyses may be made in respect of today's managerial strategies. However, it has also been suggested that the kinds of personality adjustments required of sales and interactive service workers also be developed and applied to *non-service* employees.

The quality of personal interaction has become increasingly important in all areas of work, including manufacturing, with the emphasis on positive customer relations as one of the ways to organizational success, and the improvements in production that may be gained through workforce 'empowerment', Total Quality Management (TQM), and so on. Firms compete via their services to customers, teams within firms compete via their services to other teams as the service relationship becomes generalized through the process of production itself. 'Performance' is becoming central to the work process, and increasingly, performance monitoring (made more effective as a consequence of IT developments) is becoming a major form of control over the workforce. Empirical investigations such as the Workplace Industrial Relations Survey have recorded these developments through such factors as the increase in Performance Related Pay.[4] In the language of the 1970s debates on the labour process, therefore, many commentators would seem to agree that there has been a growth in the conscious development of 'responsible autonomy', rather than 'direct control', as a means of controlling the labour process. Frenkel *et al.* (1995: 774) have described this as 'info-normative' control, that is 'control based on data objectification (performance indicators) and employee accommodation or commitment to performance standards'.

In summary, it may be argued that the shift to people-centred work as a consequence of the expansion of service employment, together with recent developments in the workplace relating to the organization and control of employees, are making it increasingly less likely that the experience of work will generate the consciousness of solidarity and cohesion with other employees which has been conventionally described as the identity of 'class'. It may be objected that workers will not find it difficult to see through these cynical attempts at manipulation, and that in essence, the purchase of a worker's personality is no different from the purchase of their labour or skills. However, as Leidner (1993) has emphasized, far from experiencing their scripts as manipulation and thus a source of resentment, interactive service workers find them very useful on the job. Similarly, it may also be objected that TQM is in essence a sham, a screen for the intensification of effort and 'downsizing' rather than a genuine attempt to improve the quality of service provision and workplace relationships (Webb, 1996). However, surveys of employees suggest that most workers *do* believe that their work has become more skilled and demanding (Gallie and White, 1993).[5]

In the contemporary employment situation, therefore, the shift towards service and related employment has been accompanied by increasing employment insecurity together with developments in technological monitoring and managerial strategy in which material rewards are becoming increasingly dependent upon individual performance. Even if employees do develop oppositional attitudes to managerial control strategies such as HRM and TQM, developments in the labour market might reduce the likelihood of a collective response. Here the increase in 'non-standard' work, increasing employment insecurity and declining establishment size, all mean that employment has become increasingly likely to generate individualistic, rather than collectivist, sentiments.

NOTES

1 There are also a range of relevant non-consumption practices which act not only as social markers, but also have material consequences. See e.g. Scott (1991) on the social practices of the upper classes.

2 The best-known sociologist to have incorporated Braverman's Marxist account of the labour process within his framework of class analysis is Wright (1989).

3 For example, the issue of May 1996 includes a leading article entitled 'Revenge of the middle manager – job insecurity: make it work for you'.

4 See Millward *et al.*, 1992. The semantic shift in the labelling of individualized pay systems – from 'piece rates' to 'performance-related pay' – is a revealing one, 'piece' referring to the product, whereas 'performance' relates to the individual.

5 The debate on 'skill' as such is not one that is being addressed directly here. One feature which makes it highly problematic is an absence of universally agreed definitions or measures of 'skill'. For example, a recent research interview established that in successive years, 28 per cent and 31 per cent of the employees of a major clearing bank had been downgraded as a consequence of reorganization (their pay would be frozen, rather than reduced, in consequence). However, they were simultaneously being exhorted to acquire selling skills.

REFERENCES

Beatson, M. (1995) 'Progress towards a flexible labour market', *Employment Gazette*, February: 55–66.

Beck, U. (1992) *Risk Society*, London: Sage.

Bourdieu, P. (1984) *Distinction: A Social Critique of the Judgement of Taste*, London/New York: Routledge & Kegan Paul.

Braverman, H. (1974) *Labor and Monopoly Capital*, New York: Monthly Review Press.

Crompton, R. and Sanderson, K. (1990) *Gendered Jobs and Social Change*, London: Unwin Hyman.

Davis, J. (1985) 'Rules not laws: outline of an ethnographic approach to economics', in B. Roberts, D. Gallie, and R. Finnegan, (eds), *New Approaches to Economic Life*, Manchester: Manchester University Press.

Du Gay, P. (1993) 'Numbers and souls: retailing and the de-differentiation of economy and culture', *British Journal of Sociology* 44: 563–587.

Frenkel, S., Korczynski, M., Donoghue, L. and Shire, K. (1995) 'Re-constituting work: trends towards knowledge work and info-normative control', *Work, Employment and Society* 9(4): 774–796.

Gabriel, Y. (1988) *Working Lives in Catering*, London: Routledge.

Gallie, D. and White, P. (1993) *Employee Commitment and the Skills Revolution*, London: Policy Studies Institute.

Goldthorpe, J.H., Lockwood, D., Bechhofer, F. and Plat, J. (1969) *The Affluent Worker in the Class Structure*, Cambridge: Cambridge University Press.

Gregg, P. and Wadsworth, J. (1995) 'A short history of labour turnover, job security and job tenure: 1975–94', *Oxford Review of Economic Policy* III (1): 73–90.

Hochschild, A. (1983) *The Managed Heart*, Berkeley: University of California Press.

Ingham, G.K. (1996) 'Some recent changes in the relationship between economics and sociology', *Cambridge Journal of Economics* 20: 243–275.

Lash, S. and Urry, J. (1994) *Economies of Signs and Space*, London: Sage.

Leidner, R. (1993) *Fast Food Fast Talk*, Los Angeles: University of California Press.

Lockwood, D. (1958, 1989) *The Black Coated Worker*, London: Allen & Unwin, Oxford: OUP.

Lockwood, D. (1966) 'Sources of variation in working-class images of society', *Sociological Review* 14(3): 244–267.

Lockwood, D. (1988) 'The weakest link in the chain', in D. Rose (ed.), *Social Stratification and Economic Change*, London: Unwin Hyman.

Mills, C.W. (1951) *White Collar*, New York: Oxford University Press.

Millward, N. *et al.* (1992) *Workplace Industrial Relations in Transition*, Aldershot: Dartmouth.

Offe, C. (1985) 'Work – a central sociological category?', in C. Offe (ed.), *Disorganized Capitalism*, Cambridge: Polity.

Riesman, D. (1961) *The Lonely Crowd*, New Haven, CT: Yale University Press.

Savage, M., Barlow, J., Dickens, A. and Fielding, T. (1992) *Property, Bureaucracy, and Culture*, London: Routledge.

Scott, J. (1991) *Who Rules Britain?*, Cambridge: Polity.
Veblen, T. (1934) *The Theory of the Leisure Class*, London: Modern Library.
Webb, J. (1996) 'Vocabularies of motive and the "New" Management', *Work, Employment and Society* 10(2): 351–371.
Wright, E. (ed.) (1989) *The Debate on Classes*, London: Verso.

The Body in Consumer Culture

Mike Featherstone

THE PERFORMING SELF

> Today's modern society puts a premium on youth and good looks, in fact *to look better, to look younger*, to *look more attractive* has become a basic need for most of us because *people who look good, are made to feel good*.
>
> (The Lasertone Beauty Therapy Treatment leaflet 1982)

A number of commentators have suggested that a new personality type has emerged in the course of the twentieth century. David Riesman (1950), for example, refers to the replacement of the 'inner-directed' by the other-directed type and Daniel Bell (1976) mentions the eclipse of the puritan by a more hedonistic type. Interest in this new personality type has been sharpened recently by discussions of narcissism (Lasch 1976, 1977b, 1979a). This new narcissistic type of individual, which it is argued has recently come into prominence, is described as 'excessively self-conscious', 'chronically uneasy about his health, afraid of ageing and death', 'constantly searching for flaws and signs of decay', 'eager to get along with others yet unable to make real friendships', 'attempts to sell his self as if his personality was a commodity', 'hungry for emotional experiences', 'haunted by fantasies of omnipotence and eternal youth'. Lasch (1979b, p. 201) argues that the culture of narcissism first took shape in the 1920s, matured in the post-war era and is now rapidly disintegrating.[1] For our purposes the interesting feature of the narcissistic type and the culture of narcissism which spawned it, is that it points to a new relationship between body and self. Within consumer culture, which approximately coincides with the culture of narcissism, the new conception of self which has emerged, which we shall refer to as the 'performing self', places greater emphasis upon appearance, display and the management of impressions.

One indication of the movement towards the performing self can be gleaned from the shift in the self-ideal proclaimed in self-help manuals from the nineteenth century to the early twentieth century. In the nineteenth century, self-help books emphasised the Protestant virtues – industry, thrift, temperance, not just as means but as valid ends in their own right (Lasch 1979a, p. 57). Achievement was measured not against others but against abstract ideals of discipline and self-denial. With the bureaucratisation of the corporate career these virtues gave way to an emphasis upon competition with one's peers, salesmanship, 'boosterism' and the development of 'personal magnetism'.

Warren Susman (1979) has characterised this shift as entailing the replacement of the nineteenth-century concern with character by a new focus upon personality in the early twentieth century. The words most frequently associated with *character* were: citizenship, democracy, duty, work, honour, reputation, morals, integrity and manhood. Locating the transition in the middle of the first decade of the twentieth century, Susman argues that subsequent advice manuals emphasised *personality* and a new set of associated adjectives came into prominence: fascinating, stunning, attractive, magnetic, glowing, masterful, creative, dominant, forceful. In his book *Personality: How to Build It* (1915) H. Laurent remarked: 'character is either good or bad, personality famous or infamous' (Susman 1979, p. 217). A comparison of two books written by O. S. Marden, separated by twenty years at the turn of the century, further illustrates the transition. His *Character: the Greatest Thing In the World* (1899) stressed the ideals of the Christian gentleman: integrity, courage, duty as well as the virtues of hard work and thrift. In 1921 he published *Masterful Personality* which emphasised a set of different virtues: now attention should be given to 'the need to attract and hold friends', 'to compel people to like you', 'personal charm', and women should develop 'fascination'. Good conversation, energy, manners, proper clothes and poise were also deemed necessary (Susman 1979, p. 220).

The new personality handbooks stressed voice control, public speaking, exercise, sound catering habits, a good complexion and grooming and beauty aids – they showed little interest in morals. Susman (1979, p. 221) remarks:

> The social role demanded of all in the new culture of personality was that of performer. Every American was to become a performing self . . . the new stress on enjoyment of life implied that true pleasure could be obtained by making oneself pleasant to others.

Individuals should attempt to develop the skills of actors, a message not just emphasised by self-help manuals, but by advertising and the popular press in the 1920s. Hollywood provided many of the models for the new ideal with stars marketed as 'personalities'. Douglas Fairbanks, the archetype 'personality star', even wrote his own kind of self-help book *Make Life Worthwhile: Laugh and Live* (Susman 1979, p. 233).

Richard Sennett's book *The Fall of Public Man* is interesting in this context because he examines the historical origins of the new belief that appearance and bodily presentation express the self. In the eighteenth century, he suggests, appearance was not regarded as a reflection of the inner self but more playfully distanced from an individual's character which was regarded as fixed at birth. The replacement of this traditional holistic world view by a more 'existentialist' view in which each individual was responsible for the development of his own personality, occurred in the nineteenth century. Following Marx's fetishism of commodities argument Sennett sees the development of the department store in the second half of the nineteenth century as crucial to the process. The department store sold the newly available cheap mass-produced clothing by using increasingly sophisticated techniques of advertising and display. Clothing which indicated a fixed social status came to be avoided and an individual's dress and demeanour came more and more to be taken as an expression of his personality: clothes in the words of Thomas Carlyle became 'emblems of the soul'. Individuals had now to decode both the appearance of others and take pains to manage the impressions they might give off, while moving through the world of strangers. This encouraged greater bodily self-consciousness and self-scrutiny in public life.[2]

The 'performing self' became more widely accepted in the inter-war years with advertising, Hollywood and the popular press legitimating the new ideal for a wider audience. Within consumer culture individuals are asked to become role players and self-consciously monitor their own performance. Appearance, gesture and bodily demeanour become taken as expressions of self, with bodily imperfections and lack of attention carrying penalties in everyday interactions. Individuals therefore become encouraged to search themselves for flaws and signs of decay: as Lasch (1979a, p. 92) remarks:

> All of us, actors and spectators alike, live surrounded by mirrors. In them, we seek reassurance of our capacity to captivate or impress others, anxiously searching out blemishes that might detract from the appearance we intend to project. The advertising industry deliberately encourages the pre-occupation with appearances.

If individuals are required to be 'on stage' all the time, it can lead to what Goffman (1969) has termed as 'bureaucratisation of the spirit', for the performing self must produce an even performance every time. The demands here are no less stringent for professional actors: White (1981) recounts the story of a promising theatre actor from New York who was interviewed for a film part in Hollywood, but was declared a non-starter by the studio after his first interview, without being given a screen test, because he lacked the stylised, off-stage actors' presentation of self which had become mandatory in Hollywood. It is not enough to have the capacity to perform within specific contexts, it becomes essential to be able to project constantly a 'winning image'.

Behind the emphasis upon performance, it can be argued, lies a deeper interest in manipulating the feelings of others. Anthropologists and ethnologists have long been interested in developing theories of non-verbal bodily communication. One offshoot in the post-war era has been the positivist study of body behaviour: kinesics (Kristeva 1978) which seeks to reconstruct the grammar of body language. Paul Ekman, a researcher in this field, has recently catalogued 7,000 facial expressions, which according to his experiments can be used to tell exactly what individuals are feeling. There has also been some interest in the practice of kinesics from the popular press and self-help literature: 'keep a controlling hand in arguments and negotiations', 'decide when the other person is lying', 'interpret gestures of friendliness and flirtation', 'detect boredom', runs the advertising blurb for a popular paperback entitled *How to Read a Person Like a Book*. Another, entitled *Kinesics: The Power of Silent Command*, tells the reader how to learn to 'project unspoken orders that must be obeyed', 'how Silent Command brings you the love and admiration of others'. Arguing that we should try to break through this type of body manipulation by attempting to produce a widespread competence in body language, Benthall (1976, p. 92) writes:

> The body as a whole is still a repressed element in our culture, we tend to *believe* (or find it hard to disbelieve) the sincerity of the politician when he looks us straight in the eye over the TV screen, or that of the actress whose flashing teeth urge us to buy her brand of toothpaste. In both cases a verbal message is lent considerable persuasiveness by the controlled use of certain tricks of bodily deportment, which work largely at an unconscious rather than a conscious level. Until we become more aware of the body's power and resourcefulness, we will not feel a sufficiently educated outrage against its manipulation and exploitation. Rather than campaigns for literacy or numeracy, we may need a campaign for corporacy.

The performing self has also gained impetus from the institutional changes which have brought about the rise of the managerial-professional middle class. One effect of the bureaucratisation of industry and the growth of bureaucratic administrative organisations has been to undermine the bourgeois achievement ideology so that there are more and more areas of work in which the precise evaluation of an individual's achievement on universalistic criteria becomes impossible. Hence 'extra-functional elements of professional roles become more and more important for conferring occupational status' (Habermas 1976, p. 81). The difficulty of evaluating an individual's competence on strictly rational criteria opens up the space for the performing self, schooled in public relations techniques, who is aware that the secret of success lies in the projection of a successful image. In the dense interpersonal environment of modern bureaucracy, individuals depend upon their ability to negotiate interactions on the basis of 'personality'. Impression management, style, panache and careful bodily presentation therefore become important.

It has also been argued that this type of individual has been furthered by the growth of the 'helping professions' which have expanded by discrediting traditional mores and family-centred remedies in favour of a new ideology of health, based upon therapy, human growth and scientism (Lasch 1977a). In education, social work, health education, marriage guidance, probation, the helping professionals have not only been able to develop careers based upon interpersonal skills but have also transferred and imposed the new modes of emotional and relational management on to their clients (de Swaan 1981, p 375). Social relations take on a veneer of informality and equality, but actually demand greater discipline and self-control as management through command gives way to management by negotiation. The 'negotiating self' is also granted legitimation outside the work sphere as the new styles of social interaction spread into family life not only through the direct intervention of experts but also through the feature articles, advice pages and problems programmes of the popular media (Hepworth and Featherstone 1982; Ehrenreich and English 1979). In effect the professional-managerial middle class, which expanded in the course of the twentieth century, are in the process of becoming 'the arbiter of contemporary lifestyles and opinions' (de Swaan 1981, p. 375).

The tendency towards narcissism, the negotiating, performing self is therefore most noticeable in the professional-managerial middle class who have both the time and money to engage in lifestyle activities and the cultivation of the persona. It is arguably spreading to sectors of the working class (Dreitzel 1977) and up the age scale to the middle-aged (Hepworth and Featherstone 1982; Featherstone and Hepworth 1982). This is not to suggest that the implications of the consumer culture imagery of the body and the performing self do not encounter resistance: groups like the Grey Panthers and the Women's Movement have mounted a strong (if as yet ineffectual) critique of 'ageism' and 'sexism'. While pockets of working-class culture clearly remain, it has been suggested that the working class increasingly draw upon the media as a source of identity models (Davis 1979). Consumer culture imagery and advertising cannot be dismissed as merely 'entertainment', something which individuals do not take seriously. In rejecting this position and its obverse, that individuals are somehow programmed to accept essentially false wants and needs, we can indicate two broad levels on which consumer culture operates: (1) it provides a multiplicity of images designed to stimulate needs and desires, (2) it is based on and helps to change the material arrangements of social space and hence the nature of social interactions. Taking the interactional level first, it can be argued that changes in the material fabric of everyday life have involved a restructuring of social space (e.g. new shopping centres, the beach, the modern pub) which provides an environment facilitating the display of the body. Individuals may of course choose to ignore or neglect their appearance and refuse to cultivate a performing self, yet if they do so they must be prepared to face the implications of this choice within social encounters.

Finally, with regard to the proliferation of images which daily assault the individual within consumer culture, it should be emphasised again that these images do not merely serve to stimulate false needs fostered on to the individual. Part of the strength of consumer culture comes from its ability to harness and channel genuine bodily needs and desires, albeit that it presents them within a form which makes their realisation dubious. The desire for health, longevity, sexual fulfilment, youth and beauty represents a reified entrapment of transhistorical human longings within distorted forms. Yet in a time of diminished economic growth, permanent inflation and shortages of raw materials the contradictions within the consumer culture values become more blatant, not only for those who are excluded – the old, unemployed, low paid – but also for those who participate most actively and experience more directly the gap between the promise of the imagery and the exigencies of everyday life.

NOTES

1 A number of commentators have criticised Lasch's periodisation. Oestereicher (1979) sees the narcissistic as merely a continuation of the inner-directed individualistic self. Wrong (1979, p. 310) and Narr (1980, p. 68) criticise the vagueness of Lasch's periodisation. Lasch (1979b) has since replied to his critics and attempted to clarify this issue. It is worth adding that Lasch (1977b, 1979a) has referred to the writings of Stuart Ewen (1976) linking together the rise of consumer culture in the 1920s with the growth of narcissism.

2 While Sennett traces the origins of the new personality structure back to the 1860s he argues that it became more noticeable in the 1890s revolt against Victorian sobriety and prudery. This brought into prominence tighter-fitting, more colourful clothes for women with a more widespread use of makeup – discreetly advertised in women's magazines (Sennett 1976, p. 190). Lasch and Susman both locate the transition from character to personality as occurring around the turn of the century.

REFERENCES

Bell, D. (1976) *Cultural Contradictions of Capitalism*, London: Heinemann.

Benthall, J. (1976) *The Body Electric: Patterns of Western Industrial Culture*, London: Thames & Hudson.

Davis, H. (1979) *Beyond Class Images*, London: Croom Helm.

Dreitzel, P. (1977) The Politics of Culture, in N. Birnbaum (ed.) *Beyond the Crisis*, New York: Oxford University Press

Ehrenreich, B. and English, D. (1979) *For Her Own Good: 150 Years of Experts Advice to Women*, London: Pluto Press.

Ewen, S. (1976) *Captains of Consciousness: Advertising and the Social Roots of the Consumer Culture*, New York: McGraw-Hill.

Featherstone, M. and Hepworth, M. (1982) Ageing and Inequality: Consumer Culture and the New Middle Age, in D. Robbins *et al.* (eds) *Rethinking Social Inequality*, London: Gower.

Goffman, E. (1969) *The Presentation of Self in Everyday Life*, London: Allen Lane.

Habermas, J. (1976) *Legitimation Crisis*, London: Heinemann.

Hepworth, M. and Featherstone, M. (1982) *Surviving Middle Age*, Oxford: Blackwell.

Kristeva, J. (1978) Gesture: Practice or Communication, in T. Polhemus (ed.) *Social Aspects of the Human Body*, Harmondsworth: Penguin.

Lasch, C. (1976) *The Narcissistic Society*, New York Review of Books.
Lasch, C. (1977a) *Haven in a Heartless World: the Family Besieged*, New York: Basic Books.
Lasch, C. (1977b) The Narcissistic Personality of Our Time, *Partisan Review*.
Lasch, C. (1979a) *The Culture of Narcissism*, New York: Norton.
Lasch, C. (1979b) Politics and Social Theory: A Reply to the Critics, *Salmagundi*, 46.
Narr, W. D. (1980) The Selling of Narcissism, *Dialectical Anthropology*, 5, 1.
Oestereicher, E. (1979) The Privatisation of the Self in Modern Society, *Social Research*, 46, 3.
Riesman, D. (1950) *The Lonely Crowd: A Study of the Changing American Character*, London: Yale University Press.
Sennett, R. (1976) *The Fall of Public Man*, Cambridge: Cambridge University Press.
Susman, W. (1979) Personality and the Making of Twentieth Century Culture, in J. Higham and P. K. Conkin (eds) *New Directions in American Intellectual History*, Baltimore: Johns Hopkins University Press.
Swaan, A. de (1981) The Politics of Agoraphobia, *Theory and Society*, 10, 3.
White, E. (1981) *States of Desire*, New York: Bantam.
Wrong, D. (1979) Bourgeois Values, No Bourgeoisie: The Cultural Criticism of Christopher Lasch, *Dissent*.

Commerce and Femininity

Rachel Bowlby

Psychoanalysis was not the only enterprise around the turn of the century to be interested in the answer to Freud's famous question, 'What does a woman want?' Women's desires and the objects of their investment were of the greatest interest and profit to the respected company Stuart Ewen dubs the 'captains of consciousness'.[1] His phrase is intended to evoke the transformation of business concerns from production to consumption – from concentration on the manufacture of goods under the management of the nineteenth-century captains of industry to the manufacture of minds disposed to buy them. The later captains of American industry, Ewen argues, were engaged in a deliberate endeavor to create the needs among potential buyers which would ensure the selling of the increasing quantities and types of commodity which their ever more efficient and productive factories were turning out. The use of distinctive brand names, display techniques and other means of advertising implied new methods of marketing aimed at selling the 'image' of a product along with, or as part of, the thing itself.

The shift in the late nineteenth and early twentieth centuries to what is now conventionally known as 'consumer society' was common to the countries of Dreiser, Gissing and Zola. Britain, France and the United States had the three highest per caput GNPs in the world at the beginning of the twentieth century; together with Germany, they were by far the most developed countries in terms of the scale of industrialization and their reliance on cheap raw materials imported from colonies. Despite individual variations in specific economic and cultural histories (American immigration, then at its height, clearly inflected developments in ways that did not apply to the European countries; France was far less affected by industrialization and urbanization in the nineteenth century than was Britain, and so on), it is reasonable to consider these states together, as occupiers of parallel positions in the world system, with cultural and ideological forms that can usefully be compared.[2]

As the proportion and volume of goods sold in stores rather than produced in the home increased, it was women, rather than men, who tended to have the job of purchasing them. Even though, particularly in the United States, large numbers of women were themselves beginning to enter the industrial wage-earning force, they also performed the services of housework and shopping for the home. More significant still, middle- and upper-class ladies were occupied with the beautification of both their homes and their own persons. The superfluous, frivolous associations of some of the new commodities, and the establishment of convenient stores that were both enticing and respectable, made shopping itself a new feminine leisure activity. On both counts – women's purchasing responsibilities and the availability of some of them for extra excursions into luxury – it follows that the organized effort of 'producers' to sell to 'consumers' would to a large measure take the form of a masculine appeal to women.

It does not matter in this regard how far individual capitalists or admen, or the fraternity as a whole, were conscious of this as deliberate strategy (though Zola's *Au Bonheur* will present evidence to suggest that they certainly were in some cases); nor does it matter whether the modification of buying habits necessarily had the quality of intentional manipulation. The essential point is that the making of willing consumers readily fitted into the available ideological paradigm of a seduction of women by men, in which women would be addressed as yielding objects to the powerful male subject forming, and informing them of, their desires. The success of the capitalist sales project rested on the passive acceptance or complicity of its would-be buyers, and neither side of the developing relationship can be thought independently of the other.

On the other hand, women's consumption could be advocated unequivocally as a means towards the easing of their domestic lot and a token of growing emancipation. Elizabeth Cady Stanton punctuated her lectures of the 1850s with exemplary tales like the one about 'The congressman's wife,' an unfortunate soul possessed of an ill-equipped kitchen and a husband who complained about her cooking. Mrs Stanton's advice to her was:

> Go out and buy a new stove! Buy what you need! Buy while he's in Washington! When he returns and flies into a rage, you sit in a corner and weep. That will soften him! Then, when he tastes his food from the new stove, he will know you did the wise thing. When he sees you so much fresher, happier in your new kitchen, he will be delighted and the bills will be paid. I repeat – GO OUT AND BUY![3]

This apparently foolproof recipe – his fury, your tears, a nice meal, a 'fresher, happier' you, then finally his conversion – is an interesting lesson in that much-vaunted nineteenth-century feminine power of 'influence.' Significantly, the injunction to buy comes from woman to woman, not from a man, and involves first bypassing and then mollifying a male authority. To 'go out' and buy invokes a relative emancipation in women's active role as consumers. Given the assignment of women to the domestic sphere, shopping did take them out of the house to downtown areas formerly out of bounds, and labor-saving equipment could make home work more manageable.[4]

At first sight there is a distinction between consumption for use and luxury consumption. A loaf of bread, in a society where bread is part of the staple diet, can hardly be considered an object of wishful desire on the part of its purchasers; a stove facilitates the performance of a socially necessary task. Women's relation to the materials of cooking and housework might, in fact, be the same as that of laborers or artisans to the tools required for their work.

Recent debates in Marxist and feminist theory, usually taking as their starting point Engels' *Origin of the Family, Private Property and the State* (1884), have addressed the issue of the ambivalent position of women in terms of social class and the relations of production. Engels argued that women's position in the modern family was analogous to that of the worker under capitalism, with the husband/father in the position of the capitalist exploiter of labor power. The theory referred to above, according to which housewives are equivalent to workers, stresses from another angle that homes perform the function of reproducing labor power: they are part of the infrastructure that maintains the male breadwinners as operative in their jobs.

But differences in the roles fulfilled by working-class and bourgeois wives – the unpaid domestic servant and the lady of leisure – immediately put in question the priority of gender to class in determining the social position of women. This can be identified either with that of the family wage-earner or with that of all women, considered as finally subject to patriarchal domination, whether their function be primarily industrious or ornamental. Women's ambivalent status is intensified by the consideration that large numbers of them became (low-paid) members of the workforce during this period. This made them on the one hand doubly exploited, contributing to capital's surplus at both labor and retail ends; and on the other, divided between the roles of masculine wage-earner and feminine housekeeper and consumer.

In a lucid exposition of the complexities of contemporary debates, Michèle Barrett concludes that it is neither possible nor useful to look for a simple formulation that would account for the place of 'woman' as a generic or homogeneous social category.[5] Rather, the various, analytical perspectives demanded by the different and often contradictory parts played by women in the society, and by the concept of 'woman' in ideological constructions, must be used and examined in conjunction. There are risks involved in easy assimilations which can only mask the problematic and inconsistent nature of the various social and ideological positions woman/women may occupy in given social formations and in different historical periods.

No more straightforward is the status of those objects which women consume. A loaf of bread always provides a given quantity of nutrition, just as a vacuum cleaner presumably gets the cleaning done. But a selling process which involves competition between different brands of bread is necessarily engaged in presenting one particular type of loaf as superior to or different from its competitors. In the post-sliced-bread era, a loaf may be marketed as being uniquely nutritious; as containing some special ingredient as the natural choice of the decent housewife ('Mother's Pride'); or as particularly good for slimmers (an interesting, inversion of its original use, which can claim the reduction of calorific content, the amount of energy-producing food, as a positive quality). Clearly, if bread was not a source of food, no loaves would ever have been baked or sold. But the values attributed and added to it, whether in actual changes of substance or in the mental associations of the name or the image 'sold' with the bread, raise it, with yeast of magical properties, to a status that exceeds (though it may still include) the functional. Need and kneading go together only up to a certain point. [. . .]

Baudrillard thus makes the argument that, in the same way that there is no nature in the codes of consumer society, but only the idea of naturalness, so there is no such thing as a person, but only the code or cult of 'personality' or 'personalization.' This is constituted through the valorization of an authentic subjectivity actually acquired only through the objects which one owns and the habits one cultivates out of a limited set of options. Uniqueness has nothing to do with anything like an original character, just as in French the word *personne* is both 'someone' and 'no one.' The boundaries of subject and object, active and passive, owner and owned, unique and general, break down in this endless reflexive interplay of consumer and consumed. One consequence is that the clear separation of masculine and feminine roles as applied to the consumer/commodity relation cannot be maintained. It would seem, in fact, that in the priority of commodities to persons, the feminine commodity is in the dominant position – though the consequences of this for female consumers are evident.

The notion of 'image' is useful for thinking about consumer forms of subjectivity. Significantly, it is a visual word harking back to the myth of Narcissus frustratedly gazing at himself in the pool. In Ovid's version of the story, recounted in the *Metamorphoses*, *imago* is the word used for the beloved reflection. In modern society, the image has other concrete and specific forms related to, but different from, the simple reflexive mirror relationship of self and self-image. Photography, cinema, billboard advertising were all being developed and coming to permeate social life in the period at the turn of the century. The photographic medium enabled a form of exact representation of places, people and things; in the multiple uses to which it was put, it both indicated and helped to promote a desire and willingness on the part of society to look at images of itself, collectively and individually – to see its own image reflected or refracted back through the technological medium.[6]

Narcissus' tragedy is that he cannot free himself from the image with which he has fallen in love, which he wishes to grasp and possess and know (the Latin *comprendere* includes all three meanings), but cannot recognize as being only a derivative reflection of his own body. He is seduced by, and wants to seduce, something which is both the same as and different from himself, something both real and unreal: there to be seen but not tangible as a substantial, other body. It is an ideal image in which he sees nothing to threaten an unquestioning love. Narcissus is fatally caught inside a trap of attraction which he does not see to be of his own making, moving according to his own movements. The consumer is equally hooked on images which s/he takes for her own identity, but does not recognize as *not* of her own making.

Freud, writing during the early stages of the consumer period of capitalism, found in the Narcissus myth an apt evocation of one of the constitutive stages in the formation of human subjectivity, figuring the ego in its initial attachment to and identification with an ideal and all-fulfilling image both separate from and an extension of itself. The narcissistic stage is chronologically and structurally prior to the socialization of the child, who moves beyond the dyadic reflexivity of relation to the mother's body and his or her own potentially satisfying image, into the rules and practices of social convention. These are experienced as a limit to the child's omnipotence and self-sufficiency, but are ultimately internalized as the superego. In psychoanalytic terms, the impetus for this development is provided by the threat of castration. While boys can eventually respond to this by internalizing an active, moral identity modeled on their father's, girls must come to terms with the fact that they are already castrated, lacking the male organ and what it represents. If they do not, they are engaged in a futile attempt to take on the functions of a masculine subjectivity not their own. Hence the tendency of women to remain closer to the narcissism of childhood and outside the arenas of public achievement.

The point has often been made that Freud's account of female subjectivity is overtly male-centred because of the significance attached to possessing the power associated with the phallus. His portrayal of women as defective males can be taken to demonstrate Freud's misogyny, considered either as a personal warp or, more generally, as a prejudice typical of his time and social class. Alternatively, the male bias of the work can be seen as a legitimate reflection of the hierarchy of sexual difference as it is in modern patriarchal society. As an interpretation of the implicit form of gender relations in Europe at the turn of the century, with more or less pertinence to other societies, Freud's description can then be allowed to stand.[7]

Such a perspective potentially offers a means of using Freud in a more historical way. Freud gives one kind of description of why it might be that women in bourgeois society are less active than men – less likely to leave the domestic sphere and, inversely, more narcissistically absorbed in themselves, their beauty, their desirability as potential objects of male love.[8] The determinants of this pattern are shown to exist within the nuclear family, which consists of a kind of cross-generational variant of the romantic love triangle – a structure which would itself, in Freud's theory, be a repetition in later life of the earlier family situation. Freud does not extensively address the relationship of the family to the social structure of which it is a part. If Freudian theory can be integrated into an understanding of the conditions of bourgeois society, there must be a relation between the passive, pre-social destiny of its anatomically female children, and wider forces governing the forms of familial and social experience in the society at large. The 'family romance' is not the whole story.

Returning, then, to the question of the relation between women and commodities, there is an obvious connection between the figure of the narcissistic woman and the fact of women as consumers. 'What does a woman want?' is a question to which the makers of marketable products from the earliest years of consumer society have sought to suggest an infinite variety of answers, appealing to her wish or need to adorn herself as an object of beauty. The dominant ideology of feminine subjectivity in the late nineteenth century perfectly fitted woman to receive the advances of the seductive commodity offering to enhance her womanly attractions. Seducer and seduced, possessor and possessed of one another, women and commodities flaunt their images at one another in an amorous regard which both extends and reinforces the classical picture of the young girl gazing into the mirror in love with herself. The private, solipsistic fascination of the lady at home in her boudoir, or Narcissus at one with his image in the lake, moves out into the worldly, public allure of *publicité*, the outside solicitations of advertising.

'Just looking': the conventional apology for hesitation before a purchase in the shop expresses also the suspended moment of contemplation before the object for sale – the pause for *reflection* in which it is looked at in terms of how it would look on the looker. Consumer culture transforms the narcissistic mirror into a shop window, the *glass* which reflects an idealized image of the woman (or man) who stands before it, in the form of the model she could buy or become. Through the glass, the woman sees what she wants and what she wants to be.[9]

As both barrier and transparent substance, representing freedom of view joined to suspension of access, the shop window figures an ambivalent, powerful union of distance and desire. Unlike Narcissus' reflection, the model in the window is something both real and other. It offers something more in the form of another, altered self, and one potentially obtainable via the payment of a stipulated price. But it also, by the same token, constitutes the looker as lacking, as being without 'what it takes'. Next year, or next door, at the superior establishment, the fashion will be different: the longing and lacking of the consumer are limitless, producing an insatiable interplay between deprivation and desire, between what the woman is (not) and what she might look like. The window smashes the illusion that there is a meaningful distinction in modern society between illusion and reality, fact and fantasy, fake and genuine images of self.[10]

NOTES

1 Stuart Ewen (1975) *Captains of Consciousness: Advertising and the Social Roots of the Consumer Culture*, New York, McGraw-Hill.

2 For further information see Daniel Chirot, *Social Change in the Twentieth Century* (1997, New York, Harcourt Brace Jovanovich). Accounts of the development of consumer capitalism in each country can be found in Dorothy Davis, *A History of Shopping* (1966, London, Routledge & Kegan Paul); Peter d'A. Jones, *The Consumer Society: A History of American Capitalism* (1965, Harmondsworth, Penguin) and Étienne Thil, *Les inventeurs du commerce moderne* (1966, Paris, Arthaud).

3 Quoted in Lloyd Wendt and Herman Kogan, *Give the Lady What She Wants* (1952, Chicago, Rand McNally), 29.

4 In addition, the proliferation of gadgets marketed for the speedier performance of previously undreamt-of culinary complexities or household chores can be seen as an

effective abolition of the distinction between necessary and superfluous domestic tasks.

5 Michèle Barrett, *Women's Oppression Today: Problems in Marxist Feminist Analysis* (1980, London, Verso).

6 On the social effects of photography and its offshoots, see Benjamin, 'A short history of photography,' Walter Benjamin, 'A short history of photography,' trans. Stanley Mitchell, *Screen* 13(1), 1972, 5–26.

7 The intention here is not to debate the vexed question of how far Freud's theories can be applied to periods other than his own, but to place his preoccupation with questions of pleasure, subjectivity and gender identity in the context of some other contemporary discourses and social forms.

8 See in particular 'On Narcissism: An Introduction' (1914), vol. XIV, pp. 73–102 in the *Complete Psychological Works, Standard Edition*, 24 vols, (trans. James Strachey, 1953–1974, London, Hogarth Press). Key texts on the development of femininity are: 'Some Psychical Consequences of the Anatomical Distinction Between the Sexes' (1925), vol. XIX; 'Female Sexuality' (1931), vol. XXI and 'Femininity' (1933), vol. XXII.

9 The reading given here is mediated by Lacan's early essay, translated as 'The Mirror Stage as formative of the function of the I' in *Écrits*, pp. 1–7 (trans. Alan Sheridan, 1976, London, Tavistock). Psychoanalytic accounts concentrate on the determining effects of general cultural structures and early family relationships rather than on the way that particular sociohistorical forms inflect the shape of early and later experiences. The concern here is to look at mirrors of identification and separation other than, though related to, those that the baby sees: to suggest how available imaginary identifications in the late nineteenth century were open to historically new forms of (ad)dress or *interpellation*, to use Louis Althusser's term.

10 It was assuredly no accident when British suffragettes spent the afternoon of 1 March 1912 systematically smashing the windows of famous London stores such as Liberty's, Marshall and Snelgrove, and Swan and Edgar. Signifying possibly irreparable cracks in bourgeois ideology, 'the argument of the broken window pane,' as George Dangerfield put it, quite properly appeared to the man in the street as rather 'unseemly outbreaks.' See Dangerfield, *The Strange Death of Liberal England* (1935, New York, Harrison Smith and Robert Haas).

23

The Three-Body Problem and the End of the World

Hillel Schwartz

HESPERIDEAN ESTATES

Our first haven is the department store. There, during the last half of the nineteenth century, just as 'overnutrition' ballooned into a significant disorder, another disorder was making a name for itself. The two disorders had more than a little in common. They were remarkably easy to detect, remarkably difficult to control. They were both caught up in the perplexing folds of an industrial abundance. They were diseases of desire and consumption, of impulse and will.[1]

The department store was a feast for the eyes and the hands. Most things lay open to view and to touch. Mirrors and cornucopia moldings on walls and ceilings did not so much deceive the senses as expand the horizons of desire. With this bazaar of riches clearly marked at fixed prices, the shopper was disengaged from the rawness of the open market and the pressures of the attentive small retailer. She was left to flow through aisles toward the finer things in life, or she was lifted from beneath by elevators rising toward the higher goods.[2]

So seductive that 'even the hand of the strong-minded instinctively grasps her pocket-book before the bewitching array,' department-store counters became the site for a new class of shoplifting where the taking, not the owning, seemed paramount. By the 1870s, this impulsive theft, this taking-without-needing, had a popular name, 'kleptomania,' which referred in particular to that department-store syndrome of stereotypically well-to-do middle-aged women who stole trivial items they could easily afford. Kleptomania was as clear a case of overnutrition as contemporaries could wish for, outside the dining room. It was not simply coincident with concerns about obesity; it was a cultural correlate.[3]

Baptized *klopémanie* in 1816, renamed in successive decades 'cleptomania,' *kleptomanie* and 'kleptomania,' the disorder saw the light of day as one of a new category of 'chronic cerebral affections' known as monomanias and characterized by 'a partial lesion of the intelligence, affections or will.' Like pyromania, kleptomania was a lesion of the will, therefore an *instinctive monomania*, according to the French physician Jean Etienne Esquirol, who described such a monomaniac as the sort of patient 'who is drawn away from his accustomed course to commit acts . . . which conscience rebukes and which the will has no longer the power to control. The actions are involuntary, instinctive, irresistible.'[4]

Critical to the diagnosis of this monomania was the prevailing assumption that a monomaniac aside from his (or her) single mania was not only physically healthy but discernibly righteous. 'I know an instance of a woman, who was exemplary in her obedience to every command of the moral law, except one,' wrote American physician Benjamin Rush in 1786. 'She could not refrain from stealing. What made this vice the more remarkable was that she was in easy circumstances, and not addicted to extravagance in any thing.' Rush had not at hand (of course) the later terms, but this hardly meant that monomaniacs in general, or kleptomaniacs more specifically, had held their manias at bay until the turn of the century. As with the new designation of any disease, there was much retrospective diagnosis of kleptomania when the monomanias became certifiable between 1825 and 1860.[5]

'Certifiable' – because monomania, or partial insanity, had legal implications. Partial insanity had been used as a defense in a case of repeated petty theft as early as 1815. Beau-Conseil (!), ex-commissioner of police in Toulouse, argued that he should not be held responsible for his admitted crimes because he was a virtuous man suffering from an indomitable propensity to steal. Subsequent cases would follow this line of reasoning, which was legally curious, since the defense would be engaged in highlighting a characterological inconsistency such that the crime(s) became proof of the disease. The sentence against a French woman convicted in 1844 of three petty thefts was overturned when a physician showed that the woman's entire character changed while she was stealing gloves, ribbons, cloth, brooches: normally she was a calm, reasonable, economical housewife and mother, but when she was in the kleptomaniac state, she was agitated, bitter, profligate and a lover of vegetable soups.[6]

The point was that the crimes – the thefts – were committed despite oneself, and apparently with no other motive than, ultimately, to spite oneself. The purloining was neither polished nor profitable; the loot was either hoarded or given away most innocently as gifts, and in either case immediately forgotten. Kleptomaniacs were fiendishly but randomly furtive, hence easily exposed, as if compelled to draw public shame upon their otherwise upstanding lives. It was, if not absolute insanity, a kind of moral insanity.

This was the term implied in the first important case of kleptomania cited in American medical jurisprudence.[7] Charles Sprague, twenty-five, at eight in the morning of an August day in 1848 in Brooklyn, had run up behind Sarah Watson, thrown her down, wrestled a shoe off one of her feet, and run away. The shoe was found dangling from Sprague's coat pocket the next morning at the printing office where he worked as a journeyman. Charles made no effort to conceal the shoe or explain its presence.

In court, his father, the Congregationalist minister Isaac N. Sprague, testified that Charles had been acting strangely for some ten years, ever since he had fallen from a second-storey balcony. Women in the family had begun to miss some of their shoes (always singletons) and these were usually to be found under his son's pillow, between the layers of his bed or in his pocket. Charles would deny the thefts, would be punished, but 'he seemed not to have a memory of the fact,' and, the Reverend added, his son's 'moral sense seemed to be somewhat blunted,' though only in respect to these crimes. After testimony from two physicians who defended the concept and relevance of monomania or partial insanity in the case of Charles Sprague, the judge told the jury that 'the peculiarity of his insanity consisted in what appears to the sane mind an objectless desire to possess himself of the shoes of females, and to hide and spoil them.' The jury soon returned a verdict of 'Not Guilty.'

This case deserves an essay of its own, but whether we might now put forward other diagnoses, for English-speaking courts in the mid-nineteenth century, *People v. Sprague* began to shape the legal and cultural perceptions of kleptomania. Even in its youth kleptomania was far more than the propensity for inconsequential thefts by well-heeled individuals. It entailed also a 'vacancy of the eye,' a hint of fetishism, a selective amnesia, an erratic furtiveness, all indicative of a dramatic if temporary change of character.

Medical historians may here accuse me of willfully and woefully confusing partial insanity with monomania, monomania with moral insanity, moral insanity with fetishism, and so on. The confusions, however, were endemic to Victorian psychiatry, especially with regard to kleptomania, since petty theft itself, unattended by other evidence of mental or physical disorder, was often acknowledged as the first and premonitory symptom of full-fledged mania or, worse yet, something called general paralysis.[8] The confusions were also popular confusions, for as kleptomania moved in its maturity to the Hesperidean Estates, inexplicable petty thefts could appear as 'the impulse of a diseased imagination,'[9] a transient failure of will or an abrupt and fugitive moral delinquency – each in its turn subtly promoted by department-store designers.

Indeed, before and beyond the end of the century, it was common for critics to blame department stores for the concurrent epidemics of shoplifting and kleptomania. These large stores with their novelties, open counters, bargain sales and lavish appointments, encouraged the sudden impulse, exploited the confusion of need and desire, sapped the will, wore probity down to its white knuckle. The stores were worlds apart, 'dream worlds,' where shoppers with vacant eyes might fall in love with things, where time might be forgotten, where a woman of means might find sanctuary in changing rooms, restrooms, tearooms. And the stores made every effort to obscure the boundaries between the utilitarian and the luxurious, the vital and the trivial.[10]

No wonder women stole. Men too, but mostly women, and not simply because women did the shopping. Department stores played on women's innate weaknesses: their delight in the senses, their daydreaminess, their somehow-glandular cravings for the exotic, most obvious during their monthly periods and their pregnancies. Women had been known to steal at such times, usually shiny things, baubles, trinkets, notions. If the department stores were suffused with a certain sensuality, as most meant to be, then they were affording women a larger arena of expression for their natural propinquities toward a trancelike theft.[11]

Trance. During an epoch when spirit mediums and somnambulists toured Europe and North America giving lectures and demonstrations, when séances were parlor games and animal magnetism was giving way to hypnotism, when photographers managed convincing pictures of floating apparitions and invalids spent long afternoons in semidarkness, trance was clearly in the air.[12] Shoplifters were distinguished from kleptomaniacs by purposiveness, directedness, the absence of trance. They were purposive, of course, because they were poor, and whatever temptations they succumbed to were by virtue of neediness rational temptations, therefore culpable. That shoplifting and kleptomania were differentiated less by the theft than by the social circumstances of the thief was noted early on, and is still to this day being pointed up.[13] I need not detail further such injustice; rather, I need to insist that shoplifting and kleptomania were also differentiated by the quality of the thieving act. If it seemed calculated, if it was a collaboration, if it had been repeated exactly in other stores, if the escape route demanded forethought, then the theft was shoplifting; if it seemed impulsive, if it was solitary, if it followed no plan, if the woman continued aimlessly about the store, then the theft was kleptomania. Shoplifters in their poverty (or their greed, or their criminal slyness) were always clearheaded; innocent kleptomaniacs were momentarily in a trance.[14]

Impulse and trance rarely go together, one would think, but from the start kleptomania had been fixed by paradox: the person of means who steals trifles – the king who takes spoons, the chevalier who takes napkins, the philanthropist who takes candy. It occurred to some observers around the turn of the century that a third paradox might also be at work: the housewife educated to domestic thrift who steals patently impractical kickshaws. Such women, wrote the superintendent of a large department store in 1915, were made thieves by their husbands, who granted them little spending money of their own. After arguments over household budgets, after pleas for pin money, such a woman would come to the department store feeling guilty, resentful, overwrought. Then, seeing this gewgaw, that bright bagatelle, she became 'like the child who wished for the moon – and almost instinctively – she reached for it.'[15]

More radical observers put this reaching for the moon within the context of the capitalist drive to acquire, the industrialist pattern of personal thrift and economic expansion – avenues open to women only by proxy in the midst of the commercial splendor of the department store. 'It seemed to me as if everything belonged to me,' the kleptomaniac was supposed to confess; 'I might have taken all.' Recently historians have emphasized that the theories of scientific management that underlay the Second Industrial Revolution had their repercussions in terms of domestic economy, and that well-off women imbued with the modern scientific spirit were as likely to have put themselves on severe budgets as factory girls or immigrant wives. Inside that infinitely blessed world of the department store, what they could least justify as a purchase – the fanciful, the frivolous, the evanescent – was what they were most prone to take. Their thefts were at once an unconscious protest against the tyrannies of a male, industrial thrift, and an instant claim to the rightness of their own desires for what was beautiful and phenomenal.[16]

The department store itself profited from paradox, despite the thefts. The large stores operated on the contradictory axioms that no desire is unfulfillable and that no desire can be fully satisfied. Store managers were reconciled to a degree of theft, especially by 'respectable' women; shoplifting had to be as chronic as shopping. Kleptomania was the faithful if furtive handmaiden of abundance.[17]

By the 1880s, kleptomania was a disorder popular enough to be used as a defense in a case of horse-stealing in Texas, familiar enough to be the pivot in an English three-act comedy prevalent enough that a lady's maid could hand calling cards to Boston shopkeepers from whom her mistress had just stolen. People in the year 2000, looking back to 1887, would recall the handling of kleptomaniacs as the sole parallel to their own handling of criminals, since in Edward Bellamy's utopia, crime would lack any rational motive.[18]

By the year 1905 there had been a Minnesota suit asking for divorce on grounds of a wife's kleptomania; a transoceanic ruckus over the trial of a wealthy San Francisco belle caught red-handed with eighteen tortoiseshell combs, seventeen fans, sixteen brooches, seven ivory-framed hand mirrors, as well as sable and chinchilla furs all stolen from London shops; cross-channel speculation about the identity of the English duchess arrested for shoplifting in Paris; secondhand news of German and Russian princesses forced to pay heavy fines for their own thievery; vaudeville skits and songs, including 'Mamie (Don't You Feel Ashamie),' in which Mamie was 'a kleptomamie, kleptomamiac'; and a quite significant film in the history of cinema, Edwin S. Porter's *The Kleptomaniac.*[19]

So much publicity meant that the criteria for a legal and medical judgment of kleptomania were common knowledge in higher social circles and among professional thieves and their respective lawyers. In fact, anyone caught stealing – above all, from department stores – might hope to beg off as a kleptomaniac, unless obviously indigent or disreputable. How a culprit reacted when caught was itself telling evidence for or against a presumption of kleptomania.

There were virtues and advantages, then, to knowing exactly how a kleptomaniac should react. Respectable, moneyed women who failed to respond in a kleptomaniacal manner might well be prosecuted as ordinary thieves. Mary Plunkett, store detective for Macy's in Manhattan, felt compelled to call the law down upon one respectable woman, 'as she was rebellious, reticent about herself, and disposed to be defiant.' Another woman was also respectable, 'but had to be made to feel the weight of the law, because she pleaded innocence and protested and threatened.' The criminal type was brazen, vehement, hard to crack.[20]

The true kleptomaniac, on the other hand, was shocked and penitent. When accosted just outside the store, she began with an indignant denial born of amnesia and surprise, in this format: 'How dare you accuse me of such a thing! I would never do such a thing! Why would I of all people do such a thing?' But soon, back in the privacy of the manager's office, there would be a naive, tearful confession and a fearful shame: 'I have no recollection of taking the articles, I did not mean to take the articles, I had no intention to steal anything, I was in a daze, I did not know what I was doing, an irresistible power prompted me, a sudden impulse, I was not myself, I couldn't help myself, I'll pay for the articles, please do not expose me, my name would be ruined, I could never live it down.'[21]

We may appreciate the popularity of the kleptomaniac confession and the notoriety of the shoplifting shame by glancing at two articles across the front page from each other in the *New York Times* for December 11, 1897. On the right, in column six, 'She Stole to Live,' a homeless woman, neither so young nor so pretty as a morning paper had led the *Times* reporter to believe, confessed to stealing four bottles of whiskey and a jar of olives from S.S. Pierce Company of Boston. In her testimony Marie Wilkins claimed to be dazed by her arrest, so dazed that she gave two different addresses and said, 'I have forgotten everything.' No, she had never stolen before, and she did not know why she had taken the whiskey, as she never drank. Guilty or not, dazed or not, here was a woman desperately trying to lay claim at once to poverty and kleptomania, a heroic and nearly impossible task.

On the left, in column two, 'Rush to See Shoplifters,' was a crowd overflowing from the spectator seats into the dock of the court where four quite respectable women of Lynn, Massachusetts, were to appear on counts of petty theft. Hours before the court opened, some of the most well-known women in town had lined up to catch a glimpse of the accused, who arrived veiled and would shortly be driven away in a closed carriage to the judge's private office.

Given such contexts, we may appreciate as well the thrust of the kleptomaniac confession, which reclaimed innocence by taking the theft first through the daze of shopping, then through the haze of a temporary amnesia, offering it up at last before the altar of an irresistible power. This ritual did not merely purify the sinner, who had taken without needing to take; the ritual shifted responsibility on to something, someone else. It was not, after all, I who did that. So women in their confessions and court cases often gave false names, due in part to fears of a husband's wrath or social disgrace, but also and equally due to a belief that someone else really was the thief.[22]

Who? Wasn't it, really, another self, active, dexterous, emotionally labile, needy and not to be denied? Wasn't it, really, a self without the moral, social and economic restraints of a sober married woman? Wasn't it, really, a younger self, childish, perhaps adolescent, open to the sensual prospects of the department-store world and given to an impulsive grabbing? Wasn't it a rapscallion self, rebellious, theatrical, like the vaudeville Mamie?

She took things so easy, she took things so nice,
People always wondered, she never took advice.
She became an actress, but the show could never start.
For Mamie took a fancy, she took a leading part.[23]

Wasn't it a daring, even arrogant self, one likely to risk total ruin?

NOTES

1 On 'overnutrition' or obesity, see my book, *Never Satisfied: A Cultural History of Diets, Fantasies and Fat* (New York, 1986). Compare also an article that appeared after my book was completed: Keith Walden, 'The Road to Fat City: An Interpretation of the Development of Weight Consciousness in Western Society,' *Historical Reflections* 12 (Fall 1985), pp. 331–373.

2 Rosalind H. Williams, *Dream Worlds: Mass Consumption in Late Nineteenth-Century France* (Berkeley, 1982), pp. 66–71 *passim*; William R. Leach, 'Transformation in a Culture of Consumption: Women and Department Stores, 1890–1925,' *Journal of American History* 71 (1984), pp. 319–342; Susan Benson, *Counter Cultures: Saleswomen, Managers and Customers in American Department Stores 1890–1940* (Urbana, 1986).

3 William R. Leach, *True Love and Perfect Union* (New York, 1980), p. 253, quoting an 1876 article from *New Century*; Elaine Susan Abelson, '"When Ladies Go A-Thieving": The Department Store, Shoplifting and the Contradictions of Consumerism, 1870–1914,' Ph.D. thesis, New York University, 1986.

4 Patricia O'Brien, 'The Kleptomania Diagnosis: Bourgeois Women and Theft in Late Nineteenth-Century France,' *Journal of Social History* 17 (Fall 1983), pp. 65–78; Raymond de Saussure, 'The Influence of the Concept of Monomania on French Medico-Legal Psychiatry (from 1825 to 1840),' *Journal of the History of Medicine and the Allied Sciences* 1 (1946), pp. 365–397, quoting Esquirol's *De la monomanie* (Paris, 1838).

5 Benjamin Rush, 'An Inquiry into the Influence of Physical Causes upon the Moral Faculty (1786),' in his *Medical Inquiries and Observations upon the Diseases of the Mind*, 4th edn (Philadelphia, 1815), vol. 1, pp. 101–102; Franz Josef Gall, *Sur les fonctions du cerveau* (Paris, 1825), vol. 4, pp. 206–222.

6 *Journal de Paris*, March 29, 1816, summarized in an important early work, Isaac Ray, *A Treatise on the Medical Jurisprudence of Insanity*, ed. Winfred Overholser (1838. Reprint: Cambridge, Mass., 1962), p. 142; M.H. Girard, 'Kleptomanie,' *Gazette médicale de Paris*, 2e sér., 13 (1845), pp. 735–737.

7 *People v. Sprague* (1849), in Amasa J. Parker, *Reports of Decisions in Criminal Cases . . . in the Courts of Oyer and Terminer of the State of New York* 2 (New York, 1869), pp. 43–48.

8 See Francis Wharton, *A Monograph on Mental Unsoundness* (Philadelphia, 1855), p. 155; John C. Bucknill and Daniel H. Tuke, *A Manual of Psychological Medicine* (Philadelphia, 1858), pp. 208–211; W. Julius Mickle, 'General Paralysis,' in *A Dictionary of Psychological Medicine*, ed. Daniel H. Tuke (London, 1892), vol. 1, p. 521.

9 Wharton, *Mental Unsoundness*, p. 157, citing *The Times* (London) of April 1855.

10 Abelson, '"When Ladies Go A-Thieving,"' pp. 62–65; Williams, *Dream Worlds*, esp. pp. 64–66; Michael B. Miller, *The Bon Marché: Bourgeois Culture and the Department Store 1869–1920* (Princeton, 1981), esp. pp. 197–206; 'Kleptomania,' *Medico-Legal Journal* 14 (1896), pp. 231–232, extracting a paper by Alexandre Lacassagne; Paul Dubuisson, 'Les Voleurs des grands magasins,' *Archives de l'anthropologie criminelle* 16 (1901), pp. 1–20, 341–370.

11 Alfred Swain Taylor, *Medical Jurisprudence*, 2nd American edn, from 3rd London edn (Philadelphia, 1850), p. 653; Thomas Byrnes, *Professional Criminals of America* (New York, 1886), p. 32; 'Nervous Disorders (Especially Kleptomania) in Women and Pelvic Disease,' *American Journal of Insanity* 53 (1897), pp. 605–606; O'Brien, 'The Kleptomania Diagnosis,' pp. 68, 75 n.13; Arthur B. Reeve, 'The Kleptomaniac,' *Hearst's International Magazine* 22 (December 1912), p. 67.

12 See esp. Logie Barrow, 'Anti-establishment Healing: Spiritualism in Britain,' in *The Church and Healing*, ed. W.J. Sheils (Oxford, 1982), pp. 225–248; Edward M. Brown, 'Neurology and Spiritualism in the 1870s,' *Bulletin of the History of Medicine* 57 (1983), pp. 563–577; Howard Kerr, *Mediums, and Spirit-Rappers, and Roaring Radicals: Spiritualism in American Literature, 1850–1900* (Urbana, Ill., 1972); R. Lawrence Moore, *In Search of White Crows: Spiritualism, Parapsychology and American Culture* (New York, 1977); Janet Oppenheim, *The Other World: Spiritualism and Psychical Research in England, 1850–1914* (Cambridge, 1985).

13 Bucknill and Tuke, *Manual*, p. 324; Juanita A.N. Dobmeyer, 'The Sociology of Shoplifting,' Ph.D. thesis, University of Minnesota, 1971; Meda Chesney-Lind, 'Women and Crime: The Female Offender,' *Signs* 12 (Autumn 1986), pp. 78–96.

14 See esp. J. Baker, 'Kleptomania,' in *A Dictionary of Psychological Medicine*, ed. Daniel H. Tuke (London, 1892), vol. 2, pp. 726–729.

15 'The Husband Who Makes His Wife a Thief: The True But Little-Known Reason Behind Many a Shoplifting Incident,' *Ladies' Home Journal* 32 (March 1915), p. 16; and cf. John C. Bucknill, 'Kleptomania,' *Journal of Mental Science* 8 (1862), p. 265.

16 'Kleptomania,' *Progressive Review* (*London*) 2 (1897), pp. 311–315; 'Kleptomania,' *Medico-Legal Journal* 14 (1896), p. 232, citing the French criminologist Motet on kleptomaniac confessions; Abelson, '"When Ladies Go A-Thieving,"' ch. 1; Mary Owen Cameron, 'An Interpretation of Shoplifting,' in *The Criminology of Deviant Women*, eds Freda Adler and Rita J. Simon (Boston, 1979), pp. 161–162.

17 On the reluctance to prosecute, see 'With a Mania for Stealing,' *New York Times*, July 1, 1883, p. 12, col. 3; 'Shoplifting in New York,' *New York Times*, January 2, 1906, p. 15, col. 1; Abelson, '"When Ladies Go A-Thieving,"' pp. 208–209.

18 *H.H. Harris v. The State* (1885), *18 Texas Court of Appeals*, pp. 287–295; and cf. *Alfred Lowe v. The State* (1902), *44 Texas Criminal Reports*, pp. 224–226; Mark Melford, 'Kleptomania: A Farcical Comedy in Three Acts,' *French's Acting Edition* 138 (London, 1888), produced in 1888, 1890 and 1908; 'She Cannot Help Stealing,' *New York Times*, December 22, 1888, p. 2, col. 5; Edward Bellamy, *Looking Backward 2000–1887* (1888. Reprint: New York, 1951), p. 164.

19 *Lewis v. Lewis* (1890), *44 Minnesota Reports*, pp. 124–126; *New York Times*, articles on Mrs Castle, the San Francisco belle, from October 8 through December 31, 1896, esp. October 14, p. 9, col. 1, and October 18, p. 1, col. 6; 'Kleptomania as a Disease and a Defense,' *American Lawyer* 4 (1896), p. 533; 'Kleptomania,' *Law Times* 102, November 14, 1896, p. 28; 'Mrs Castle's Case,' *Medical Magazine* (*London*) 5 (1896), pp. 1211–1213; Shobal v. Clevinger, *Medical Jurisprudence of Insanity* (Rochester, NY, 1898), vol. 2, pp. 848–849, citing the *Chicago Record*, September 26, 1895, on the English duchess, with references also to Germans and Russians; Abelson, '"When Ladies Go A-Thieving,"' p. 317, n.39; Gus Edwards and Will D. Cobb, 'Mamie (Don't You Feel Ashamie),' in *Song Hits from the Turn of the Century*, eds Paul Charosh and Robert A. Fremont (New York, 1975), pp. 162–166.

20 'Mrs Martin's Dual Role,' *New York Times*, April 11, 1895, p. 8, col. 1.

21 Ibid.; M. Motet, 'French Retrospect: Shoplifting,' *Journal of Mental Science* 26 (1880), pp. 625–629, reviewing the work of Lasègue; Dubuisson, 'Les voleurs des grands magasins,' p. 348; Abelson, '"When Ladies Go A-Thieving,"' pp. 265, 268, 269, 270, 271, 94; and cf. 'A Self-Accused Thief,' *New York Times*, May 9, 1877, p. 8, col. 3.

22 On false names, see Abelson, '"When Ladies Go A-Thieving,"' pp. 265–268.

23 Edwards and Cobb, in *Song Hits*, p. 166.

PART FOUR

Objects and Technology

INTRODUCTION TO PART FOUR

Objects rarely come alone. They invariably travel in packs. Indeed, there is something unnerving about an object that appears to be detached. For example, think of objects such as urinals, basketballs and house bricks that occasionally find themselves wrenched from their everyday banality in order to be recontextualized in art galleries. Not only can one not help but imagine them in relation to other objects, but the very act of isolation is precisely what draws one's attention to those usually taken-for-granted linkages. Thus a house brick anticipates a house. A house anticipates a garden. A garden anticipates a lawn. A lawn anticipates a lawn-mower. A lawn-mower anticipates a shed. A shed anticipates a lock. A lock anticipates a key. A key anticipates a key-ring or a hook or a shelf or a drawer or a pocket, and so on. While this 'calculus of objects' may lay a trail that is more or less functional, the essential issue is that each object always refers to *other objects*. They identify with one another (like a lock and a key) – and by the same token they differentiate themselves from one another (like a hover-mower and a rotary-mower). By playing upon a potentially infinite matrix of identity and difference, objects communicate with one another. Karl Marx famously characterized these *social relations* among *things* as the 'fetishism of commodities'. Ever since, countless authors have put forward various accounts of the relationship between objects, the most interesting of which are presented in this part of the Reader and find their echo in Part 5. What they all share is an insistence that the relationship between objects has a significance, specificity and coherence all of its own, and that this relationship enters into dialogue with human beings. Thus the extracts in Part 4 of the Reader draw out different facets of what one might call the ongoing 'negotiations' between the subjects and objects of consumer society. Which side will have had the upper hand remains to be seen.

Pasi Falk's account of 'the genealogy of advertising' describes a fundamental shift in the relationship among objects. In the early twentieth century, 'announcements' gave way to 'advertisements'. In so doing, objects stopped appearing as mere *tautologies* in order to become *representatives* of whatever was said to be necessary, desirable and missing from one's life. Above all, Falk argues that they ceased to be just objects in order to become *good* objects – especially for *you*. He makes the important point that necessity and usefulness are not given to us by nature, but are actively solicited and created. The consumer is *made* to have needs, and objects are presented as things that can satisfy those needs. In this way, products come to complement what is lacking in the consumer. At the limit, they proffer *pure* satisfaction and *complete* fulfilment. To illustrate how this happens he considers the advertising of breakfast cereals and healthcare products, which promises that you can be absolved of past, present and future negativity through the exercise of good choice. In short, *mass* production and *mass* consumption revolve around you. As a recent advertising slogan for the retailer Marks & Spencer put it, modern advertising promotes products that are 'exclusively for everyone'.

While Falk draws our attention to the language of modern advertising which solicits our individual attention and cultivates our personal desires (Part 3), **Bruno Latour** is more concerned with the social *work* that objects perform. Indeed, Latour's take on things is a salutary reminder that consumption is first and foremost a *material practice*, and that our society consists of human *and non-human* actors: plants, animals, objects, and so on. So, rather than fixate on what objects *mean* for different kinds of people, Latour focuses on what objects *do*. This is a radical departure from mainstream social science. For when social scientists and others think about society, social relations and social action, they almost always turn their attention to people. However, in making this turn, social scientists have erected what are arguably the two most problematic and contested divides in the social sciences: an *ontological apartheid* between humans and non-humans; and the privileging of *either* human agency *or* social structures in the explanation of social life. Typically, people are treated either as autonomous individuals who strategize within the bounds of family, community and the various strata of social life (voluntarism and humanism) or else as automata who are programmed to follow the norms, rules and laws of enduring and impersonal social structures (determinism and structuralism). Accordingly, consumers tend to be presented as either *rational actors* or *zombified dupes* (**Part 3**). More enlightened social scientists have tried to reconcile these two extremes, more often than not by making the durability and efficacy of social structures dependent on the continued practice of knowledgeable and purposive human beings (see **Bourdieu**). By contrast, Latour argues that social scientists will never be able to provide a satisfactory explanation of society unless and until they broaden their interests to include the social life of *things* as well as the social life of people. These are our 'missing masses', whose social work enables us to square the durability of social life with soft humans and weak moralities.

When Latour begins to take the social work of things seriously, four issues become rapidly apparent. First, objects do a lot of work in modern societies. Huge amounts of agency have been *delegated* to all kinds of objects. Latour is not especially thinking here of things such as Artificial Intelligence, but of much more mundane objects (for example, traffic-lights, keys and clocks). Second, the effort expended by people in order to keep our world going is exceeded many times over by the agency of things. Think, for example, of what our world would be like without synchronized clocks or the humble screw. Third, just as the social relations among humans are politicized, so too with the social relations among non-humans. Finally, just as elites have always believed that the masses should know their place, we humans tend to believe that objects should also know their place: to respond obediently to our demands. What Latour demonstrates so well is the fact that here, as elsewhere, things rarely – if ever – live up to our repressive expectations. For Latour, the come-uppance is simple. We need to take the agency of things seriously, and explore the often fraught 'negotiations' between human and non-human actors, which invariably lead our expectations, our demands and our world astray. The body of work that has taken up this challenge is usually referred to as 'actor-network theory', a phrase that is meant to foreground the dispersal of agency.

Christina Kiaer's consideration of Aleksandr Rodchenko's contribution to the 1925 Paris Exposition is a useful complement to Latour's suggestion that we need to stop treating non-humans as if they were nothing more than *passive* objects and start treating them as a vital part of our society. Although Latour's democratic disposition towards the 'non-human masses' chimes with a certain kind of Marxism, he calls for them to be accommodated and assimilated into the fabric of social theory, rather than for the formation of a truly revolutionary class-consciousness among non-humans, something which **Baudrillard** has taken seriously in his recent work on objects. Rodchenko, a Russian Constructivist, is interesting to the extent that he counterposed a *socialist object* to the *bourgeois object*. He was revolted by the miserable life of bourgeois objects, which existed as *slaves* and *prostitutes* to be used and abused by a culture of economic exploitation and voyeuristic display. Commodities were born into servitude, hurled into a world of indifferent exchange, and manipulated by perverted desires. The extract by Kiaer hints at another way of relating to objects. Specifically, the socialist objects brought by Rodchenko to Paris – the spiritual home of bourgeois objects (cf. **Schivelbusch**) – were conceived of as *comrades* and *co-workers*, fully engaged in the utopian construction of a just society. Yet while most Marxists continue to *recoil* from the fetishism of commodities – in the vain hope of restoring properly *social* relationships between people, and *material* relationships between things; thus reinforcing the 'ontological apartheid' that so troubled Latour – the Constructivists tried to move in the opposite direction by harnessing the fetishism of commodities – the 'uncanny of the commodity,' as Kiaer puts it – for socialist ends.

In his consideration of toys, **Roland Barthes** also has the social work of objects in mind. On his reading, the ensemble of toys made available to French children do little more than socialize them into the adult world of alienated work, private property and domestic banality. Dolls, garages, soldiers and train sets '*literally* prefigure the world of adult functions'. The child is presented with a ready-made world that appears as a timeless gift of nature, from which all traces of how that world was constructed and struggled over, and how it might have been otherwise, are erased. 'The child can only identify himself as user, never creator,' writes Barthes. 'He does not invent the world, he uses it.' Once again, the world of toys functions as a constellation of *bourgeois* objects: only this time it is the creative evolution of *both* humans *and* non-humans that is passified and enslaved in the process. As with adult life, the bourgeois object proffers only two roles: either one is a *user* of things (a worker) or one is an *owner* of things (a capitalist). In this way, modern toys naturalize and dramatize the fundamental schism of class society. To that extent, they exist *as if* they were dead – 'and once dead, they have no posthumous life for the child'. Meanwhile, Barthes finds his own *socialist* objects hidden in the margins of the modern world of toys. 'The merest set of blocks . . . implies a very different learning of the world', writes Barthes. The actions the child 'performs are not those of a user but those of a demiurge. He creates forms which walk, which roll, he creates life, not property: objects now act by themselves.' Perhaps this is why the famous analytical psychiatrist Carl Jung set aside some time each day to play with toys such as building blocks in order to keep the other world – the world of the consumer – at bay.

The 'negotiations' between humans and non-humans is especially apparent when one considers technologies. **Roger Miller** focuses on the changing relationship between middle-class women, domestic labour and city living in America between 1850 and 1920. The introduction of services such as domestic sewerage; reliable supplies of gas, electricity and water; and a whole new range of domestic technologies, such as refrigerators, washing machines and vacuum cleaners, promised to liberate middle-class women from household labour. The work that was delegated formerly to domestic servants could now be delegated to objects, although Miller makes the important point that in many cases new domestic technologies were actually *more* labour-intensive than earlier ones, which tended to mean that they were adopted initially by wealthier households with access to even greater amounts of domestic labour. Consequently, the work of middle-class women shifted from a manager of domestic labour to a user of domestic technologies. 'Women had to rationalize their schedules with those of the new machines they tended.' Needless to say, this change of profession not only required the guidance of advertisers, publicists and expert advice-givers, it also served the interests of manufacturers well. As the home became a 'machine for living', it was suburban middle-class women who found this new regime most constraining. In addition to devoting themselves to domestic technologies in order to maintain an ideal – and increasingly hygienic – home, they were denied the various forms of collective provision that were a feature of urban apartment blocks and urban neighbourhoods. They were also at a considerable remove from the main sources of casual domestic labour, and so had to become even more self-reliant. Finally, the professionalization of philanthropy and social welfare, the refocusing of everyday life around the nuclear family, and the diffuse nature of the suburbs, all mitigated against the circle of friends and culture of visiting that had been the mainstay of middle-class women's social lives in the nineteenth-century city. Between 1850 and 1920, what was once anathema to middle-class women had become their principal occupation: domestic labour. (See also **Redfern**.)

While Miller provides an insight into how the relationship between middle-class women and the suburban home was renegotiated via new domestic technologies, **Elaine Abelson** considers the technological mediation between middle-class women and department stores (see **Bowlby**). By the 1870s, there was a widespread moral panic in Europe and America about department store kleptomania (see **Schwartz**). Faced with the absolute novelty of a profusion of aestheticized commodities in what Émile Zola famously characterized as 'brothels of modern commerce', middle-class women were apparently so overcome by a desire to take possession of these commodities that they were unable to prevent themselves from stealing them. What worried the department store owners was not only the fact that huge quantities of stock were being stolen (especially ribbons, handkerchiefs, hats and hairbrushes), but that the store's most valued customers were often responsible for the theft. In this extract, Abelson considers the ambivalence of department stores when it came to mediating the relationship between consumers and commodities. While one set of technologies was employed in order to *solicit* the consumers' desire for things, another set was used to *frustrate* consumers in their ability to act impulsively on those desires. Suddenly, the brothels of modern commerce were possessed with a new security consciousness, complete with an entire arsenal of fixtures and fittings for the protection of commodities that functioned as 'silent and unnoticed detectives'. From the 1890s, mirrors, artificial light, glass and store detectives were being used both to *seduce* consumers and to *safeguard* commodities. Yet in the struggle between the two regimes, seduction invariably took precedence over protection (see **Schivelbusch**). Although Abelson does not address it directly in this extract, the actual 'two-way mirror' – used by department stores, police forces, advertising agencies and TV shows – is a famous *misnomer*. The so-called 'two-way mirror' is really a one-way surveillance window, which just happens to have a reflective surface on its other side. As Abelson's titular use of the term implies, this 'accidental' property turns out to be as important to the consumer society as its 'functionally determined' side (see **Bauman**).

The extract by **Jon Goss** considers the nature of consumer surveillance at the close of the twentieth century, with particular reference to computer systems that deal with geodemographic data (i.e. demographic and psychographic data on consumers that are spatially referenced, usually to their place of residence). Responsible for a veritable revolution in the direct-marketing industry, these systems do much more than keep track of the voluminous public and private-sector data trail left by individual consumers. Through a variety of statistical and modelling techniques, such as cluster analysis and factor analysis, they generate what are taken to be more or less stable and coherent consumer segment schemes. Goss aptly refers to these segments as 'data narratives' and 'statistical fictions'. Not only do marketers use them to try to regulate and discipline consumers, but they are also beginning to substitute *simulation* for statistical *abstraction* in the generation of segmentation schemes and abstract types. Goss is concerned about how people are reduced to – and judged against – an array of abstract types, often through spatial referencing or statistical resemblance. He is also troubled by the fact that the spatial referencing of geodemographic systems reifies one's home address into the principal unit of consumption – the household – and thereby 'anthropomorphizes ZIP [i.e. post] codes, speaking of them as if they were possessed of specific tastes and purchasing power'. Taken as a whole, this spatial fetishization of consumption effectively 'partitions social life into life-style areas'. Finally, the 'statistical subjects' that crystallize out of geodemographic data are a perfect illustration of how the distinction between humans and non-humans has become increasingly tenuous. Data subjects are not just *representations* that need to be evaluated in terms of how well they *resemble* a real flesh-and-blood subject. More importantly, data subjects are our *representatives* who need to be evaluated in terms of how well they *perform* in another dimension.

The Genealogy of Advertising[1]

Pasi Falk

Advertisement, in its modern and proper meaning (Leiss *et al.*, 1986), originated with the massive breakthrough around the turn of the century of consumer society and its huge markets for consumer goods in the major European centres and particularly in the United States (Fraser, 1981; Hayes, 1941). The step from *announcement* to *advertisement* came with the recognition that making the product known to people formed an integral part of sales; or, to paraphrase Clausewitz, when this was recognised as a continuation of sales by other means. An early formulation of the idea was presented in 1904 by American advertising guru John E. Kennedy, whose simple but ingenious thesis was: 'advertising is salesmanship in print' (Pope, 1983, 238).

As well as a particular mode of production (Marx), modern advertising required a particular 'mode of information' (Poster, 1990) that made possible the transformation of concrete products into representations, into complex meanings carried by words and images.

This connection between mass markets and mass communication is crystallised in modern advertising. In the mass production of consumer goods, the target, on the one hand, is an anonymous 'mass'; on the other hand, that mass is recognised as consisting of individual buyers. The same duality is repeated in the sender–recipient logic of the mode of information: the message is sent out equally to the whole body of recipients, but it is received individually by each one of them.

This paradoxical duality helps to explain the image of a Janus-faced consumer that unfolds not only in marketing philosophies but also in theories of modern consumption since the late nineteenth century: the mass of consumers was represented as more or less irrational 'adult children' – as it was put in the *Printer's Ink* magazine in 1897 (Lears, 1984, 376) – and on the other hand as consisting of individuals who were capable of making sensible decisions and choices (even though the conduct would not be considered as strictly rational in economic terms).

What really lies behind this dual consumer role is a confrontation of two antithetical perspectives. On the one hand, the producer's concern is to realise the material mass of similarity (mass product) on what is regarded as a homogeneous consumer market. This is supposed to happen with the help of marketing to make sure that the correspondence between the two masses is established. On the other hand, the producer and the marketing apparatus must recognise the individual consumer's inalienable, freedom of choice.

NEED, DESIRE, WILL

Even though the (somewhat problematic) distinction between *needs* and *desires* (see Falk, 1994, 93–150, 182–3, n. 1) is crucial in the deconstruction of the Janus-faced consumer, this distinction appears to have only marginal relevance from a marketing and advertising point of view. At the point where a willingness or readiness to sell is transformed into an active intention to sell, every product that is for sale becomes necessary ('you need this'), desirable ('this is what you desire'), missing ('you still lack this'); in a word, it becomes something that is *good* ('for you'). As far as the intention of selling is concerned it is of course a basic condition that the product is realised in the transaction of exchange, which from the buyer's (consumer's) point of view is defined as a choice, as a realisation of a definite *will* to buy a certain product.

Because of its strategic significance, will becomes a key concept that supersedes both needs and desires. This reformulation is also found in the neoclassical economics of the late nineteenth century: the discourse on needs and desires is replaced by theorising of the consumer's 'wants' and 'preferences' (cf. Falk, 1994, Chapter 5). As far as the business transaction is concerned, the most important thing is what the buyer wants, regardless of how the underlying motives are described.

However, the situation appears in a somewhat different light when it is approached from the marketing theorist's or the advertising psychologist's point of view. In an attempt to establish what sort of representation most effectively attracts the will and attention of the consumer to the product, the advertising expert will need to go beyond the surface and probe into the possible motives that lead to the act of will to buy. At the same time, the advertising expert is confronted, time and time again, with the problem of the multitude of these motives. The solution to that problem lies (seemingly) in the Janus-faced consumer, in its different variants. One of the most interesting versions was presented back in 1926 in the American journal *Advertising and Selling*, which said that 'consumers made purchases on emotional impulse and then justified them with "reason-why" rationalisations' (Marchand, 1986, 153). Later a different approach was adopted to the problem of the Janus-faced consumer: the analysis was now based on a classification of different types of consumer goods – a theme that nowadays occurs mainly in the discussions on the dimensions of 'high and low involvement' in consumer behaviour (for example, Rajaniemi, 1992) – and on different types and groups of consumers ('market segmentation'; see Pope, 1983, 291ff.).

THE GOOD THINGS OF LIFE

But no matter what approach they adopted to arranging and classifying motives, marketing theorists and advertising psychologists still remained captive to the seller's point of view. That is, no matter what words or images beyond the simple statement of availability were attached to the product, they had to promote a positive expression of want, a positive decision to buy. There are various different ways to argue in favour of a product; you may say it is 'useful', 'comfortable', 'healthy', that it brings 'social prestige' or simply that it 'makes you feel good'. The crucial thing is that an image is created of a 'good object' – to use the psychoanalytical concept introduced by Melanie Klein (1932) – that *you* do not yet have.

Some of these arguments may of course appear as more 'rational' than others, but even so their function remains the same. Even the rational argument serves to provide a representation of the (good) thing you are lacking, and in this capacity it often boils down to straightforward rationalisation (in the psychological sense of the term). An example is provided by an American advert where the argumentation for a 'completely new type' of deodorant bottle says it saves time, thanks to the new wider ball; you need just one stroke instead of two (Wills, 1989). Gary Wills demonstrates in his analysis that this 'advantage' – 'rolls on fast dries fast!', as the slogan goes – is in fact an imaginary product quality, the 'rationality' of which lies in your saving a few minutes of time per year.

Arguments attached to the product and/or its consumption may even exploit the register of negative images, but the overall effect must be positive, otherwise people are not going to come in and buy the product. The idea that both the positive and negative register could be used was formulated by Roy Johnson, father of the 'impressionistic principle' of advertising, in 1911 as follows:

> we suggest the comfort or profit which results from the use of the product, or the dissatisfaction, embarrassment, or loss which follows from its absence.
>
> (*Printer's Ink*, 75 (25 May 1911), 10–11; cited in Lears, 1984, 382)

In other words, even where indirect or inverted means are applied, these must help to transform the product and its context into a representation of 'good'. The negative register is used for depicting the state of deficiency that follows with the absence of the product and/or its use.

First, a negative image may be linked up with an identifiable moment in the present time, in which case it will appeal to the buyer's actual experience of deficit (the classical 'before–after' scheme); alternatively, it may interpret the consumer's current situation as 'relative deprivation' in comparison with the better situation that will (should) follow with the purchase and/or use of the product. This type of advert played an important role not only in the advertising of patent medicines towards the end of the nineteenth century, but also in the socially stigmatising, 'anxiety format' adverts of the 1920s and 1930s (see Leiss *et al.*, 1986, 52–3; Marchand, 1986, 18–21). The latter appealed to the glances of reproof from neighbours and significant others and in general depicted the evil, the deficit that the absence and non-use of the product would cause to social relations, career prospects, and so on.

This type of advertisement was used in the marketing of mouthwash, for instance, a product which promised to resolve once and for all the problem of bad breath and at once the distress, discrimination and loneliness that followed. It promised a completely different world of happiness, a world that was to be later outlined by Dale Carnegie in his famous guidebook for those who wanted to win friends and money and influence people. It is symptomatic that the marketing of this particular product named not only the product itself (*Listerine*) but also the deficit it promised to do away with. Discovered in an old medical dictionary, 'halitosis' was defined as a condition that could be cured by *Listerine*, opening the doors to a happy (social) life: 'For Halitosis use Listerine!' (Marchand, 1986, 19).

In modern advertising one is hard put to find such uses of the negative register, although it may sometimes occur in the form of self-irony, with parodic repetition of 'outdated' advertising jargon and 'old' styles and patterns appealing to a sense of anxiety and guilt. Modern advertising operates almost exclusively with the positive register, depicting the happy and content soap user for whom there is always room even in a cramped lift rather than the distressed non-user who is left out.

Secondly, a negative image may be projected into a conceivable future as a threat or as an otherwise undesirable state of affairs – loss of health, loss of face – that the product promises to keep at bay. Whereas negative images referring to the present time have more or less disappeared from advertising, images of threats projected into the future continue to occupy a quite firm position in present-day advertising jargon. This is most particularly true of adverts for beauty and health products, which do not portray horror scenes but rather hint at the prospect. Here, too, the negative is excluded from the frame of representation.

The strategy is applied most notably to selling today's 'patent medicines'; that is, the various vitamin products, the consumption of which has been increasing significantly in the Western world during the last two decades (Klaukka, 1989). Unlike their precursors in the nineteenth century, these products do not need to make promises that they have curative effects because their 'use value' is located in a possible future – and in the threat contained therein – rather than in the present time and its verifiable effects. In this sense these products make for an ideal marketing item: no one can say they don't work because there is no argument that the effects are visible here and now. The promises of future health and longevity cannot be falsified unless one lives one's life all over again in a 'control group'. This has to do with more than just the placebo character of a product (cf. Richards, 1990, 193): the utility of this product is equally verified by the presence and absence of any effects; after all things could have been worse if. . . .

MODERN ADVERTISING: DIMENSIONS OF CHANGE

Whether the advertisement uses the positive or the negative register of representation, the outcome must establish a positive link between the identified product and the 'good' that characterises it. The building of this link implies a metamorphosis in which the product transforms into a representation – and it is this that modern advertising is basically about. The basic pattern has remained unchanged through the century-long history of modern advertising, but the modes and methods of creating representations of 'good' have changed. These changes can be roughly outlined in three stages of development:

1. The shift from product-centred argumentation and representation to a thematisation of the product–user relationship and further to the depiction of scenes of consumption which emphasise the *experiential* aspect of consumption.
2. The shift from the emphatically rational mode of argumentation supported by essentially falsifiable 'evidence' of product utility towards representations of the satisfaction that comes with using the product – again emphasising the experiential aspect of consumption.
3. The shift in communication from verbal and literary means to audio-visual means, based on the development of communications technology that is closely related to the two former dimensions. Pictures were introduced to printed advertising in 1880s, photographs in 1890s and the next century offered the new powerful medias of cinema, radio and television.

These trends in development would seem to be heading towards a form of representation which has increasing independence *vis-à-vis* its point of reference, that is, the product, and which increasingly operates with expressions of the positive register. In other words, the language and argumentation of advertising is moving towards a pure 'good', towards a true positive experience. Of course, this argument needs to be backed up with a closer definition of what is meant by representation of 'good', and this is in fact one of the chief concerns of the discussion that follows. Suffice it to note at this point that the shifts identified above characterise not only the history of modern advertising, but more generally the completion of the modern world of consumption, particularly in its emphasis on the *experiential nature of consumption*.

The story to follow is primarily about how this world of goods becomes visible to the consumer and how this visibility constitutes a direct consumer–product relationship.

NAMING THE NAMELESS

Modern advertising identifies and singles out products as representations which are intended to appeal to the consumer at once as an individual and as a mass. Or, as Leo Lowenthal puts it: the individualising 'for you' actually addresses 'all of you' (Marchand, 1986, 108). On the other hand, each argument and quality that is attached to the product identifies it precisely as a positive thing or 'good object' which in one way or another promises to fill in the 'empty space' that consumers feel is there, *even though they do not know how to name it*.

What could the identification of this unnameable deficit mean? If hunger were plain hunger and bread just plain bread, then the whole problem of anonymity wouldn't even exist. The naturalist theory of needs has a name both for the deficit (hunger) and for whatever it is that fills the stomach (bread). However, a fundamental thesis of modern advertising is that 'bread is not only bread', which necessarily means that the same applies to the other side of the equation: 'hunger is not only hunger'. In other words, both are (also) something else and more; but what? This leads us inevitably to the problem of namelessness, which has two addresses – but which as we shall soon discover are just different entrances to the same house.

First, on the one hand there is the problem of the nameless *product*, which is resolved by naming and individualising the product and by setting it apart (in positive terms) from other products by means of marketing (packaging, adverts, and so on). A paradigmatic case of product identification is the pioneering work that was done by the American Quaker Oats Company (renamed as The American Cereal Company in the early twentieth century) towards the end of the nineteenth century (Marquette, 1967). In 1880, an oat producer by the name of Schumacher started packing his produce – which so far had been known simply as 'oats' – into sacks that had the producer's name printed on them (Marquette, 1967, 16). This was the first step in creating a brandname, which was soon to be followed by a technological innovation (steam mill) that made possible the processing of oats into a specific, identifiable form (rolled oats) and finally by the packaging and naming of this product: Quaker Oats.

Hence the transformation of plain oats into a potential *brand*; a potential that in this case was well exploited. The relationship of people to oats was of course nothing new, but the individualised (albeit imaginary) relationship of the (mass) consumer to Quaker Oats very definitely was. It was in this particular packaged, identified form – as a representation – that the product became more than just 'plain oats'.[2] The making of the product did of course require some real effort in terms of product development (from flour into flakes), which changes the sensory qualities of the raw material. However, the crucial thing here is that all the characteristics that emphasise the distinctive identity of the product are packaged together in the form of a representation. And if the physical characteristics no longer suffice to support the unique identity of the product, the other elements of the representation, the surplus 'goods' will have to take their place.

No product can expect to have the market all to itself for very long. Quaker Oats was soon followed by Kellogg's Company, Postum Cereal Company and many others who went after the consumer with cornflakes and rice krispies, which by now have become the staple food of breakfasts the world over. Therefore a positive character must be continuously created and re-created for a named product.

Second, on the other hand, as a message to the potential consumer, the building of a positive product identity implies a representation of that product as a *complement*. It is in this function that the advertisement has to name something that is fundamentally nameless, the negative form of which is the consumer's *deficit* (which is eliminated by the complement) and the positive form of which is the *wholeness* that buying the product and/or using it promises to bring. 'It' is fundamentally nameless, not representable, and that is why it is always given new names, over and over again.

But regardless of whether the naming of 'it' focuses on the negative (deficit) or on the positive (wholeness), the role of the product as a *complement* is always and necessarily positive. As a representation appealing to the potential consumer, it promises something good, either in terms of eliminating the evil or in terms of offering surplus good; in this latter case the state preceding the surplus good is necessarily redefined as a deficit. In other words the duality of deficit (negative) and wholeness (positive) is present in the representation either explicitly or implicitly, on the reverse side of either deficit or wholeness. In any event the role of the product as a complement remains positive, as Roy Johnson (see p. 186) clearly understood.

So in the end the two addresses of identification or naming in modern advertising – the singularisation of the product and its representation as a complement to an identified deficit and/or as a surplus which produces the wholeness identified – lead to exactly the same place. Transformed into a representation, the product must stand clearly apart from other similar products, in which case it will also be individualised to the potential consumer as a party to a bilateral (albeit imaginary) relationship. All the attributes of 'good' attached to the product in its representation fulfil both of these functions. In fact the full range of 'goods' – which has since been repeated in various combinations by modern advertising – was already in use in the marketing of selected pioneer products (that is cereals) in the late nineteenth century:

> Every device of the Advertiser's art was used by *American Cereal Company* between 1890 and 1896. Techniques later men claim to have innovated were tested by Crowell as early as 1893. In his ads he appealed to love, pride, cosmetic satisfactions, sex, marriage, good health, cleanliness, safety, labour-saving, and status-seeking. His boldness, at the height of prudish Victorianism, reached its peak in 1899 in an advertisement in *Birds* magazine and several other periodicals of the day. The illustration was a voluptuous, bare-breasted girl, her torso draped in Roman style, sitting on a *Quaker Oats* box.
> (Marquette, 1967, 51)

If for reasons of distinction or competition advertising has to resort to the positive register ('our' product is better than 'theirs', or 'better value for money'), the situation is somewhat more complex as far as the building of a product–consumer relationship is concerned. This is because the naming may concern both the deficit that the product 'promises' to make disappear and the 'good' that is secured by ownership and use of the product. It is on this dimension of duality that we can follow the development of the modern advert towards the positive. This process is made visible by the history of patent medicines, which in effect lies at the very root of modern advertising. Indeed, the thematic shift from the novelty products of the foodstuffs industry to patent medicines is not all that dramatic; most of the former were introduced to the general public precisely as 'health' products (see, for example, Levenstein, 1988; Young, 1967; Porter, 1989), such as one particular successor of patent medicines called Coca-Cola.

NOTES

1 A more extensive version of this article was published with the title 'Selling Good(s) – on the genealogy of modern advertising', in Falk (1994), *The Consuming Body.* Sage Publications, Theory, Culture & Society book series, London, 1994, ch. 6, pp.151–185, by permission of Sage Publications Ltd.

2 Haug (1980); the 'commodity aesthetics' theorist would say that this represented a transition from real use values to empty promises of use value. However, the problem cannot be brushed aside as easily as that, as I have tried to demonstrate elsewhere (Falk, 1982). It is also interesting to note that in her critique of the world of goods, Susan Willis (1991) uses precisely the case of Quaker Oats as an example of how marketing (naming and packaging) distorts people's 'natural' relationship to oats.

REFERENCES

Falk, P., 'Tavarametafysiikkaa' (Commodity Metaphysics), *Tiedotustutkimus*, 5 (3) (1982), 78–84.

Falk, P., *The Consuming Body* (London: Sage TCS, 1994).

Fraser, W. H., *The Coming of the Mass Market, 1850–1914* (London: Archon, 1981).

Haug, W. F., *Warenästhetik und kapitalistische Massenkultur: Systematische Einführung in die Warenästhetik* (Berlin: Argument Verlag, 1980).

Hayes, C. J. H., *A Generation of Materialism, 1871–1900* (New York: Harper & Bros, 1941).

Klaukka, T., *Lääkkeiden käyttö ja käyttäjät Suomessa* (*The Use and Users of Medicine in Finland*) (Helsinki: Kansaneläkelaitos, 1989).

Klein, M., *The Psycho-Analysis of Children* (London: Hogarth Press, 1932).

Lears, J. T. J., *Some Versions of Fantasy: Toward a Cultural History of American Advertising, 1880–1930, Vol. 9. Prospects*, in Salzman, J. (ed.), *The Annual of American Cultural Studies* (New York: Cambridge University Press, 1984).

Leiss, W., Kline, S. and Jhally, S., *Social Communication in Advertising* (Toronto: Methuen, 1986).

Levenstein, H., *Revolution at the Table* (New York: Oxford University Press, 1988).

Marchand, R., *Advertising the American Dream* (Berkeley: University of California Press, 1986).

Marquette, A., *Brands, Trademarks and Good Will: The Story of the Quaker Oats Company* (New York: McGraw-Hill, 1967).

Pope, D., *The Making of Modern Advertising* (New York: Basic Books, 1983).

Porter, R., *Health for Sale* (Manchester: Manchester University Press, 1989).

Poster, M., 'Words without things: the mode of information', *October*, 53 (1990), 63–77.

Rajaniemi, P., 'Conceptualization of product involvement as a property of a cognitive structure', Vol. 29, in Mikkonen, K. (ed.), *Acta Wasaensia* (Vaasa: University of Vaasa, 1992).

Richards, T., *The Commodity Culture in Victorian England: Advertising and Spectacle, 1851–1914* (Stanford: Stanford University Press, 1990).

Willis, S., *A Primer for Daily Life* (London: Routledge, 1991).

Wills, G., 'Message in the deodorant bottle: inventing time', *Critical Inquiry*, 15 (3) (1989), 497–509.

Young, J. H., *The Medical Messiahs: A Social History of Health Quackery in Twentieth-Century America* (Princeton: Princeton University Press, 1967).

Where Are the Missing Masses?

The Sociology of a Few Mundane Artifacts

Bruno Latour

To Robert Fox

> Again, might not the glory of the machines consist in their being without this same boasted gift of language? 'Silence,' it has been said by one writer, 'is a virtue which render us agreeable to our fellow-creatures.'
>
> Samuel Butler (*Erewhon*, ch. 23)

Early this morning, I was in a bad mood and decided to break a law and start my car without buckling my seat belt. My car usually does not want to start before I buckle the belt. It first flashes a red light 'FASTEN YOUR SEAT BELT!', then an alarm sounds; it is so high pitched, so relentless, so repetitive, that I cannot stand it. After ten seconds I swear and put on the belt. This time, I stood the alarm for twenty seconds and then gave in. My mood had worsened quite a bit, but I was at peace with the law – at least with that law. I wished to break it, but I could not. Where is the morality? In me, a human driver, dominated by the mindless power of an artifact? Or in the artifact forcing me, a mindless human, to obey the law that I freely accepted when I got my driver's license? Of course, I could have put on my seat belt before the light flashed and the alarm sounded, incorporating in my own self the good behavior that everyone – the car, the law, the police – expected of me. Or else, some devious engineer could have linked the engine ignition to an electric sensor in the seat belt, so that I could not even have started the car before having put it on. Where would the morality be in those two extreme cases? In the electric currents flowing in the machine between the switch and the sensor? Or in the electric currents flowing down my spine in the automatism of my routinized behavior? In both cases the result would be the same from an outside observer – say a watchful policeman: this assembly of a driver and a car obeys the law in such a way that it is impossible for a car to be at the same time moving AND to have the driver without the belt on. *A law of the excluded middle* has been built, rendering logically inconceivable as well as morally unbearable a driver without a seat belt. Not quite. Because I feel so irritated to be forced to behave well that I instruct my garage mechanics to unlink the switch and the sensor. The excluded middle is back in! There is at least one car that is both on the move and without a seat belt on its driver – mine. This was without counting on the cleverness of engineers. They now invent a seat belt that politely makes way for me when I open the door and then straps me as politely but very tightly when I close the door. Now there is no escape. The only way not to have the seat belt on is to leave the door wide open, which is rather dangerous at high speed. Exit the excluded middle. The program of action[1] 'IF a car is moving, THEN the driver has a seat belt' is enforced. It has become logically – no, it has become sociologically – impossible to drive without wearing the belt. I cannot be bad anymore. I, plus the car, plus the dozens of patented engineers, plus the police are making me be moral (Figure 25.1).

According to some physicists, there is not enough mass in the universe to balance the accounts that cosmologists make of it. They are looking everywhere for the 'missing mass' that could add up to the nice expected total. It is the same with sociologists. They are constantly looking, somewhat desperately, for social links sturdy enough to tie all of us together or for moral laws that would be inflexible enough to make us behave properly. When adding up social ties, all

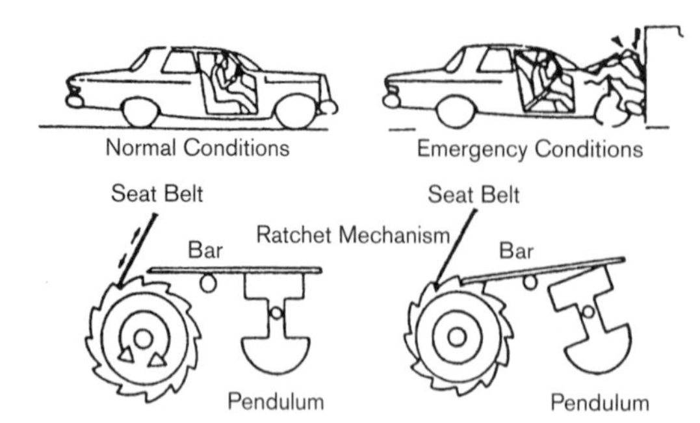

Figure 25.1 The designers of the seat belt take on themselves and then shift back to the belt contradictory programs: the belt should be lenient and firm, easy to put on and solidly fastened while ready to be unbuckled in a fraction of a second; it should be unobtrusive and strap in the whole body. The object does not reflect the social. It does more. It transcribes and displaces the contradictory interests of people and things.

does not balance. Soft humans and weak moralities are all sociologists can get. The society they try to recompose with bodies and norms constantly crumbles. Something is missing, something that should be strongly social and highly moral. Where can they find it? Everywhere, but they too often refuse to see it in spite of much new work in the sociology of artifacts.[2]

I expect sociologists to be much more fortunate than cosmologists, because they will soon discover their missing mass. To balance our accounts of society, we simply have to turn our exclusive attention away from humans and look also at nonhumans. Here they are, the hidden and despised social masses who make up our morality. They knock at the door of sociology, requesting a place in the accounts of society as stubbornly as the human masses did in the nineteenth century. What our ancestors, the founders of sociology, did a century ago to house the human masses in the fabric of social theory, we should do now to find a place in a new social theory for the non-human masses that beg us for understanding.

NOTES

1 The program of action is the set of written instructions that can be substituted by the analyst to any artifact. Now that computers exist, we are able to conceive of a text (a programming language) that is at once words and actions. How to do things with words and then turn words into things is now clear to any programmer. A program of action is thus close to what Pinch *et al.* (1992) call 'a social technology,' except that all techniques may be made to be a program of action.

2 In spite of the crucial work of Diderot and Marx, careful description of techniques is absent from most classic sociologists – apart from the 'impact of technology on society' type of study – and is simply black-boxed in too many economists' accounts. Modern writers like Leroi-Gourhan (1964) are not often used. Contemporary work is only beginning to offer us a more balanced account. For a Reader, see MacKenzie and Wajcman 1985; for a good overview of recent developments, see Bijker *et al.* (1987). A remarkable essay on how to describe artifacts – an iron bridge compared to a Picasso portrait – is offered by Baxandall (1985). For a recent essay by a pioneer of the field, see Noble (1984). For a remarkable and hilarious description of a list of artifacts, see Baker (1988).

REFERENCES

Baker, N. (1988) *The Mezzanine.* New York: Weidenfeld & Nicolson.

Baxandall, Michael (1985) *Patterns of Intention. On the Historical Explanation of Pictures.* New Haven, CT: Yale University Press.

Bijker, Wiebe E., Hughes, Thomas P. and Pinch, Trevor J. (eds) (1987) *The Social Construction of Technological Systems.* Cambridge, MA: MIT Press.

Butler, Samuel (1872) (paperback edition 1970) *Erewhon.* Harmondsworth: Penguin.

Leroi-Gourhan, A. (1964) *Le Geste et la Parole.* Paris: Albin-Michel.

MacKenzie, Donald and Wajcman, Judy (eds) (1985) *The Social Shaping of Technology: a Reader.* Milton Keynes: Open University Press.

Noble, David (1984) *Forces of Production: A Social History of Industrial Automation.* New York: Knopf.

Rodchenko in Paris

Christina Kiaer

'The light from the East is not only the liberation of workers,' the Russian Constructivist Aleksandr Rodchenko wrote in a letter home from Paris in 1925, 'the light from the East is in the new relation to the person, to woman, to things. Our things in our hands must be equals, comrades, and not these black and mournful slaves, as they are here.'[1] Rodchenko was in Paris on his first and only trip abroad to arrange the Soviet section of the Exposition Internationale des Arts Décoratifs et Industriels, for which he built his most famous Constructivist 'thing,' the Workers' Club interior. His lucidly spare, geometric club embodies the rationalized utilitarian object of everyday life proposed at this time by Russian Constructivist artists and *Lef* theorists such as Boris Arvatov. And Rodchenko's invocation of the socialist object from the East as a 'comrade' corresponds to Arvatov's theory of the new industrial object as an active 'co-worker' in the construction of socialist life, in contrast to the passive capitalist commodity – Rodchenko's 'black and mournful slaves' – oriented toward display and exchange.[2] Yet there is something uncanny about the stark, constrained order of the Workers' Club that exceeds Arvatov's theory, a visual uncanny that corresponds to the curious intensity and pathos of Rodchenko's verbal plea for 'our things in our hands.'

In this essay I want to propose that the language of Rodchenko's letters and the visual forms of his club elaborate, in a more subjective register, upon the Constructivist theory of the object – an elaboration that endows that object with a body and places it within the field of desire that is organized, under capitalism, by the commodity form. Rodchenko offers this elaboration, on the one hand, precisely as a response to the psychic and sensory overload of the Parisian commodity world; his idea of the object from the East must somehow cogently *respond* to his new, intimate knowledge of the Western commodity and its extraordinary power to organize desire and construct identities. On the other hand, despite the rhetoric of his letters, Rodchenko knew very well that 'East' and 'West' were not quite so cleanly opposed in 1925. The West had industrial technology, while Russia was only beginning to industrialize, but at the same time Moscow was no haven from the commodity: the New Economic Policy (NEP) in the Soviet Union had unleashed a vital if idiosyncratic commercial culture. The Constructivist theory of the object had been developed within the conflicted context of the mixed Soviet economy, and therefore already encompassed an acknowledgment of the phantasmatic component of consumption and the role of objects in negotiating it. The evidence we have from Rodchenko's encounter with Parisian consumer culture in 1925 offers an especially vivid articulation of a texture of desire that I want to claim is integral to the conflicted, utopian Constructivist object. It attempts to encompass, rather than repress, the desires organized by the Western commodity fetish, even as its goal is to construct new, transparent relations between subject and object that will lead to the collective ideal of social utopia illuminated by 'the light from the East.'

THE TRANSPARENCY OF THE CONSTRUCTIVIST OBJECT

Constructing the modular, movable furnishings of the club interior out of cheap, lightweight wood, and using open-frame construction, Rodchenko was intent on conserving materials and eliminating excess weight or bulk. While the objects in the club have the social function of materially organizing the leisure time in the everyday lives of workers, they are related formally to the nonutilitarian sculptural constructions of early Constructivism, such as Rodchenko's *Spatial Construction No. 9* (*Suspended*) of 1921.[3] Made of minimal material elements – plywood painted the color of metal – the *Spatial Construction* begins its life as a flat, two-dimensional circular form with a series of concentric circles carved straight through its surface. Aleksandr Lavrent'ev's reconstruction of the hexagonal *Spatial Construction No. 10* gives an idea of how the similar circular construction might have looked before it was opened up. When each concentric section is opened out to a different point in space and the structure is suspended from above, it is infinitely transformable within the logic of its own system. Precisely this formal, functional logic reappears in the dismountable orator stand for the Workers' Club, where these expanding and collapsing elements reappear in the fold-out screen for projecting slides and the contractible bench and speaker's platform. Other objects in the Workers' Club also operate like the orator stand: the side flaps of the table can be raised or lowered, depending on the activity of the club member; the chess ensemble in the back of the room, under the poster of Lenin, consists of two chairs separated by a nifty revolving chessboard on hinges above it, the case for the 'wall newspaper' allows for daily changes.

Rodchenko's club may resemble a standard piece of interior design in the utilitarian style of international modernism, but the extravagant premise of Constructivism is that it is not interior design at all, but rather an entirely new kind of *art* object.[4] Generations of critics have doubted the theoretical feasibility, or even the political integrity, of this Constructivist attempt to take the self-referential, systemic structures that were so revelatory as modern art and harness them for utilitarian tasks in transforming everyday life. The contemporary Soviet version of this critique of Constructivism was made forcefully by the critic Iakov Tugendkhol'd in his review of the Paris Exposition, in which he lumped the 'spiritless geometry' of the Russian Constructivist exhibits together with the rationalized geometry of those of the Esprit Nouveau group in France, exemplified by Le Corbusier's exhibit of a starkly furnished house as a 'machine for living,' complete with a maid's room. Tugendkhol'd had little patience for utopian technicism, whether from the left or the right: 'The fetishism of the machine, the worship of industry – here is the pathos of this group of artists, serving in essence as ideologues of the large-scale capitalism flourishing in France.'[5] He warned the Constructivists against their participation in this 'new style,' because helping to align people with the products of modern industry most often simply facilitates their subjection to its (capitalist) logic. This critique reappears in Manfredo Tafuri's dark vision of modernist utopianism, for example, as well as in Jean Baudrillard's postmodern critique of modern design.[6]

Most recently, Hubertus Gassner has offered a provocative analysis of Rodchenko's series of hanging constructions of 1921 as transparent systems that metaphorize and organize both the body and the unconscious – only to claim that the utilitarian turn in Constructivism destroyed the purity (and interest) of these systemic forms by harnessing them to the service of Soviet modernization and industrialization.[7] According to Gassner, the hanging construction, in allowing for nothing that exceeds determination by the system, permits the Constructivist artist-engineer to achieve organized self-consciousness through the very process of making it. 'If the structure is completely systematic in its inner logic and entirely transparent in its making or functional modes, i.e., if the object is "constructed throughout,"' Gassner writes, 'it appears as a homologous model of the producer's unconscious of which he has become fully aware. The artistic subject becomes as transparent as his creation. The previously impenetrable dark of his subconscious and body is illuminated and rendered transparent through the exposure of the logic of their functional modes.'[8]

Gassner's confidence that the conscious subject can become 'fully aware' of her unconscious desires in this way may fly in the face of a more sober, psychoanalytically described reality, but precisely this *fantasy* of a transparent relay between the consciousness of the maker and the consciousness of the object fuels the most utopian ideal of the Constructivist object, such as Arvatov's notion of the object as a conscious 'co-worker' formed within the active, dynamic, and conscious processes of industrial production. Gassner further identifies a compelling homology between the Constructivist object (the co-worker) and the human *body*:

> In the constructivist universe, objects exist solely as organs of human activity. They adjust to people's actions, expand and die with them, while constantly renewing their own shape and function. The constructivist objects are congruent counterparts of the subject. Therein lies their utopian potential.

> Ideally, they would have transformed material reality into an unrestricted space in which free people could act.[9]

The Constructivist object as a 'congruent counterpart' of the human subject, an object that 'expands and dies' with the human body, brings us close to what Rodchenko might mean when he calls the object a 'comrade.' Yet Gassner claims that the displacement of this homology between the body and the object on to utilitarian tasks – the transition, that is, from the *Hanging Construction* to the Workers' Club interior – would lead only to the subjection of human bodies to the forces of industrialism. Gassner thus brings us back to the dystopian conclusions of Tugendkhol'd and Tafuri. The Constructivists failed to transform reality into a space of freedom, he concludes, because the moment of perfect transparency, which is also a fleeting moment of pure autonomy for the art object (it is responsible only to itself, to its own coherent system), is destroyed once it is brought into contact with history – when the self-referential and 'non-utilitarian' structures have utilitarian imperatives imposed on them from outside the system.[10]

I want to suggest, however, that Gassner's insights into the bodily and unconscious functioning of the nonutilitarian Constructivist object can actually be used, instead, to support a claim for its utopian potential precisely in its utilitarian form. For Gassner offers the first useful analysis we have seen of the uncanny *content* of the Constructivist object: its doubling of the human body. In Marx's definition of commodity fetishism, the system of exchange inverts social relations, resulting in 'material [*dinglich*] relations between persons and social relations between things.'[11] Hal Foster has recently suggested that in this trading of semblances between producers and products, 'the commodity becomes our uncanny double, evermore vital as we are evermore inert.'[12] In contrast to Marx, whose entire critique of political economy is aimed at restoring that lost set of social relations between producers, Arvatov's theory of the Constructivist object attempts to recuperate for proletarian culture this notion of thing-like relations between producers and of social relations between newly active and materially appropriate things.[13] Constructivism aims, in effect, to remake or harness the uncanny of the commodity – its ability to act as the *Doppelgänger* for the human producer – for socialist ends. The uncanny effect of an object stems from its evocation of a repressed desire; the uncanny (*das Unheimliche*), Freud says, 'can be shown to come from something repressed which *recurs*.'[14] In the uncanny this recurrence provokes anxiety but the socialist object would make a space within the uncanny (a home within the *Unheimliche*) that could also be the site of release from or acknowledgment of the repressed desire. For the 'secret nature' of the uncanny is that this recurrence is 'in reality nothing new or foreign' (nothing *unheimlich*), 'but something familiar and old-established in the mind' (something *heimlich*), which is why Freud insists that *das Heimliche* cannot be differentiated from *das Unheimliche*.[15] In its uncanny animation, the Constructivist object will be the figure of the automaton, working to align human subjects with the modernizing 'light from the East,' but in its very embodiedness it will also mark out a homely space for the potential humanizing of the un-homely products of industrial culture, bringing those products into the human field of desire.

NOTES

1 Aleksandr Rodchenko, 'Rodchenko v Parizhe. Iz pisem domoi,' *Novyi Lef* 2 (1927), p. 20 (letter of May 4, 1925). A substantially different version of the letters, containing less of his personal or subjective commentary, was published in *A. M. Rodchenko: Stat'i, vospominaniia, avtobiograficheskie zapiski, pis'ma*, ed. V. A. Rodchenko (Moscow: Sovetskii Khudozhnik, 1982). This and other translations from the Russian are my own.

2 See Boris Arvatov, 'Byt i kul'tura veshchi' (Everyday Life and the Culture of the Thing), in *Al'manakh Proletkul'ta* (Moscow, 1925), p. 79.

3 The term 'nonutilitarian' (*vneutilitarnyi*) is used by Boris Arvatov in *Iskusstvo i klassy* (Gosudarstvennoe Izdatel'stvo: Moscow–Petrograd, 1923), p. 40. While descriptively helpful, the term has promoted a certain teleology toward utilitarianism in early Constructivism. See, for example, Christina Lodder's use of the term in *Russian Constructivism* (New Haven: Yale University Press, 1983).

4 Soviet scholars grouped around Selim O. Khan-Magomedov and the journal *Tekhnicheskaia Estetika* have, since the 1960s, defended the significance of Constructivism by insisting precisely that it initiated the modern medium of design (*dizain*). The need to take this approach was partly conditioned by political circumstances in the USSR in the 1960s and 1970s that discouraged the study of Russian Constructivism as a viable form of modern art, but the consequence has been

an unfortunately narrow view of Constructivism's ambitions in Soviet and now Russian scholarship.

5 Iakov Tugendkhol'd, 'Stil' 1925 Goda. (Mezhdunarodnaia Vystavka v. Parizhe),' *Pechat' i Revoliutsiia* 7 (1925), p. 42.

6 See Manfredo Tafuri, 'USSR–Berlin, 1922: From Populism to "Constructivist International,"' in *Architecture, Criticism, Ideology* (Princeton: Princeton Architectural Press, 1985), pp. 121–181. Tafuri concludes that the utopian vision of Russian Constructivism led, in both the USSR and in Europe, to an apolitical ideology of technicism and total organization that was completely available to capitalism. For Baudrillard's critique of the elitism and repressiveness of so-called utilitarian modern design, see *For a Critique of the Political Economy of the Sign*, trans. Charles Levin (St Louis: Telos Press, 1981), especially his critique of the Bauhaus in chapter 10, 'Design and Environment.'

7 Hubertus Gassner, 'The Constructivists: Modernism on the Way to Modernization,' in *The Great Utopia: The Russian and Soviet Avant-Garde, 1915–1932* (New York: The Guggenheim Museum, 1992). Gassner's reading of the hanging constructions forms part of a larger argument about the change in Rodchenko's work from an emphasis on *faktura* (in his painting up until around 1920), which Gassner associates with intuition and feeling, to an interest in nonmaterial line and system, associated with logical planning and anti-individualist making. Gassner sees the uniformly smooth, shiny silver paint surface of the hanging constructions as a negation of the materiality of paint and canvas in favor of pure, systemic construction. While this transition does exist in Rodchenko's work, he also went on to make all kinds of other objects after 1921 that contradict the declared pre-eminence of line over *faktura*. Gassner's definition of *faktura* as tied to intuition and feeling is also idiosyncratic; other definitions stress the *antisubjective* nature of an interest in material surface, tying *faktura* to a materialist, workmanlike relation to art making. See, for example, Benjamin Buchloh, 'From *Faktura* to Factography,' *October* 30 (Fall 1984).

8 Gassner, 'Constructivists,' p. 317.

9 Ibid., p. 318.

10 Ibid., p. 314. Gassner claims that the Constructivists' decision to throw in their lot with the Soviet campaign for industrialization stemmed not from a genuine commitment to socialism, but from their self-interested struggle to maintain a power base within the Soviet system after their initial success at taking over organizational posts in museums and art education was curtailed around 1920. However, he offers no concrete evidence to support this depoliticized reading. See pp. 315–316.

11 Karl Marx, *Capital*, vol. 1. trans. Ben Fowkes (New York: Vintage, 1977), p. 166.

12 See Hal Foster, *Compulsive Beauty* (Cambridge: MIT Press, 1993), p. 129. My thinking about the Constructivist theory of the object in relation to the commodity is indebted to Foster's insightful suggestion that the uncanny as a concept may be historically dependent on the rise of the mass-produced 'mechanical commodified' object.

13 Arvatov's attempted recuperation of the agency of things assumes as a necessary prerequisite the elimination of capitalist exploitation of labor and commodity exchange. I discuss Arvatov's theory of the object in 'Constructivism and Bolshevik Business: Theory and Practice of the Socialist Commodity,' chapter 2 in 'The Russian Constructivist "Object" and the Revolutionizing of Everyday Life, 1921–1929' (Ph.D. diss., University of California, Berkeley, 1995).

14 Sigmund Freud, 'The "Uncanny"' (1919), in *Studies in Parapsychology*, ed. Philip Rieff (New York: Collier Books, 1963), p. 47. While Foster emphasizes the ways that the uncanny recalls infantile anxieties of blindness, castration, and death, for my purposes here I want to underscore that Freud specifies that any kind of emotional affect, once repressed, can be the source of the 'uncanny.'

15 Ibid.

Toys

Roland Barthes

French toys: one could not find a better illustration of the fact that the adult Frenchman sees the child as another self. All the toys one commonly sees are essentially a microcosm of the adult world; they are all reduced copies of human objects, as if in the eyes of the public the child was, all told, nothing but a smaller man, a homunculus to whom must be supplied objects of his own size.

Invented forms are very rare: a few sets of blocks, which appeal to the spirit of do-it-yourself, are the only ones which offer dynamic forms. As for the others, French toys *always mean something*, and this something is always entirely socialized, constituted by the myths or the techniques of modern adult life: the Army, Broadcasting, the Post Office, Medicine (miniature instrument-cases, operating theatres for dolls), School, Hair-Styling (driers for permanent-waving), the Air Force (parachutists), Transport (trains, Citroëns, Vedettes, Vespas, petrol-stations), Science (Martian toys).

The fact that French toys *literally* prefigure the world of adult functions obviously cannot but prepare the child to accept them all, by constituting for him, even before he can think about it, the alibi of a Nature which has at all times created soldiers, postmen and Vespas. Toys here reveal the list of all the things the adult does not find unusual: war, bureaucracy, ugliness, Martians, etc. It is not so much, in fact, the imitation which is the sign of an abdication, as its literalness: French toys are like a Jivaro head, in which one recognizes, shrunken to the size of an apple, the wrinkles and hair of an adult. There exist, for instance, dolls which urinate; they have an oesophagus, one gives them a bottle, they wet their nappies; soon, no doubt, milk will turn to water in their stomachs. This is meant to prepare the little girl for the causality of house-keeping, to 'condition' her to her future role as mother. However, faced with this world of faithful and complicated objects, the child can only identify himself as owner, as user, never as creator; he does not invent the world, he uses it: there are, prepared for him, actions without adventure, without wonder, without joy. He is turned into a little stay-at-home householder who does not even have to invent the mainsprings of adult causality; they are supplied to him ready-made: he has only to help himself, he is never allowed to discover anything from start to finish. The merest set of blocks, provided it is not too refined, implies a very different learning of the world: then, the child does not in any way create meaningful objects, it matters little to him whether they have an adult name; the actions he performs are not those of a user but those of a demiurge. He creates forms which walk, which roll, he creates life, not property: objects now act by themselves, they are no longer an inert and complicated material in the palm of his hand. But such toys are rather rare: French toys are usually based on imitation, they are meant to produce children who are users, not creators.

The bourgeois status of toys can be recognized not only in their forms, which are all functional, but also in their substances. Current toys are made of a graceless material, the product of chemistry, not of nature. Many are now moulded from complicated mixtures; the plastic material of which they are made has an appearance at once gross and hygienic, it destroys all the pleasure, the sweetness, the humanity of touch. A sign which fills one with consternation is the gradual disappearance of wood, in spite of its being an ideal material because of its firmness and its softness, and the natural warmth of its touch. Wood removes, from all the forms which it supports, the wounding quality of angles which are too sharp, the chemical coldness of metal. When the child handles it and knocks it, it neither vibrates nor grates, it has a sound at once muffled and sharp. It is a familiar and poetic substance, which does not sever the child from close contact with the tree, the table, the floor. Wood does not wound or break down, it does not shatter; it wears out, it can last a long time, live with the child, alter little by little the relations between the object and the hand. If it dies, it is in dwindling, not in swelling out like those mechanical toys which disappear behind the hernia of a broken spring. Wood makes essential objects, objects for all time. Yet there hardly remain any of these wooden toys from the Vosges, these fretwork farms with their animals, which were only possible, it is true, in the days of the craftsman. Henceforth, toys are chemical in substance and colour; their very material introduces one to a coenaesthesis of use, not pleasure. These toys die in fact very quickly, and once dead, they have no posthumous life for the child.

28

The Hoover® in the Garden

Middle-class Women and Suburbanization, 1850–1920†

Roger Miller

TECHNOLOGICAL INNOVATION AND DOMESTIC CONSUMPTION PATTERNS

One important set of circumstances that contributed to women's constrained daily lives revolved around major changes in domestic technology between 1850 and 1920. In fact, reduction of the number of (and in many cases elimination of) servants in the household was ultimately made possible by the development of new technologies and products which greatly reduced the time required for household chores and activities. Some of these technological innovations were associated with the municipal provision of services. The connecting of houses to supplies of gas, water, and electricity and the provision of domestic sewerage eliminated many tasks. Related innovations, such as the provision of indoor plumbing, central heating, and the development of more efficient lighting fixtures and stoves for cooking, also cut down on the amount of hard manual labor required for maintaining the household (Strasser, 1982; Wright, 1981). Perhaps the most important innovations in domestic technology were the refrigerator, the washing machine, and the vacuum cleaner. Improvements in food storage technology, the refrigerator in particular, eliminated the need for daily shopping. The washing machine cut the most arduous of the weekly tasks down to reasonable size. The vacuum cleaner, an important innovation after the introduction of electricity into households, shortened cleaning time and made it possible for housewives to keep their homes cleaner (Strasser, 1982).[1]

Nevertheless, it would be wrong to think that the introduction of these domestic labor-saving devices, in and of itself, provided more leisure time for women. Although they cut down the time for many household tasks, they also encouraged the middle-class housewife to assume more direct responsibility for household maintenance tasks that previously would have been delegated to servants. A housewife's work was as often as not managerial when there were a number of servants or even hired helpers in the household. But with the reduction in the number of domestic laborers in the household and the introduction of housekeeping technology, housewives tended to be directly engaged in a form of 'factory life in the home' (Cowan, 1976a, 1976b; Ehrenreich and English, 1975; Strasser, 1982).

This increased reliance on domestic labor-saving devices was vitally important to the economy at large. Increasingly, suburbanization began to mean the decentralization of consumption (Douglass, 1925; Walker, 1977). A home-buying, home-owning middle class represented the major market for manufactured goods in the USA. Buying the home was only the first stage in consumer development. Homes, once bought, had to be furnished, and the individual most

† Hoover is a registered trademark. The author would like to emphasize the contradictions inherent in the use of a registered trademark in the title of an essay on the development of ideological control in the advertising of commodities for domestic work.

likely to direct further purchases was the housewife. Advertisements from this period reflect the growing awareness of women's buying power and their control over purchases in many homes. Increasingly, women were told by advertisers that they, individually, were responsible for their family's health, welfare, and safety. Advertisements for everything from cereals to laundry detergents stressed the central role of the housewife as guardian of the family (Cowan, 1976a; Ewen, 1976). Housewives were depicted as doing the washing, fixing meals, and cleaning the house in these advertisements, whereas they earlier had been portrayed as instructing maids and houseworkers in the use of household products.

Changes in daily life

The combination of the processes of suburbanization, growth of social mobility, and changing domestic consumption led to a considerably more constrained life for many women living in twentieth-century suburbs than what had been the norm even in the supposedly rigid Victorian era. Although middle-class women had been put upon ideological pedestals in the mid-nineteenth century, they were forced to develop sophisticated managerial skills in order to run large and complicated households. The presence of servants meant that many tasks could be delegated, which gave the housewife significant periods of free time to carry out social duties and take part in discretionary activities. By 1920, however, domestic servants had ceased to be the major sign of achieved middle-class status in suburban areas. Rather, home ownership had come to signal social arrival. In order to buy even a modest house in a distant suburban tract, some households, particularly those that had barely attained middle-class incomes, had to make compromises, and chief among these was the substitution of domestic technology and casual domestic labor for live-in domestic help. As a result, the housewife had to take on more responsibility for housework herself.

Although the decreasing numbers of available servants and the substitution of domestic technology were problems for all middle-class households, the ultimate effects on the lives of female homemakers were different in suburban and urban settings. Suburban wives tended to be severely constrained in their daily activities, especially when compared with their counterparts in urban apartment settings. In urban apartments it was possible to manage with fewer servants, for several reasons. First, many of the household maintenance tasks were performed centrally in the apartment building. Heat was supplied, grounds were kept up by a maintenance staff, and, in some rare cases, food was provided from central kitchens (Wright, 1981). Second, hourly help was more likely to be available in neighborhoods near the center of the city than in a distant suburb (thus many city apartments were built without servants' rooms). Third, the central neighborhoods had the density to attract support services such as bakeries and laundries. In contrast, the suburban home required the expenditure of time on gardening and the upkeep of grounds, on shovelling snow, and on maintaining the structural integrity of the home. Fewer services were provided close to the home and hence the need for domestic servants was correspondingly greater. When servants became harder to find, a larger share of the increased amount of work required to maintain the household fell to the housewife.

There was also a change in the location of discretionary activities between 1850 and 1920, as well as a change in the unwritten rules and customs governing participation in these activities. Earlier suburbs were closer to the center of the city and initial suburbanites were wealthier as a group than their later counterparts. Hence, more city-centered activities were accessible *and* affordable to middle-class patrons in 1850 than in 1920. Women were tied more closely to the home both for discretionary and for nondiscretionary activities by 1920, especially if they were unable to hire daily help because of limited financial resources or the unavailability of a non-live-in domestic labor pool. As the individuals responsible for actually carrying out domestic tasks, they had to remain in the home for longer hours, despite the supposed time-savings represented by labor-saving devices. In part this was because the tasks to be accomplished had been redefined. Although women were responsible for the health, safety, and moral welfare of their families both in 1850 and in 1920, the tasks associated with these goals had increased dramatically by 1920 (Ehrenreich and English, 1975). Women were bombarded with information about the importance of domestic hygiene, which advertisers couched in terms of blatant appeals to guilt. In short, standards became more stringent as the means for meeting them became generally available. Thus, the ability to wash clothes more efficiently did not mean that the wash would be done in less time; it meant that the same clothes could be washed more frequently. The Hoover® brought the rug out of the attic, where it was stored between social events, and put it down to be trampled and vacuumed every week. Women had to rationalize their schedules with those of the new machines they tended. Although a housewife might not be constantly engaged in doing laundry on wash days, she could not leave her washing machine unattended for long periods because manual labor was still required at almost every stage, including loading, adding detergent and chemicals, unloading, wringing, hanging out to dry, and ironing. As Palm and Pred (1974) have shown for contemporary women, it is necessary to consider both the duration and *distribution* of periods of free time in order to determine potential access to activities.

The increasing distance between city and suburb would not have been so deleterious to suburban women had there been adequate suburban activities to replace those that were no longer accessible in the city. Activities outside the home had been class specific in 1850, but, owing to the smaller proportion of the population that had attained middle- or upper-class status, the class distinctions had been broader and somewhat more inclusive. Also, the city provided a greater variety of class-specific opportunities, so if one activity was foreclosed, others were potentially available. Women could engage in philanthropic and church-related activities in 1850, as well as a wide range of social and recreational pursuits. Even when discretionary activities were confined to the home, they were likely to include individuals from beyond the immediate household because of the prevalence of frequent visiting and the maintenance of extended family ties (Smith-Rosenberg, 1975).

Many of these possibilities were no longer available by 1920. The professionalization of philanthropy and the rise of social welfare work removed the outlets of 'good works' and 'friendly visits.' The dispersed residences of the suburbs were not very conducive to formal visiting, one of the mainstays of city social life. The greater focus on the nuclear family meant that home-bound activities were likely to be solitary activities. Opportunities in residential suburbs were limited and many of them, if not most of them, focused attention on women's domestic roles. Suburban activities also tended to be quite class specific, reflecting the suburban residential landscape itself, which had been built to reify class distinctions through spatial homogeneity.

CONCLUSIONS

As we have seen, women became more directly involved in domestic tasks for a variety of reasons, many of which were related to larger social processes. Demographic and social change led to a declining number of females going into domestic service, creating a labor shortage problem which was exacerbated for women in suburban settings. Innovations in domestic technology allowed the preferences for suburban residences to be maintained, but at a cost to some middle-class housewives.

A key point which has not yet been addressed concerns the change in attitude which accompanied these shifts in activities. Why was it acceptable, and even encouraged, for middle-class women to do domestic work in 1920, when the opposite was true in 1850? First, this attitudinal shift was not a sudden one; it occurred over several decades. As some middle-class wives found themselves in new living situations and were faced with new difficulties, they had to respond to their changed conditions. Professional advice-givers, from members of the American Home Economics Association to the advertisers in the magazine *House Beautiful*, steered the responses these women made. They provided the validation for the new modes of behavior as they promoted new forms of domestic technology and household products. Several researchers have questioned which came first – the public validation of new forms of behavior or the new responses themselves (Cowan, 1976a; Ewen, 1976). The question possibly bypasses the major point of the process. New attitudes grew out of the activities in which these women engaged. The activities were both constrained by and enabled by social processes, and the latter were themselves the expressions of other individuals facing different structural and structurating situations. Ideology in the domestic sphere was not separate from ideology in other areas: women were absorbing codes and ideas for appropriate behavior from their past experiences as well as from contemporaneous attempts to mold their opinions. The attitudinal shifts that were apparent in the advertising of household products between 1895 and 1920 both reflected and shaped social practices. The processes related to suburbanization, growth of social mobility, and domestic technological innovation did not occur in a strictly linear fashion or at a constant rate. Coping behaviors were thus responses to shifting situations.

NOTE

1 It should be noted that innovations in household technology were not always adopted because they were perceived as time-saving devices. In a number of instances new technology required more labor than earlier methods. For instance, the vacuum cleaner initially was a large and cumbersome device. Early vacuum cleaners required two persons for operation – one to crank or pump to create the vacuum and another to clean. Such innovations were often adopted by wealthier households first, since they were expensive and often labor intensive. As production continued, new models utilizing electric power and embodying fewer decorative elements could be made less expensively. At this point the technology was adopted by middle-class families.

REFERENCES

Cowan, R. S. (1976a) 'The "industrial revolution" in the home: household technology and social change in the twentieth century.' *Technology and Culture* **17**: 1–23.

Cowan, R. S. (1976b) 'Two washes in the morning and a bridge party at night: the American housewife between the wars.' *Women's Studies* 3: 147–171.

Douglass, H. P. (1925) *The Suburban Trend* (Appleton-Century-Crofts, New York); reprinted 1970 (Johnson Reprint Corporation, New York).

Ehrenreich, B. and English, D. (1975) 'The manufacture of housework.' *Socialist Revolution* 5 (4): 5–40.

Ewen, S. (1976) *Captains of Consciousness: Advertising and the Social Roots of the Consumer Culture* (McGraw-Hill, New York).

Palm, R. and Pred, A. (1974) 'A time-geographic perspective on problems of inequality for women.' WP-236, Institute of Urban and Regional Development, University of California, Berkeley, CA.

Smith-Rosenberg, C. (1975) 'The female world of love and ritual: relations between women in nineteenth-century America.' *Signs* **1**: 1–29.

Strasser, S. (1982) *Never Done: A History of American Housework* (Pantheon Books, New York).

Walker, R. A. (1977) *The Suburban Solution: Urban Geography and Urban Reform in the Capitalist Development of the United States.* Ph.D. dissertation, Department of Geography, The Johns Hopkins University, Baltimore, MD.

Wright, G. (1981) *Building the Dream: A Social History of Housing in America* (Pantheon Books, New York).

29

The Two-Way Mirror

Elaine S. Abelson

As more goods and a greater variety of goods came on to the market, the need to protect them became increasingly urgent. Shoplifting was already a critical problem in the stores by the 1890s; 'a growing evil,' wrote one reporter. Initially, the new fixtures seemed to solve the dilemma of the greater availability of merchandise and the increased incidence of casual theft. Descriptions of different display cases for individual departments and particular types of merchandise, each one promising to protect the goods from theft, seemed to flood the pages of the store journals in the early 1890s: leather goods, toilet articles, umbrellas, parasols, jewelry, 'fancy' goods, gloves, handkerchiefs, shirts, suits, men's furnishings – suddenly, all required protective glass cases, allowing the customer to closely inspect but not to handle the merchandise – in sight but enclosed and definitely out of reach.[1]

George G. Hammond, a salesman in the black silk department of the R. H. Macy Company in the late 1880s, related the following incident during an interview many years later. His testimony seems to reinforce the claim that unprotected goods were at risk. 'The buyer went downtown and bought a bolt of grosgrain silk which was put out on the counter at about 10:30,' Hammond recalled, 'and at noon [it] had disappeared, evidently stolen, as no one could tell what had become [of] it.'[2] A *Dry Goods Reporter* [*DGR*] 'interview' with an unnamed Chicago merchant sought to dramatize the same problem for its readers and point out the appropriate moral: The merchant related how he had placed a display of shoe findings ('blocking trees, shoe horns and the like') on a table in the front of the department, but found that, instead of the expected profit from the table, he 'had to contend with a loss by persons helping themselves to these articles. Being on a table, many persons, not genuine thieves, of course, but absent minded, helped themselves to polish, laces, etc., on leaving the store. I saw after a time that the gain did not make up for the loss and I bought a show case . . . I placed it on the table and put my findings inside, so that people could not get at them readily. My findings sales increased, as many persons would stop and look into the case.'[3]

Ribbons had their special cases by 1892: a glass front cabinet with shelves that slid out to one side, providing easy access for the saleswoman, but not for the customer. Before the end of the decade, advertisements for a more elaborate, revolving ribbon case assured the merchant that 'no ribbons can be removed from [the] case except by [a] clerk.'[4] An obvious nuisance, but an important fashion accessory, ribbons were a 'must' for the stores: a valuable and easy-to-hide commodity, they were also a magnet for shoplifters. The problem seemed never to have resolved itself; for ten years later, fixture companies were still claiming that their particular ribbon cabinets were indispensable for 'convenience in handling goods' and for 'protection and attractive display of stock.'[5] Handkerchiefs, too, were suddenly more effectively displayed in showcases. 'The nature of their borders, whether self or fancy colored, can be seen through the glass, and when the customer desires to acquaint herself with the texture, the boxes may be taken out for inspection.'[6] The implication was clear.

A *Dry Goods Economist* [*DGE*] article on 'The Millinery Department' presented a particularly interesting view of protection. 'Everything,' the article began, 'must be displayed openly.' But 'openly' meant something quite different for differing grades of merchandise. The carefully drawn illustration showed trimmed hats (and it implied that they were expensive) exhibited in rather elegant, free-standing glass cases; untrimmed hats (obviously much less costly) lay about on a table top, unprotected.[7] Trimmings followed hats.

Ordinary merchandise was often easily and openly available, while 'a glass case . . . contained extreme and delicate novelties in this line.'[8]

Even the commonplace hairbrush came in for its share of concern as trade journals advised merchants that 'the modern woman brushes her hair every day and has several different brushes' for the ritual.[9] 'The Improved Graves Brush Rack' eliminated 'hair brushes thrown in a heap with marred backs and broken bristles' from constant 'pawing' over. The Graves Rack promised to display 'each one of 8 to 10 dozen hair brushes' and guaranteed to 'do away with shoplifting, increase your sales, save your time and add to your profits.'[10]

There was, suddenly, a whole industry centered on department store fixtures and the requirements of a new security consciousness. The advertisements of the Norwich Nickel and Brass Works of Norwich, Connecticut, illustrate the heightened interest in store interiors and the competing demands of visibility and protection. The company published an eighty-page illustrated catalog in 1891, and advertised it as 'an arsenal' of display fixtures, counters, mirrors, and window dressing aids of every possible combination.[11] During portions of 1892 and 1893, Norwich ran a full-page advertisement every week in the *DGE*, each one differing from the previous week, and each one highlighting a particular store fixture. There was little subtlety in the Norwich copy; shoplifters were a given, and the copywriters simply pointed out the dual uses of the products, the opportunity offered both for profit and protection.

Émile Zola fully understood the connection between department store display and shoplifting. His 1883 novel, *Au bonheur des dames*, was an attempt to dramatize the materialism of late nineteenth-century France and the moral implications of mass consumption as they were manifested in its quintessential institution, the large dry-goods bazaar. The power of the big store and the power of women within the store were the twin foci of the novel and the conflicting poles around which all action revolved. Not surprisingly, neither the tempo of life nor the realities of commercial seduction in the Parisian Ladies' Paradise were very different from those of a comparable American department store. The pressures to sell and to buy were equally intense. When an 'aristocratic-looking' customer intimidated the salesclerk at the lace counter into showing ever more expensive samples, Zola captured the tension of the scene and the inevitable outcome: The counter covered with a fortune in Alençon lace, Mme. de Boves demanding to be shown still another sample, the clerk afraid to resist, yet hesitating, 'for salesmen were cautioned against heaping up these precious fabrics, and he had allowed himself to be robbed of ten yards of Malines the week before.' The salesman, Deloche, yielded, of course, and 'abandoned the Alençon point for a moment to take the lace asked for from a drawer.' A sample disappeared up the sleeve of Mme. de Boves, and Deloche was once again the ready victim.[12]

One doubts that many department store owners and managers read Zola, but one can bet they read C. G. Phillips, author of the influential 'Wide-Awake Retailing' column in the *Dry Goods Economist*. Throughout the 1890s, Phillips pressed for acceptance of glass cases and counters to prevent such incidents of shoplifting. Although new showcases were a capital expenditure like any other, stores were slow to convert. Some stores in the major cities replaced many of the older wooden counters with the new glass combinations, but even here there was an obvious reluctance to introduce a change that transformed the look of the traditional store interior. Again and again Phillips' articles touted modern fixtures as both a valuable asset and an indispensable protector of merchandise. The copy changed, but the message was always the same:

> Up to date showcases are worth their weight in gold . . . they sell goods, they preserve goods from injury, they prevent lots of mark downs, they are effectual protection against shoplifters, they give an air of brightness and general up-to-dateness to the store that makes business.[13]

Glass became an attraction in its own right and offered the merchant the money-saving advantage of a large quantity of goods in full view, theoretically safe from the ever-lurking shoplifter. 'You save money by securing your goods against thieves, who are constantly operating in stores where the large quantities of easily concealed goods are lying about loose on the counters, not one in a hundred of these "light fingered" people ever being discovered.'[14] Phillips wrote this in May 1899; in July he upped the estimate: 'For one so detected there are probably two hundred who go uncaught, the safe side is worth the necessary expense of showcases and glass counters.'[15] 'More money is lost annually,' the Wide-Awake Retailer continued, 'by having stock spread in open boxes and on counters than would pay for glass tops or glass show cases twice over.'[16] Wide-Awake's final word for 1899 was marked by cautious resignation: 'If you do not feel disposed to do anything just now, the *best plan* is to keep your counters as clear of goods as possible.'[17]

As in the early days of retailing, the counter again became a sales mediator, a barrier between the customer and the merchandise. Merchants attempted to find a median between glassed-in stock, which meant slower service but a greater defense against theft, and open selling, which fostered impulse buying but also easier access and theft on a much broader scale. They were never very successful. There was no entirely satisfactory resolution of the problem, as the *DGE* recognized: 'Either the custom of displaying goods promiscuously on table and counters without an adequate amount of help to watch them will have to be modified or the merchandise will have to be kept out of the reach of dishonest fingers in the boxes or in cases.'[18] Sixty years later, a report by the National Retail Dry-Goods Association, 'Controlling Shortages and Improving Protection,' demonstrated how little progress had been made; the dialogue still revolved around the critical question of well-designed fixtures. 'Merchandise on counter tops,' the report contended, 'was at risk'; only those counters free of merchandise were safe. The monumental problem of stock protection remained just that.[19]

By 1910 most department store displays were fully under or behind glass, and, as in the days of A. T. Stewart, customers were again forced to ask to be shown specific things. Management hoped the inoffensive glass barrier would stimulate purchases as it discouraged both 'the general, almost unconscious habit of handling goods unnecessarily' and the tendency of far too many customers to pocket something on the sly.[20] But the great revolution in display seemed to have gone a full 360 degrees.

The mixed possibilities inherent in glass counters and showcases were duplicated in the newly extended use of mirrors. Although they had been used in stores since the 1860s, large mirrors were not manufactured in the United States before the 1880s and were still something of a novelty in the early 1890s.[21] A. T. Stewart had imported the first full-length mirrors from France in 1852 for his Marble Palace, and for the rest of the century 'French plates' continued to be the quality reflectors.[22]

'Mirror, Mirror on the Wall' became more than an incantation; it expressed the newly discovered, essential role of reflection in the dry-goods bazaars. In the 1890s mirrors sprouted on walls and on columns, in showcases and in windows, and wherever customers might stop and look, indulging the sort of fantasies the stores sought to encourage. 'Where is the woman,' the *DGE* asked, 'who is not satisfied on finding herself before a mirror?'[23] Seemingly no function was too farfetched. A 1902 article suggested that 'one of the best features of mirrors is the influence they exert on busy days. They put waiting customers in a more satisfied frame of mind and induce them to wait without complaint longer than they would ordinarily.'[24]

As the psychological aspect of selling became more important, mirrors assumed a complex function. More than anything else, they changed the sense of dimension, 'concentrating and heightening the light, sharply defining the articles displayed,' making the window or display case seem larger and the material abundance even greater.[25] Mirrors doubled the stock without increasing the investment, one practical merchant pointed out. 'Not an adornment,' mirrors were an 'absolute necessity.'[26] In Zola's Ladies' Paradise, 'mirrors, cleverly arranged on each side of the window, reflected and multiplied the forms without end, peopling the street with these beautiful women for sale.'[27] They became an inseparable adjunct to the merchandise, a powerful 'silent suggestor' which reflected the goods from a variety of angles and added to the illusion of unlimited availability of things.

The *DGE* maintained, however, that store personnel were unaware of the manifold possibilities of mirrors, and took upon itself the job of educating them. The journal argued that mirrors could be both 'bait' to bring a woman into a store, and a toy to 'entertain her' while she was there. Instructions told up-to-date merchants to place them – particularly the novel duplicate and triplicate ones – in strategic areas to enhance the merchandise while 'they please the customers.' They pay for themselves 'over and over again during a season,' the copy promised.[28] That a modern store could not have too many of 'these look-at-yourself affairs about' became a commonplace.[29] With the increased availability of ready-to-wear, 'our mirror sells more cloaks than a couple of salesmen,' one mirror company advertised. 'A customer can't doubt its reflection.'[30]

What was implicit in this ongoing discussion was the corresponding use of these same mirrors to provide a 'judicious system of reflections' and scrutiny in an 'unlimited multiplication of areas, objects and stock.'[31]

In the 1890s the Lazarus dry-goods store in Columbus, Ohio, had a mirror arrangement in the ceiling that could be observed from the elevated first-floor office of the owners, Fred and Ralph Lazarus. The mirrors 'were at different angles . . . you really could see through brick and plaster walls by seeing around them, reflection after reflection,' Fred Lazarus, Jr., recalled years later. Lazarus thought that shoplifting was not much of a problem in the 1890s but acknowledged that the mirrors were there to both 'watch customers and see how salespeople were waiting on them, and they could see them.'[32]

Some store architects made liberal use of mirrors at the back of shelving by the late 1890s. The object was far more than ornament: it was security.

> They serve as silent and unnoticed detectives. The clerk whose back is turned to a shopper while taking articles from the shelves or putting them back can observe every motion of the visitor. Many an act of shoplifting can thus be avoided and in the most graceful way. 'Did you wish to take that article, Madame?' is quite sufficient to make the person examining it drop it like the proverbial hot potato. And there is no scene, no police station, no story in the papers.[33]

As with glass counters and cases, mirrors were charged with a twofold mission. While they reflected and enhanced the merchandise and the store itself, encouraging the fantasies of consumption, they simultaneously served for surveillance, reflecting the surreptitious and the illegal activities of a great many customers.

> One firm in Washington, D.C., protects what would otherwise be a most dangerous corner by a full-length mirror. It is in the book department where customers and shoppers naturally expect to rove unmolested from counter to counter and where an obtrusive watching of their movements would be impolitic. The mirror tells its silent tale: the visitor is watched but is all unconscious of the surveillance.[34]

'Sales people can watch the customers even when their backs are turned,' became a vital selling point.[35] However, an article that appeared in the *Brooklyn Eagle* in the summer of 1901 was more persuasive than weeks of reasoned argument. Headlined 'Alleged Shoplifters Caught,' the piece described how a store detective 'was looking at a mirror with his back turned to two shoppers when his attention was drawn to the movements' of one. He saw the customer 'open one of the cases and take from it a garment.' Until that chance look in the mirror, he said, the work 'was so adroitly done that no one but an expert would have noticed them.'[36] Far more than alert salesclerks or detectives, glass in its various manifestations became the primary safeguard for the merchandise.

The new uses of lighting furnish still another view of the continual dichotomy between display and protection. Proper illumination did not become an issue until technology provided the possibility of greater light at reasonable cost, and until the stores became large enough to require more than a minimum of interior illumination. By the end of the nineteenth century, 'poorly lighted' stores came to mean 'uninviting,' and gloomy establishments drew only those women who entered out of necessity, not those who came to 'shop.'[37] Zola returned repeatedly to descriptions of the half-light in the old-fashioned shop of the dry-goods merchant Baudu: the funereal appearance of the store in which the goods were hardly visible and the customers only occasional visitors.[38] The failing Baudu was a relic of the past who exemplified everything the new bazaars sought to change.

Kerosene and gas were early methods of introducing light into the dry-goods bazaars, but both had the multiple defects of heating up the atmosphere, poor combustion, yellowish color, and limited capacity. Once electric light became feasible for large interior spaces, these other forms of illumination lost ground. Often a combination of the new and the old appeared side by side in the 1890s. Descriptions of stores of this period regularly mention the supercharged atmosphere, particularly in the afternoons when the growing crowds, the 'glaring intensity of the light,' and the heat of many hundreds of gas jets and electric arc lights made the stores stifling and even breathing difficult.[39]

Huge windows and a momumental center rotunda capped by a skylight characterized the Astor Place dry-goods palace of A. T. Stewart in 1870. Reflecting the best architectural and technological capability of its day, Stewart's used electricity, but only to automatically light the gas fixtures at a predetermined hour.[40] The result, however, was transforming. A description of the sudden metamorphosis of the store on a December afternoon when 'suddenly, as if by magic, this vast pile became irradiated with a sea of light,' allows us to appreciate the drawing power of illumination in the nineteenth century.[41]

The first stores to use electricity for light were, of course, the big, popular, highly competitive institutions. The R. H. Macy Company provides a good example of the progression from gas to electricity. A Macy's Christmas advertisement in the *New York Herald* in December 1875 mentions a 'store beautifully illuminated outside and in.' The illumination was gas. The first electric light, an electric arc light, appeared in 1878 outside of the 14th Street building, and although gas fixtures predominated for many more years, arc lights were used in the display windows and inside the store during the 1879 holiday season.[42]

A letter from Isidore Straus to the Manhattan Electric Light Company written in September 1888 indicated that Macy's wanted to fully modernize the lighting system at that time, but for some reason, probably the limited capacity of the lighting company, the store could not or did not. Only when they moved from 14th Street to 34th Street in 1902 was Macy's fully electrified.[43] The plant for lighting this extraordinary new store was said to be 'the largest isolated lighting plant in New York.'[44]

The pioneer decade for electric lighting in the big bazaars was the 1880s. 'Well lighted buildings . . . always give people a sense of comfort, safety and security,' Macy's advertised in 1881, and merchants and customers alike knew it to be true.[45] Light was still a beacon that represented safety and life and added to the already immense attraction of the dry-goods bazaars. Zola's descriptions of the 'furnace-like brilliancy' of the Ladies' Paradise [which] 'shone out like a lighthouse, and seemed to be of itself the life and light of the city' vied with *Dry Goods Economist* reports of Macy's ingenious use of 'concealed incandescent lights and a powerful electric force above,' which made the windows a 'mass of sparkling jewels.'[46] Nighttime window illumination produced spectacular results. Macy's brightly lit Christmas windows, 'for the benefit of all New Yorkers,' were an 'instant, crowd-pulling sensation.'[47] Diarist Clara Pardee was probably among the dense crowds, as she invariably noted 'going downtown to see Christmas.'[48] Although still only a seasonal extravaganza, the illuminated holiday windows were powerful symbols, which competitors quickly imitated. An article in *Outlook* in 1895 described New York shop windows 'ablaze behind invisible plate glass barriers.'[49]

However insufficient it was as a source of illumination, natural light had its adherents; there was a longstanding prejudice for daylight, which in the past had occasionally served merchants' interests. It was not unknown for merchants to pass on shoddy merchandise in the half-light of traditional dry-goods stores, and early discussions of good lighting often hinged on the problem of getting sufficient daylight into the cavernous interiors of the new bazaars.[50] Throughout the late nineteenth century, people still widely believed that natural light was healthier than any form of artificial illumination. Large windows and light-diffusing prisms, which magnified and refracted daylight but were insufficient for the acres of floor space in the larger emporiums, were compared favorably to the arc lamp systems, which theoretically produced a steady, even distribution of light but in fact often flickered and surged in intensity and were widely believed to be responsible for a variety of nervous disorders, headaches, and eye problems.[51] Many shoppers complained of the overheated atmosphere in stores. In *Au bonheur des dames* Zola linked the heavy, still air, the intensity of hundreds of gas jets, and the buying frenzy of the overheated crowds to an image of sensuality, covetousness, and moral flexibility.[52]

From the owners' and managers' vantage point these contradictions were only marginally important. A well-lighted store was both a 'continual and most effective advertisement,' and a store that provided a positive deterrent to the unyielding problem of customer theft. Incidents of customers' going to the door with merchandise 'for better light' were commonly reported. In August 1887, Mrs Julia Hershey was arrested in O'Neill's, a well-known Sixth Avenue dry-goods store, on charges of stealing an umbrella. Claiming that she was nearsighted and only wanted 'to examine the silk of the umbrella more closely' in better light, Mrs Hershey had walked out with the umbrella.[53]

Adequate illumination minimized shadows and diffused light 'in all dark corners . . . behind posts, shelf fixtures and all other obstructions.'[54] Fully lighted counters and showcases enhanced the merchandise but obviously made observation easier as well. Managers assumed that with good lighting the fear of detection would inhibit many instances of impulsive shoplifting.

The Detective, the monthly 'Official Journal of the Police Authorities and Sheriffs of the U.S.,' described the effect of light on a potential shoplifting incident in its August 1899 issue. The woman in question used the excuse of poor light to move expensive material away from the direct supervision of the salesperson. The customer had repeatedly taken bolts of silk to the end of the counter, 'ostensibly for the purpose of examining them in the light.' When she had a number of such bolts at the far end of the counter and other women moved down to look at them, the shoplifter 'slipped one whole piece into the pocket in her skirt.'[55]

The *Brooklyn Eagle* reported a similar case. A Mrs Cornelius Wigham was arrested in the Abraham and Straus store after six pairs of hose, a number of handkerchiefs, and an umbrella were found in her possession. 'Liberated from jail' by her husband, Mrs Wigham denied the charge of shoplifting and explained that she went to the umbrella counter to 'make a purchase for her servant'; she took the umbrella to the door merely to see the material in a better light and was not aware that there were hose or handkerchiefs in its folds.[56]

The importance of light as a psychological factor in the attempt to prevent theft was unstated but implicit. What happened, however, was unpredictable. Illumination often aroused unmet desires and created the very conditions for shoplifting that management hoped it had eliminated. The 'delight afforded to eyes eager to behold the new fabrics which glow under the brilliance of the electric pageant' was not simply advertising copy but a new reality.[57] Margarete Bohme's portrayal of a jewelry sale in Mullenmeister's Department Store on a Saturday evening near Christmas highlights the complexity of such stimulation, often amounting to a physical sensation.

> Electric reflectors drew colored rays from the jewelry that lay on black velvet cushions . . . the gazers seemed to be in a state of excitement verging on ecstasy. A fat lady in furs was almost gasping.
>
> 'If I only had credit till the end of January! I should so like to have those emerald buttons,' she said to her neighbor. 'But at Christmas time . . . !'[58]

Frequently, temptation was a benign 'silent salesman,' but temptation of this sort was subversive to both the store and the woman customer.

By the early 1890s security was embedded in a variety of technological responses, all of which promised to protect the store and its contents. Security came to mean light and mirrors and glass. In these guises it functioned to invite customers into the store even as it provided a subtle barrier to the merchandise itself.

Unconscious surveillance, clear visibility without handling, illuminating engineering, an inoffensive barrier – these were the key phrases that pointed to a problem of wide dimensions, but one the stores took great pains not to publicize. Rarely was there direct or public comment on shoplifting, even as the owners and managers took concrete steps to keep the situation within tolerable limits. What these limits were was never clear, probably not even to the merchants themselves, but certainly the defensive mentality that was so evident was rife with ambivalence.

Store technology became a complex interrelationship between architecture and fixtures, and the imperatives of display, visibility, accessibility, and protection. There was never any doubt, however, that in the confrontation between the free entry principle, the democratization of luxury that the stores worked so hard to foster, and the exigencies of security, the former often took precedence. A degree of loss could be sustained if the aisles were filled with shoppers and the profits steadily increased.

Glass and light intensified the spectacle in the dry-goods bazaars and supported the carefully crafted illusion that anything was possible. By the late 1880s display was as important to sales as the merchandise itself. 'Businessmen know,' the *DGE* claimed with authority, 'that the channel to the people's pocket is through the eye . . . no sane man will deny the selling power of display.'[59] But the insoluble paradox of compelling display and the need to protect persisted.

Recognizing the limits of technology, merchants sought to exert more pervasive internal control over the employees, particularly those on the selling floor. Perhaps organization and technology could together sustain a fantasy of consumption without, as it was euphemistically called in the trade, 'shrinkage.'

NOTES

1 Lew Hahn and Percival White, *The Merchants' Manual* (New York: McGraw Hill, 1924), 52. Almost any issue of the *Dry Goods Economist* in the 1890s carried advertisements for the new glass-enclosed counters and showcases.

2 Interview with George B. Hammond, RG 10, Box 1, Macy Archives. Hammond entered the employ of the Macy store on June 17, 1886, when Charles Webster and Jeremiah Wheeler ran the firm.

3 *Dry Goods Reporter* (April 8, 1905), 39.

4 *Dry Goods Economist* (September 10, 1898), 51.

5 *Dry Goods Reporter* (July 7, 1906), 6.
6 *Dry Goods Economist* (July 30, 1892), 11.
7 Ibid. (March 18, 1893), 32.
8 Ibid. (August 3, 1893), 10.
9 Ibid. (September 28, 1901), 53.
10 Ibid. (March 4, 1893), 48, advertisement.
11 Ibid. (February 6, 1892), 115.
12 Émile Zola, *Au bonheur des dames*, trans, John Stirling (Pliladelphia: T. B. Peterson & Bros., 1873) 371.
13 *Dry Goods Economist* (November 26, 1898), 45.
14 Ibid. (May 27, 1899), 51.
15 Ibid. (July 22, 1899), 53.
16 Ibid. (August 26, 1899), 61. Newspapers were full of reports of middle-class women claiming they were overcome by the opportunity. See the *Brooklyn Eagle* (December 21, 1897), 2:6, (October 27, 1896), 16:2, and (October 12, 1897), 28:7.
17 *Dry Goods Economist* (August 12, 1899), 73.
18 Ibid. (April 4, 1891), 4.
19 'Controlling Shortages and Improving Protection,' National Retail Dry-Goods Association Store Management Group (New York, October 1953).
20 Pittsburgh Plate Glass, *Glass: History, Manufacture and its Universal Application*, (Pittsburgh, 1923), 105.
21 Robert Twyman (1954) *A History of Marshall Field & Company, 1852–1906* (University of Pennsylvania Press, Philadelphia), 5; Lloyd Wendt and Herman Kogan (1952) *Give the Lady What She Wants! The Story of Marshall Field & Company* (Rand McNally, New York), 92.
22 Winston Weisman, 'Commercial Palaces of New York: 1846–1875,' *The Art Bulletin* XXXVI (December 1954), 289. According to the *New York Evening Post* of September 21, 1846, the mirrors were installed in 1846. Cited in Mary Ann Smith, 'John Snook and the Design for A. T. Stewart's Store,' *The New-York Historical Society Quaterly*, LVIII (Jan. 1974) 25.
23 *Dry Goods Economist* (September 24, 1898), 9.
24 *Dry Goods Reporter* (August 16, 1902), 11.
25 *Dry Goods Economist* (February 11, 1893), 57.
26 Ibid. (July 4, 1896), 81.
27 Zola, *Au bonheur des dames*, 4.
28 *Dry Goods Economist* (February 15, 1896), 62.
29 *Dry Goods Reporter* (February 6, 1904), 43.
30 *Dry Goods Economist* (July 4, 1896), 81, and (July 25, 1896), 62.
31 Ibid. (August 22, 1891), 19.
32 Interview with Fred Lazarus, Jr., February 1, 1965, Federated Stores Oral History, Columbia University, New York; in Whiteley's, a large English department store, 'there were mirrors on all sides, in which every movement was reflected,' of clerks as well as customers. R. S. Lambert, *The Universal Provider: A Study of William Whiteley and the Rise of the London Department Store* (London: George G. Harrap & Co., 1939), 142.
33 *Dry Goods Economist* (September 24, 1898), 9.
34 Ibid.
35 Ibid. (April 1, 1899), 17. Queried about how to deal with shoplifting women, the editor suggested the use of mirrors built into shelving: 'The salespeople can then watch the customers, even when their backs are turned.'
36 *Brooklyn Eagle* (August 10, 1901), 2:6.
37 *Dry Goods Reporter* (December 13, 1902), 17; *Dry Goods Economist* (February 2, 1895), 67. Light was a major concern of the trade journals. Almost any issue in the 1890s had an article about illumination.
38 Zola, *Au bonheur des dames*, 5, 12.
39 Margarete Bohme (1912) *The Department Store: A Novel Today* (D. Appleton and Co., New York), 78.
40 D.J.K. (1869) 'Shopping at Stewarts', *Hearth and Home I* (January 9, 1869), 43.
41 Ibid. See discussion of the light and fantasy at Coney Island in the early twentieth century in John F. Kasson, *Amusing the Million: Coney Island at the Turn of the Century* (New York: Hill & Wang, 1978), 82–85.
42 Webster to Mr Kline, May 22, 1919, 'Personnel,' Carton 2, Macy Collection; *New York Herald* (December 12, 1875), 16:1, RG 10, Box 2, #1998, and (November 30, 1879), 20:3, RG 10, Box 2, #2014, Macy Archives.
43 See correspondence from the United States Electric Lighting Co. to Isidore Straus, June 1, 1888, Box 8, Macy Archives; letter from Isidore Straus to Manhattan Electric Light Co., September 22, 'Store,' Carton 2, Macy Collection.
44 'Real Estate,' Carton 2, Macy Collection, from *New York Times* (January 16, 1902); Macy store biographer Ralph Hower said that the Macy power plant was the largest private one in the city. Ralph M. Hower, *History of Macy's of New York*, 1858–1919: *Chapters in the Evolution of the Department Store* (Cambridge University Press, 1943), 325.

Marshall Field & Co. demonstrated a similarly timed progression. The gaslight of 1868 was amplified by a large glass-domed rotunda and skylight in 1873; electric arc lights supplemented the gas lights in 1882. Illumination in the new Marshall Field store in 1906 was all electric. Twyman, *History of Marshall Field*, 48, 61.

The new Siegel-Cooper store advertised at its opening that it had seven thousand incandescent lamps and eight hundred big arc lights, all of which were powered by nine dynamos in its own engine room. Siegel-Cooper also boasted of having 'the largest light ever made,' a huge light on its tower that not only promised to annihilate distance within the metropolitan shopping territory but supposedly had a 75-mile range. *New York Times* (September 6, 1896), 10:4.

45 *New York Herald* (December 18, 1881), 1:2, RG 10, Box 2, #2025, Macy Archives.

46 Zola, *Au bonheur des dames*, 24; *Dry Goods Economist* (December 5, 1891), 21.

47 Lambert, *Universal Provider*, 99; Tudor Jenks, 'Before Shop Windows,' *Outlook* 51 (April 1895), 688.

48 Clara Burton Pardee, *Diaries*, 1883–1938, (N-YHS), entries December 16, 1885, and December 18, 1893.

49 Jenks, 'Before Shop Windows,' 688.

50 Dorothy Davis (1966) *A History of Shopping* (Routledge and Kegan Paul, London), 106–107; Daniel Defoe (1745) *The Complete English Tradesman I* (J. Rivington, London), 270–271.

51 *Dry Goods Reporter* (December 26, 1903), 42. 'Daylight is essential in the cloak department . . . it is well known that artificial light affects colors.' *Dry Goods Economist* (March 8, 1902), 89.

52 *Dry Goods Economist* (March 22, 1902), 19. Along with real problems with lights heating up the atmosphere, there were problems with ventilation in all the large stores.

53 *New York Times* (April 11, 1888), 8:2, and (September 6, 1888), 8:4. Describing the grand opening of the new Field, Leiter store in 1868, the *Chicago Tribune* discussed the importance of the daylight the new windows provided: 'It is possible to see absolutely the color of any article in any part of the rooms, and no clerk is compelled to take a piece of goods and carry it from his desk to a better lighted portion of the room.' *Chicago Tribune* (October 13, 1868), 4:3.

54 *Dry Goods Reporter* (December 26, 1903), 42.

55 *The Detective* XV (August 1899).

56 *Brooklyn Eagle* (August 29, 1899), 2:4.

57 *Dry Goods Economist* (February 15, 1896), 79.

58 Bohme, *Department Store*, 75.

59 *Dry Goods Economist* (January 28, 1893), 57.

30

'We Know Who You Are and We Know Where You Live'

The Instrumental Rationality of Geodemographic Systems

Jon Goss

GEODEMOGRAPHICS AND THE CONSTRUCTION OF CONSUMER IDENTITY

> The internal logic of these processes (statistics, probabilities, operational cybernetics) is certainly rigorous and 'scientific,' yet it somehow doesn't get any purchase on anything, it is a fabulous fiction whose index of refraction in (true or false) reality is zero. This condition is all that gives these *models* any force, but the only truth it leaves them comes from paranoid projection tests of a caste or group, undecideability dreaming of a miraculous adequation between the real and their own models, and therefore an absolute manipulation.
>
> (Baudrillard 1988, 56)

Whatever the rhetorical claims of the promotional discourse of geodemographics, the identity, or character, of individuals is reduced to their measurable characteristics as defined by secondary data sources, unobtrusive observation of consumer behavior, and specially commissioned attitude surveys. Segmentation analysis of demographic and psychographic variables is an attempt to reduce the complex nature of consumer motivations and predispositions to patterns in digital data. It is a part of the impossible strategy of economic reason – 'to exteriorize interiority, or objectivize the subjective,' to reduce the subjectivity of the individual to the social identity with which she or he is provided (Gorz 1989, 176–177).

In geodemographics data are aggregated, correlated, and collapsed into a number of statistical types that adequately summarize patterns in quantitative data (typically capturing about 85 percent of the variance). The 'black box' mechanics of the analysis are probably little understood by most marketing executives, and the resulting abstractions or clusters are not consistent with their commonsense understanding of consumers, so geodemographics provides them with coherent and consistent identities – or 'soap operatic' characters (Goss 1994) – that fit with their own stereotypes. There is, of course, always a subjective element in the identification or naming of the groupings produced by cluster analysis, but in geodemographic systems great liberties are taken with the methodology to provide for the marketer elaborate consumer identities complete with first names, fictional slices of family life, personal dreams, and social weaknesses. Dressed as a precise science – the promotional discourse of geodemographics is replete with metaphors of precision (see Goss 1994) – the methodology constructs consumer identity as a highly subjective marketers' abstraction. The detailed 'portraits' of consumer characters created by market researchers are apparently compelling to marketers precisely because they conceive of social identity as a consumption life-style that is vulnerable to the manipulations of the industry. NDL (1993a, 2), for example, nicely suggests the marketers' vision of social life when it claims that 'as you get to know [cohorts], you'll discover these are all people you know: your friends, neighbors, relatives, and – most importantly – your customers.'

This is not the place to undertake an extensive critical review of the concept of life-style, nor to document how the sociological concept of 'style of life' after Max Weber, among other social science theorists, has been (mis)appropriated by marketing research (see Wrong 1990; Giddens 1991, 81). The geodemographic literature, however, reveals a problematic conception of social identity as a coherent and relatively stable life-style project organized around a particular mode of consumption. The statistical and stereotypical fictions of segmentation belie the contingent, fragmentary process of contemporary identity formation (Featherstone 1991; see also Brown 1990, 75), or rather they present a normative alternative, a model of consumer identity consistent with the productivist organization of the social relations of consumption.

First is the conception of consumer identity as synchronically and diachronically coherent. The consumer is presumed to lead a particular life-style and to consume a complementary set of goods, which conforms more to the desire of marketers to sell particular 'constellations of commodities' (McCracken 1988) and to organize consumption's field of meaning efficiently than to the reality of unpredictable and inconsistent purchase decisions, that has been described as 'consumer schizophrenia' (Kardon, cited in Sampson 1992, 240). Consumer identity is presumed to possess a coherent narrativity characterized by functional development over time, with the consumption of an increasing quantity and diversity of complementary goods. For example, the widely used VALS system presumes the development of consumer identity through discrete stages, from the psychological immaturity associated with 'Survivors' to full psychological maturity of the 'Integrateds,' by either a traditional outer-directed (status-oriented) development route or a contemporary inner-directed (self-actualizing) route (Mitchell 1983, 26). Passage from one stage to another in this 'double hierarchy' of values occurs as individuals satisfy their previous life-style needs (Mitchell 1983, 27). In the more recent VALS2 + system, the progression is from 'Strugglers' to 'Actualizers' through three routes – principle-oriented, status-oriented, or action-oriented – as these consumers accumulate monetary resources. The story of the modern life-style is evidently one of increasing wealth, with the narrative regularly punctuated by predictable events in the life course that change consumer status (births, marriages, home acquisition). This conception has even led to the new marketing 'science' of synchographics (Larson 1992, 79–80), the targeting of consumers on the basis of predictive models that allow marketers to preempt life-style decisions, even perhaps to encourage certain preferred outcomes.

Second is the reduction of life-style to consumption behavior or the presumption that 'you are what you buy' (Piirto 1991, 233). The social identities created in the geodemographic analysis are defined by the amount and type of goods and services consumed – that is, in terms of their functionality to expanded consumption. In SRI International's classification system, for example, 'Actualizers' possessions and recreation reflect a cultivated taste for the finer things in life. . . . Experiencers are avid consumers and spend much of their income on clothing, fast food, music, movies, and videos. . . . [Makers] are unimpressed by material possessions other than those with a practical or functional purpose (such as tools, pickup trucks, or fishing equipment). . . . Strugglers are cautious consumers . . . a modest market for most products . . . but are loyal to favorite brands' (SRI International n.d.). In the vignettes and cluster descriptions, 'positional goods' (Hirsch 1976) clearly distinguish consumer types; for example, ESOMAR, a European market research company, is developing a cross-cultural clustering scheme that measures 'economic status' using ownership of selected consumer durables (Quatresooz and Vancraeynets 1992). Marketers naturally have an interest in defining the principal divisions of contemporary society in terms of commodity consumption.

Marketers are more interested in amount of money spent and products consumed than in income earned, and geodemographics is based on the notion that the realms of production, consumption, and distribution have become relatively independent (see Bell 1973), or therefore that life-style is increasingly a matter of choice based on individual values (Mitchell 1983, viii). Strangely, however, residential location – for example, measures of 'urbanicity' (Strategic Mapping 1994, vii) – and education, occupation, and income, taken as measures of the relative affluence or 'purchasing power' of neighborhoods, all of which are fundamentally constrained by relations of production (to say nothing of race and gender) are privileged variables in the statistical models of segmentation analysis.

Geodemographics is then guilty of the 'fetishization of life-style,' or the representation of consumer life-styles as if they were independent of socioeconomic relations, consistent with a dominant ideology that persistently denies the relationality of the spheres of social life (Laclau and Mouffe 1985).

Moreover, although analysts explicitly argue that their segmentation schemes do not imply any social judgment (Michman 1991, 113), there is in *every* scheme an explicit ranking of segments, where the lowest ranks or highest cumulative scores identify the most affluent consumers. Most of the systems have developed additional means to identify high-spending consumers, by 'affluence weighting' their segments; DMIS, for example, uses its 'Socio-Economic Status Indicator' as the basis of its Affluence Model, a multiple regression analysis that estimates net worth of any geographically defined area in the United States, based on the census, the Survey of Consumer Finances, and private financial data bases. Greater premiums must be paid for the names of high-spending categories of consumers; NDL, for example, charges premiums for the names of consumers with real estate interests and high incomes, those who are frequent flyers and casino gamblers, those who donate to charities, and those who own personal computers, cellular phones, camcorders, and CD players (NDL 1993b). Marketers are obviously not equally interested in all segments, and they design strategies explicitly to exclude the segments 'that do not fit [their] consumer profile' (NDL 1993c, 11). Geodemographics is used to isolate 'deviant' consumers who do not fit the normative model of consumer behavior and to eliminate them from solicitations or service provision.[1] The focus of geodemographics is generally on those 'in the fullness of their needs' (Baudrillard 1988, 52). This means, ironically, that less affluent consumers are perhaps less vulnerable to the solicitations of the marketing industry and the invasions of privacy and attempts to control consumption identity that this brings, but of course this is not attained through choice.

As previously noted, geodemographics presumes that social life is spatially sorted by consumption characteristics, or, in the words of Jonathan Robbin, that 'humans group themselves into natural areas. . . . They create or choose established neighborhoods that conform to their life-style of the moment' (cited in Burnham 1983, 92). This allows the classification of all individuals in society by spatial inference, achieving a total ordering of social life from which no consumer escapes. Such inference, however, is based on the assumptions that households are the basic unit of consumption and that neighbors are more likely to share a set of consumption characteristics than non-neighbors. As a result, geodemographics reifies the address into the primary social unit of consumption; hence 'current resident' (Crt) is usually an appropriate recipient for direct mail and the goal is simply 'to predict the potential responsiveness of *any address* to future mailings' (Beaumont and Inglis 1989, 601, emphasis added). According to the laws of geodemographics: 'You are where you live.'[2] The address is critical because the home is the terminus of the circuits of micropowers of consumption; it is the node where the individual is plugged into the commercial communication networks that carry the messages of the marketing industry. The home is a 'consumption cottage,' or the place where the flows of energy, information, and commodities that disperse the work of consumption converge and penetrate directly into the domestic sphere (see also Putnam 1993, 156).

Ironically, although rapidly developing communication and information-technologies are said to be responsible for an 'emergent placelessness' (Borland 1988, 147) or loss of the 'sense of place' (Meyrowitz 1985), they are employed to locate households within spatialized communities of consumption. Geodemographics partitions social life into life-style areas based on the aggregation of individual characteristics, and marketers target their messages at the resulting statistical identities. Geodemographic discourse consequently also anthropomorphizes ZIP codes, or ZIP code-based segments, speaking of them as if they were possessed of specific tastes and purchasing power. Furthermore, as retailing, financial services, real estate, and media planning increasingly rely on geodemographics to inform their decision making, so an abstract territorialization designed to increase the efficiency of a specific service is rapidly being reified into the fundamental spatial unit of social life.

THE 'TRUTH-EFFECT' OF GEODEMOGRAPHICS

> The reified models of the [social] sciences migrate into the sociocultural life world and gain objective power over the latter's self-understanding.
>
> (Habermas 1971, 113)

As Giddens (1991, 81) points out, the individual in late-modern society is in effect forced to construct a coherent life-style in order to constitute an identity, and literally to *place* herself or himself within the social order, for 'spatially located activity becomes more and more bound up with the reflexive project of the self' (Giddens 1991, 147). In this society, marketing is an organized moral institution whose function is to promote social integration and control through the production of a system of coded values that are employed in the construction of these identities (Baudrillard 1988, 49), and it is increasingly responsible for the spatial ordering of life-styles. It presents a hierarchy of socially sanctioned life-styles and residential locations as models for individuals to select from, consistent within their performative capacities, structuring the possibilities in terms of limited consumption choices. Geodemographics assists in this process by preconstructing social identities, creating coherence by eliminating individual idiosyncrasies and offering a limited number of spatially defined aggregate models of identity. The genius of geodemographics is that it systematically produces such life-styles both from us and for us: it presents descriptions of our consuming selves that are actually normative models, or mean characteristics of our consumption (stereo) type to which we are exhorted to conform. Geodemographics enables marketers to make, within known levels of statistical confidence, that most psychologically effective of marketing pitches, that beginning with 'People like you. . . .'

The spatial corollary of this is that through its systematic application geodemographics both segments and potentially segregates social life, effectively producing the conditions of its own reproduction. Geodemographic marketing thus involves an effective tautology: the marketer purchases detailed psychodemographic data about consumer identity in digital form and sells these identities back to consumers in material form as consumer goods and services. Purchases by consumers generate more detailed information, updating statistical identity and providing a better understanding of the developing needs for further goods and services, and so on (see also Luke 1989, 110). The application of geodemographics to political campaigning potentially has a similar effect on the social reproduction of political values and behavior.

Although the power of geodemographics ultimately depends on an ability to persuade consumers to purchase particular commodity life-styles, consumers can make choices only if they are adequately informed, and geodemographics purposively controls the distribution of information in order to increase marketing efficiency. As geodemographics defines and *addresses* spatially specific modes of consumption, so individuals may be persuaded to adopt the statistical life-style identities that are created from them and for them, while others may be systematically prevented from doing so. Although discrimination on the basis of consumer life-styles is based on apparently less immutable categories than race, ethnicity, sex, or gender, the targeting of specific types of neighborhood may effect a de facto redlining of social life (see Larson 1992, 56; Flowerdew 1991, 9). As geodemographics is used to inform the marketer of particular 'constellations of commodities' for particular neighborhoods, so the statistical identities that they produce may have an instrumentalizing effect, (re)producing the sociospatial differentiation of society and reinforcing past behaviors in particular places.

NOTES

1 For example, the Delinquency Alert System service (provided by Credit Bureau Inc.) maintains records on consumers who have defaulted on loans and predicts further risks of delinquency on loans based on the socio-economic and psychodemographic profile of applicants or their addresses (Gandy 1989, 65).

2 This is the catchy title of Claritas, Inc.'s ongoing display in the 'Information Age' exhibit at the Smithsonian Institution's National Museum of American History.

REFERENCES

Baudrillard, J. (1988) Consumer society. In *Selected writings*, ed. M. Poster, 29–56. Stanford, Calif.: Stanford University Press.

Beaumont, J. R. and Inglis, K. (1989) Geodemographics in practice: Developments in Britain and Europe. *Environment and Planning A* 21: 587–604.

Bell, D. (1973) *The coming of post-industrial society: A venture in social forecasting*. New York: Basic Books.

Borland, J. (1988) Placing television. *New Formations* 4: 145–154.

Brown, G. (1990) *The information game: Ethical issues in a microchip world*. New York: Humanities Press.

Burnham, D. (1983) *The rise of the computer state.* New York: Random House.

Featherstone, M. (1991) *Consumer culture and postmodernism.* Newbury Park, Calif.: Sage.

Flowerdew, R. (1991) *Classified residential area profiles and beyond.* Research Report No. 18. Lancaster: North West Regional Research Laboratory, Lancaster University.

Gandy, O. H. (1989) The surveillance society: Information technology and bureaucratic social control. *Journal of Communication* 39(3): 61–76.

Giddens, A. (1991) *Modernity and self-identity: Self and society in the late modern age.* Stanford: Stanford University Press.

Gorz, A. (1989) *A critique of economic reason.* London: Verso.

Goss, J. D. (1994) Marketing the new marketing: The strategic discourse of Geodemographic Information Systems. In *Ground truth: The social implications of geographic information systems,* ed. J. Pickles, 130–170. New York: Guilford Press.

Habermas, J. (1971) *Knowledge and human interests.* Boston: Beacon Press.

Hirsch, F. (1976) *The social limits to growth.* Cambridge, Mass.: Harvard University Press.

Laclau, E. and Mouffe, C. (1985) *Hegemony and socialist strategy: Towards a radical democratic politics.* London: Verso.

Larson, E. (1992) *The naked consumer: How our private lives become public commodities.* New York: Henry Holt.

Luke, T. W. (1989) *Screens of power: Ideology, domination, and resistance in informational society.* Urbana: University of Illinois Press.

McCracken, G. (1988) *Culture and consumption: New approaches to the symbolic character of consumer goods and activities.* Bloomington: Indiana University Press.

Meyrowitz, J. (1985) *No sense of place.* New York: Oxford University Press.

Michman, R. D. (1991) *Lifestyle market segmentation.* New York: Praeger.

Mitchell, A. (1983) *The nine American life-styles: Who are we and where we're going.* New York: Macmillan.

National Demographics & Lifestyles (NDL) (1993a) *NDL FOCUS: Knowing your customers.* Brochure. Denver.

—— (1993b) *The lifestyle selector: The list that lasts.* Brochure. Denver.

—— (1993c) *Customer database development program. Database marketing services.* Denver.

Piirto, R. (1991) *Beyond mind games: The marketing power of psychographics.* Ithaca, N.Y.: American Demographics.

Putnam, T. (1993) Beyond the modern home: Shifting the parameters of residence. In *Mapping the futures: Local cultures, global change,* ed. J. Bird, B. Curtis, T. Putnam, G. Robertson and L.Tickener, 150–165. New York: Routledge.

Quatresooz, J. and Vancraeynets, D. (1992) Part 2: Using the ESOMAR harmonised demographics: External and internal validation of the results of the EUROBAROMETER test. *Marketing and Research Today,* March, 41–46.

Sampson, P. (1992) People are people the world over: The case for psychological market segmentation. *Marketing and Research Today* 20: 236–244.

SRI International (n.d.) *VALS2+: Psychographic segmentation system.* Menlo Park, Calif.: SRI International.

Strategic Mapping (1994) *ClusterPLUS 2000: The lifestyle segmentation system for today and tomorrow. Enhanced cluster description guide.* Santa Clara, Calif.: Strategic Mapping Inc.

Wrong, D. H. (1990) The influence of sociological ideas on American culture. In *Sociology in America,* ed. H. J. Gans, 19–30. Beverly Hills: Sage.

FOUR

PART FIVE

Theory

INTRODUCTION TO PART FIVE

What marks out the extracts in this part of the Reader as worthy of the accolade 'theory' is simply the *scope* of their attempts to unthink and rethink the world of consumption. The work of **Marcel Mauss** provides an excellent example of what we mean. He raises some profound questions about a subject we normally take for granted: the economy. 'Apparently there has never existed . . . anything that might resemble what is called a "natural" economy' notes Mauss. 'One hardly ever finds a simple exchange of goods, wealth, and products in transactions concluded by individuals.' With these words, Mauss challenges a great deal of received wisdom about the essential features of economic life. While it is hard to imagine an alternative economic system to our own, seemingly natural, fixation on money, capital, wage-labour, self-interest and more-or-less rational economic calculation, Mauss attempted to do precisely that. He succeeded in conveying the historical and geographical specificity of Western economic exchange by describing a very different arrangement that occurred in other places and at other times. To our impersonal circulation of money and commodities, Mauss counterposed societies organized through the circulation of gifts. Receiving a gift is very different to receiving money. Accepting a gift reinforces a social bond between the giver and the receiver. It also places one under an *obligation* to give something back in return (a counter-gift) – if only an acknowledgement of the other's generosity. So, unlike the supposed *equivalence* of monetary exchange – which is aptly expressed by the adage that 'fair exchange is no robbery' – knowing how *much* to give back, *what* to give back, and *where* and *when* to do so is a complex and delicate matter that remains haunted by the possibility of getting it wrong in the eyes of the other. Unlike money, then, the circulation of gifts is wedded to a profoundly *moral* economy. Gifts are always given and taken *personally*. They are inalienable – unlike commodities. So, like it or not, the giving and receiving of gifts is far from innocent. Gifts establish asymmetrical relations of power, authority and prestige. For this very reason, the logic of the gift can span the full spectrum of social relationships: from individuals to nations.

In an attempt to distil the social logic of the gift, Mauss focused on what he called 'archaic societies' and what others have called 'primitive', 'traditional' or 'pre-modern' societies. These are societies where gift exchange was the primary means for establishing and sustaining both social relationships and the circulation of things. Mauss showed that archaic gift exchange did not amount simply to a case of *barter*. It was not the embryonic form of the fully fledged system of commercial exchange that developed in the modern West. It was an entirely different system underpinned by an entirely different logic. In this extract, Mauss is especially concerned with how some social formations confront and oppose one another through an *aggressive* circulation of gifts. For as everyone knows, gifts are ambivalent. They can serve as an expression of intimacy and generosity, but they can also function as an act of rivalry and hostility. Indeed, in a quest for honour and prestige (see **Veblen**), some social formations 'even go as far as the purely sumptuary destruction of wealth that has been accumulated in order to outdo the rival chief'. Mauss reserved the term 'potlatch' (to feed or to consume) for this type of *agonistic* relation, and it has remained a key term in many studies of consumer culture ever since.

Inspired by Mauss' work on the social logic of the gift, **Georges Bataille** takes up his insights and extends them to reach far more radical conclusions. Given our concern with the consumer society, all this would be fairly academic were it not for the fact that gift exchange remains surprisingly important even today. For while at first sight it may appear to survive only on the margins of economic life – birthdays and marriages – one can actually find its logic at work in virtually all areas of life once one knows how to recognize it. While modern societies imagine that they have replaced the primitive ideas of gift exchange with the rational and impersonal calculations of money, Bataille claimed that they remain subject to the logic of the gift nonetheless. In marked contrast to economists' fixation on *scarcity*, Bataille's point of departure is *excess* (see **Appleby**). For him, the most crucial problem that all societies face is not the miserly allocation of scarce resources, but rather how to deal with all of those resources – time, effort, goods, people, energy and wealth – that are *not* required for mere subsistence. This is a pressing problem in our affluent societies, not least because the excess of resources over subsistence is astronomical. Basically, Bataille sought to overcome the restricted purview of the 'dismal science' of economics by focusing on what he called 'general' economy. This shift of focus underpins what has become known as poststructuralism.

By engaging with excess rather than with either spontaneous needs or manufactured scarcity, Bataille turns received economic wisdom on its head. Specifically, he maintains that most societies have consumed their excesses 'to no purpose'. In this way, Bataille presents Mauss' notion of the *squandering* of wealth – the potlatch – as the *general rule*. He calls those social formations that subscribe to this rule 'societies of consumption', and he presents the ancient Aztecs as an exemplary case. Bataille claims that consumption and sacrifice were as important to the Aztecs as production and work are to us. In contrast to the general rule of expenditure, he characterizes our own economic system as very much an *exception*. Rather than squandering our wealth for honour and prestige – in order to 'feel alive' as Baudrillard once put it – our resources are 'deliberately reserved for growth'. Accordingly, Bataille calls social formations like ours 'societies of growth'.

As principles that govern society, Bataille argues that scarcity, growth and conservation came *long after* the rule of excess, squandering and consumption. Consumer society is not a point of arrival for Bataille – like that elusive end of history, which may turn out to be utopian or dystopian depending on your point of view: it is a point of *departure*. In effect, Bataille pre-empts **Campbell**'s and **McKendrick**'s attempts (see **Part 1**) to find a convincing solution to the age-old puzzle of how the asceticism and self-sacrifice of the Protestants, whose work ethic defined the spirit of capitalism and made the Industrial Revolution possible, unintentionally mutated into the modern form of hedonism that underpins consumerism. Bataille simply reverses the terms of the puzzle: How did the general rule of squandering and excess, whose ethic of consumption defined the spirit of sociability and made the circulation of things possible, unintentionally mutate into the modern form of hoarding that underpins growth?

While we tend to believe that growth is inherently good and essentially progressive, Bataille has profound misgivings about this evaluation for two crucial reasons. First, societies of growth are plagued by the social and environmental problems that are caused by the hoarding of wealth: greed, envy, exploitation, pauperization, and so on. This worry resonates with the Marxist concern for the antagonistic class conflicts that inevitably accompany the securing of private property and the process of capital accumulation. For Bataille, one might say that the enormous amount of diffuse violence and ambient fear that pervades modern societies of growth is far more injurious than the ritual violence and sacrifice practised in archaic societies of consumption. Second, societies of growth organize themselves around the amassing of *things* rather than the living of life. Consequently, everything tends to be put into the service of accumulation and hoarding. Everything becomes enslaved to growth. So, whereas everything in societies of consumption is geared towards the maintenance of social relationships, everything in societies of growth tends to be 'relegated to the level of *things*'. Even people become considered as either instruments of growth (workers, labourers, managers) or impediments to growth (dependants, idlers, the unemployed). For Bataille, this instrumental mind-set is effectively an attempt to liquidate the social. 'Servile use has made a *thing* (an *object*) of that which, in a deep sense, is of the same nature as the *subject*' (see **Kiaer**). To put it another way, it would be like treating a heartfelt gift as if it were nothing more than a thing that may or may not be of much use.

Given our descent into servitude, the logical response for Bataille – and especially for Baudrillard – is to negate this negation: to restore the social significance of our world through the destruction of those 'things' that have alienated us from it. One must *sacrifice* – that is to say destroy – the 'accursed share', since only 'sacrifice restores to the sacred world that which servile use has degraded'. Quite simply, Bataille ups the ante on Weber's wager that the true spirit of capitalism is to be found in the self-sacrificing Protestant work ethic. For Bataille, it is precisely this form of servile self-sacrifice which must be sacrificed in its turn. Only destruction can release people from 'the order of *things*'. However, sacrifice need not destroy the thing itself – only the functional *ties* that bind it to servitude, usefulness and profitability. This is exactly what one does when one gives or receives a gift. Properly speaking, a gift is not a thing that one can possess, but the expression of a profound social relationship. Perhaps the closest our consumer societies come to Aztec sacrifice is the potlatch of Christmas, even though it is clearly enslaved to the instrumental forces of growth. Nevertheless, it still provides a great deal of scope for the guerrilla tactics of the weak (see **de Certeau**).

As well as reversing economic, historical and sociological wisdom by putting the sacrificial logic of societies of consumption first, Bataille also makes much of the *asymmetrical* relations of power, authority and prestige that Mauss identified in the logic of the gift. These do not tend towards equilibrium, but follow a process of one-upmanship. So, while orthodox economic thinking uses the calculating logic of equivalence to explain the seemingly relentless growth of production and consumption, Bataille draws upon the unruly logic of the counter-gift to explain the seemingly relentless escalation of one-upmanship and conflict. Societies of consumption are never ahistorical, 'static' societies. The come-uppance of this for Bataille and those who have been influenced by him, most notably Baudrillard, is an insistence that the *antagonistic* nature of society will never be overcome, either by an ever-increasing amount of 'affluence' provided by the growth machine or by an end to the class struggle as envisioned by Marxists. There will never be enough to satisfy the escalating wagers of the counter-gift. Facing up to this realization is all the more urgent if Bauman is right to suggest that contemporary Western societies are rapidly reverting back to the general rule of consumption.

Like Mauss and Bataille, **Thorstein Veblen** is also interested in the social logic of consumption. Once again he emphasizes the importance of social difference and rivalry that is *antagonistic* in nature. However, rather than confining this to archaic societies or opposing societies of consumption to societies of growth, Veblen claims that *all* societies have been motivated to possess and accumulate wealth – women, slaves, inferiors and things – for *honorific* rather than utilitarian reasons, and he offers a brief history of the changing forms and functions of 'ownership' in order to support his case. So, rather than marking a rupture between societies dominated by subsistence and societies devoted to affluence, the Industrial Revolution simply replaced 'physical aggression' and 'war trophies' with 'industrial aggression' and 'consumer goods' in the age-old struggle for social prestige. For Veblen, like so many other writers on consumption, the struggle for property and comfort is essentially a struggle for social recognition (see **Taylor**).

With regard to modern societies, Veblen argues that the social pecking order is established through the *conspicuous display* of wealth, and that social status has come to be measured solely in *monetary* (pecuniary) terms. Given the anonymity of modern metropolitan life, people are judged not with regard to their characters and deeds, but almost entirely in terms of what they *appear* to be worth: seeing is believing (see **Featherstone**). This not only fuels the conspicuous display of the *signs* of wealth, power and prestige, it also fuels the creation of all manner of public *spaces* to serve as 'predatory stages' upon which this tournament of value can be played out: pleasure gardens, cafés, theatres, restaurants, and so on (see **Schivelbusch** and **Zukin**).

Veblen claims that in every society there is a 'standard of wealth' which confers *merit* to those who surpass it and *shame* upon those who fall short. Given the complexity of modern societies and the fine grading of their social hierarchies, this standard of wealth is differentiated into 'pecuniary intervals' that serve to separate out the various classes. These are 'invidious' to the extent that they rate, grade and filter individuals and classes. In a similar way to McKendrick, Veblen argues that those lower down the social scale have no means of obtaining either prestige or social advancement other than by trying to *emulate* the ways of life exhibited by those higher up the pecking order. He calls this 'pecuniary emulation'. Needless to say, those further up the social ladder must continuously find new ways to *differentiate* themselves from those attempting to copy their style, lest they become absorbed into the mass rising up from below. Such is the *escalation* of the standard of wealth, the pecuniary intervals and the amount of conspicuous consumption, which once again will never be satisfied by the spread of affluence and the so-called 'democratization of luxury'. Finally, while Veblen envisaged a *single* pecking order governed by pecuniary status and shaped by those at the top of the social hierarchy (the so-called 'leisure class'), many subsequent writers have demonstrated that different social groups structure themselves around different signs of worth. The work on subcultures, neo-tribes, taste and lifestyles is perhaps the most important in this regard (see **Part 3** and **Bourdieu**). Nevertheless, despite the very particular context within which he worked (late nineteenth-century America), his relative insensitivity to social difference (especially gender and ethnicity), and his voluntaristic disposition (people are assumed to test *themselves* against the standard of wealth), Veblen's basic argument remains widely influential.

In keeping with Veblen's stress on pecuniary (monetary) emulation, **Georg Simmel** is most famous for his work on 'the philosophy of money' and its impact on the nature and experience of modern urban society. In the reading that we have included here, however, Simmel explores what he considers to be the eternal tension between *freedom* and *dependency*, a theme that is particularly central to many studies of consumer culture. Other writers have characterized this tension in terms of the struggle between individual and society, self-interest and sociability, freedom and necessity, autonomy and alienation, agency and structure, and voluntarism and determinism, and most have sought to resolve these dualisms theoretically (often dialectically). In stark contrast, Simmel argues not only that the tension between freedom and dependency is intractable, but also that it can only be resolved in *practice*, a task that each period, each class and each person must negotiate for themselves. This emphasis on *practice* should caution us against the widespread assumption that consumption is essentially communicative. Simmel reminds us that it is first and foremost embodied and performative.

In the modern context, Simmel sees *fashion* as one of the key forms through which people can 'redistribute' the quanta of dependency and freedom to their own advantage. Fashion, for Simmel, is a remarkably modern form because it gives a strong sense of the passing of the present, the fragility of established practices, beliefs and traditions, and the precariousness of individual existence. As a social form, fashion proffers an unrealizable ideal, facilitates social differentiation, and offers a road that all may travel as 'slaves to fashion'. Fundamentally, fashion is a democratic institution: anyone can elect to share an affinity with it (see **Bennett**). Indeed, fashion embodies a whole series of modern traits: equality, individuality, imitation and conspicuousness, to name but a few. However, it is the social *logic* of fashion that is of greatest interest to Simmel, especially with regard to how fashion may be used tactically by individuals to maximize their inner freedom. Fashion provides a wonderful *refuge* for the weak – 'a sphere of general imitation' – where one can hide from 'the responsibilities and the necessity of defending oneself unaided'. The weak can depend upon fashion to affirm their social belonging (compare **Comaroff**). Given the period within which Simmel made these observations, it is hardly surprising that he should explicitly link the refuge of fashion with the tactics of women subjugated by patriarchal power. By contrast, Simmel notes how 'the emancipated woman . . . lays particular stress on her indifference to fashion'. Yet there is more to fashion than a refuge for the weak. For the strong are also drawn to fashion, wherein they can display their social distinction and radical autonomy on the predatory stage. Thus fashion, for Simmel, is a profoundly ambivalent form. It is therefore the perfect *mask* for 'refined and special persons' to negotiate the 'painful oscillation' between 'the impulse towards individualization' *and* 'the drive for immersion'. In short, fashion is one of those modern forms through which people seek to 'save their inner freedom all the more completely by sacrificing externals to enslavement by the general public'.

By considering how social distinctions are expressed through taste, **Pierre Bourdieu** is often regarded as having inherited the mantle of both Veblen and Simmel. Like Veblen, Bourdieu considers social life to be antagonistic, and, like Simmel, he believes that practice is what mediates the tension between social dependency and individual freedom. Consequently, his attempt to understand how social structures are reproduced over time tries to keep to what he calls the 'historical ground of action' and 'habitude'. In addition, there is probably an even more direct lineage to **Mauss**' work on archaic societies, which he develops with an eye towards the tradition of French sociology established by Mauss' uncle, Émile Durkheim.

In an extension of Veblen's argument outlined above, Bourdieu claims that social status is no longer merely a function of one's ability to conspicuously display one's 'pecuniary (monetary) strength' in an inflationary tournament of value determined solely by culturally specific 'standards of wealth' and subsidiary 'pecuniary intervals'. On Bourdieu's reading, the struggle for social prestige is played out across many different forms of 'symbolic capital'. In addition to *economic* capital (pecuniary resources), the notion of 'symbolic capital' also embraces *cultural* capital (embodied in distinctive tastes and dispositions), and *social* capital (defined in terms of social relationships and affiliations). In a further act of variegation, some writers now even speak of *subcultural* capital. Consequently, it is the combination and permutation of these various forms of symbolic capital that determines the hierarchical ordering of society and the gestation of a variety of distinctive lifestyles.

Writers on consumption have been especially impressed with what Bourdieu says about cultural capital, which is itself a heterogeneous category. It incorporates subjective knowledge, dispositions and competences; 'cultural' objects (which effectively *objectify* subjective knowledge, thus allowing the cultural competence of the owner or consumer to be inferred); and the institutional recognition of particular aptitudes (exemplified, for instance, in educational qualifications). While cultural capital is articulated in different ways in different social fields (religion, politics, academe and so forth), in terms of consumption, broadly speaking, it translates into *taste*. Much like the gift for Bataille, and fashion for Simmel, taste does double duty for Bourdieu. It is both a 'structuring' and a 'structured' device. On the one hand, taste guides particular consumption practices: it leads different kinds of people to do and value different kinds of things. On the other hand, those self-same consumption practices also express cultural capital, conjugate with other forms of symbolic capital, and thereby express social distinctions while reinforcing class boundaries. Shared tastes promote a sense of social affiliation, while differences of taste provoke social antagonism. To borrow a phrase from Bourdieu, 'taste classifies the classifier'. People distinguish themselves by the taste distinctions they make.

In his own empirical work, Bourdieu sought to map the social space of distinction as expressed through the judgement of taste. In so doing, he was especially interested in how the two main forms of symbolic capital – *economic* capital and *cultural* capital – fracture the dominant class, and thereby open up new lines of antagonism *within* the dominant culture of the ruling class. Those who commanded economic capital represented the *dominant* fraction of the dominant class (business, financial and management groups), and their lifestyles valued practices of conspicuous consumption that demonstrated pecuniary strength. Those who commanded cultural capital represented the *dominated* fraction of the dominant class (educational, artistic and media groups), and their lifestyles valued non-utilitarian practices.

By taking the social struggle over taste in the sphere of consumption seriously, treating it as the necessary corollary of the class struggle within the world of production, Bourdieu has done much to shake off the notion of a grey, homogeneous, conformist mass culture in which consumers can only ever figure as passive dupes who need to be jolted out of their somnambulism and zombification by some revolutionary moment of disalienation. Yet he does not move especially far in the other direction, since he does not subscribe to the widely held view that individuals have effectively become free agents who are now able to fabricate an identity for themselves in a manner of their own choosing (see **Part 3**). This is because Bourdieu is keen to demonstrate that the antagonistic expression of social distinctions through the judgement of taste is almost always *unconsciously* motivated, and therefore strangely impersonal. Haunted by structuralism, he maintains that 'personal' style is never more than a deviation in relation to the common style of a period or a class. Taste is so deeply ingrained in the practices of everyday life that it would be foolish to believe that one could simply elect to exchange one persona, lifestyle, subculture or neo-tribe for another, as if one were merely faced with a series of masks that could be donned without effort or consequence. It is very difficult to come to terms with something that is not to one's taste, and even harder to come to embody it unselfconsciously in one's own repertoire of practice. Hence the importance of the notion of habit, out of which 'values are made flesh'. According to Bourdieu, habits constitute 'a grammar of actions that differentiate classes'. They provide an 'unconscious set of dispositions common to a class'. In short, society is structured through constellations of habit into which members of the various classes are socialized, particularly through one's family, peer group and formal education. Thus the social space of a society is not simply striated by its culturally specific signs of distinction. It is regionalized by habit. Each class occupies its own region of habitude: its own *habitus*.

Although he is concerned with an enormous range of practices, Bourdieu's focus is very much on *class* society. As such, his work is very strongly influenced by **Karl Marx**'s insistence that the capitalist mode of production and its various social fractions are the result of the history of class struggle. Bourdieu is often seen as providing an analysis of the *social* reproduction of class society, which complements and completes Marx's *economic* analysis of the forces and relations of the capitalist mode of production. Where Marx concerned himself with the antagonistic circulation and accumulation of *economic* capital, Bourdieu offers a parallel analysis of the antagonistic circulation and accumulation of *symbolic* capital. Indeed, Marx has frequently been regarded as offering a one-sided analysis of capitalism, centred squarely on *production* and neglectful of a full and wide-ranging consideration of consumption. He is usually regarded as taking a view fundamentally opposed to Adam Smith, who essentially held that 'consumption determines production'. Smith wrote: 'Consumption is the sole end and purpose of all production and the interest of the producer ought to be attended to, only so far as it may be necessary for promoting that of the consumer.' Marx believed that this view of the power of consumers and the subservience of producers was at best naive, and at worst an ideological justification for a system that abuses billions of people in the sphere of production on a global scale (see **Mohun**). He held that the exploitation of labour by capital in the process of production provided the key to understanding the nature, dynamism and contradictions of capitalism. His critique of political economy is therefore built on notions such as concrete and abstract labour, and the analysis of use-value, exchange-value, and, most importantly for Marx, *surplus*-value (which brings him into the proximity of Bataille's concern for excess). Yet despite appearances, Marx's emphasis on the 'mode of production' does not mean that he simply reversed Adam Smith's account. For all of the caricatures suggesting that Marx took the opposite line to Smith, arguing that 'production determines consumption', Marx actually held that 'production, distribution, exchange and consumption . . . all form members of a totality, distinctions within a unity'. In other words, he saw that consumption required production as much as production required consumption; that the two were essentially different 'moments' of the general circulation of capital. Marx's most candid remarks on consumption reveal his insight into its relation to the general circulation process, demonstrating the sophistication of his understanding of the dialectical relationship between 'consumption' and 'production'.

While many theorists of consumption have followed in Marx's footsteps, **Jean Baudrillard** is most famous for parting company with him. While broadly agreeing with Marx's critique of the political economy of capitalism, Baudrillard pulls the rug out from under this critique by highlighting what might appear to be a rather trivial oversight. Baudrillard's accusation is that Marx never got around to critiquing the category of use-value: the usefulness, utility and purpose of things. By effectively *naturalizing* use-value in this way, by regarding the usefulness of objects and commodities as a quality inherent to them, Marx could only *assume* that capitalism was a perversion of the natural order of things, and that an effective critique of political economy could restore a mutual fit between the real needs of society and the true usefulness of people and things: 'from each according to its abilities; to each according to its needs.' Marx reserved his most vehement critique for exchange-value, the mechanism for extracting surplus-value, which he saw as an artificial – and ultimately vampiric – add-on to the natural situation defined by use-values. Baudrillard, however, demonstrates that use-value is not natural: it is as much a product of the capitalist system as exchange-value. The fact that exchange-values *appear* artificial tends to promote the view that use-values are natural. But appearances are deceptive, and Marx fell headlong into this trap. For Baudrillard, use-values are very much part and parcel of the capitalist system, and not something that preceded it. Similarly, 'use' is an ideological imperative rather than a natural kind of connection to the world (see **Barthes**). In both producer and consumer societies, one *must* make use. Individuals are obliged to register themselves on the 'scale of status' (see **Veblen**), while the dictates of fashion (see **Simmel**) ensure that consumption becomes *internalized* and the individual acquiesces to social norms far more effectively than earlier forms of repressive social control could ever achieve (see **Bauman**). Consumption does not, therefore, rely on the direct manipulation of passive subjects. It ensures an *active* compliance to social norms.

The most far-reaching implication of Baudrillard's analysis, however, is that 'needs' are not natural either. Like uses, needs are socially constructed rather than gifts of nature. To put it bluntly, *all* uses and *all* needs are in a sense false and artificial. Needless to say, this counter-intuitive understanding of both the subjects and the objects of consumption invariably arouses controversy, not least because most critics of capitalism and the consumer society prefer to *limit* the extent to which uses and needs may be considered unnatural. What good is a critique if there are no higher values held in reserve? Little wonder, then, that Baudrillard should have abandoned criticism – but not radicalism – long ago. Among others, he has drawn on the ambivalence of Mauss, Bataille and Veblen to pursue his case, especially by way of symbolic exchange and the art of seduction.

In the extract reproduced here, Baudrillard inveighs against the ideological view that consumption is nothing but the satisfaction of the needs of consuming subjects by the use-values of objects. He is insistent on an *unconscious social logic* that underpins consumption. Consumption is not a rational, individualized search for satisfaction. While it presents itself as such, this 'manifest discourse' merely camouflages its latent content. This becomes clearer from Baudrillard's analysis of the object. The accepted view is simply that objects are *useful* (although see **Part 4**). While one can certainly think of objects as functional devices – a refrigerator, for instance, as 'an object that refrigerates' – this hardly gets us very far in understanding the specific character of 'objects of consumption'. Borrowing from structuralism and semiology, Baudrillard's chief theoretical innovation is to see that objects of consumption consist not only of use-values and exchange-values but also of *sign-values*. Consumer objects operate as *signs*: they come to mean something – prestige, fashionability, convenience, luxury – and do so primarily through their *difference* from other objects. Such is the 'calculus of objects'. It forms a matrix of combinations and permutations out of which consumers are called upon to forge some meaning for themselves. Thus, for example, a cup of cheap instant coffee, a 'skinny latte' and a high-quality ground coffee all differ from one another, and can be made to signify those differences. This means that consumers – intentionally or unintentionally, consciously or unconsciously – relate to objects in a way that invariably says something about their identity. Objects of consumption, then, are defined *primarily* in terms of sign-value, which should not be confused with meaning. Sign-value is what turns them into objects of consumption *as such*. Since objects of consumption are ordered into a 'hierarchical code of significations', this means that a refrigerator, a vase, a coffee, a car, and so on, can all have something fundamentally in common, which has nothing to do with use and everything to do with signification. A high-performance car and a high-quality coffee are eminently substitutable as objects of consumption (objects capable of signifying prestige or social standing, for example) – though they are clearly not substitutable in terms of use-value or exchange-value. Baudrillard's key insight is that consumption should be understood in terms of the insatiable play of sign-value, rather than in terms of individuals endowed with needs securing some final satisfaction from useful things.

Rather than try to counter the semiotic by insisting on a return to the natural order of things, Baudrillard follows Mauss and Bataille in opposing the semiotic to the symbolic. For example, although there is nothing natural about the need to wear a ring, there is nevertheless a profound difference between a wedding ring and a ring that is simply a piece of costume jewellery. A wedding ring works as a *symbol*, while an ordinary ring works as a *sign*. Where the symbolic-object takes its meaning from a 'concrete relationship between two people', the sign-object (or object of consumption) assumes its meaning 'in its differential relation to other signs'. This means that social relations are rendered opaque: they are determined by objects of consumption functioning as signs. Once again, it is this opacity that Baudrillard – like Marx and Bataille before him – sets out to destroy. 'Even signs must burn.'

Michel de Certeau, like Baudrillard, sees consumerism as an increasingly dominant aspect of the social order. Unlike Baudrillard, however, he is much more concerned with the ordinary person's engagement with that system. Specifically, he is concerned to demonstrate that everyday actions are not simply determined or constrained by the dominant social order. On the contrary, everyday actions frequently serve to *deflect* and *short-circuit* its power. Consumers, or 'users', as de Certeau prefers to call them, have a distinct habit of making unforeseeable moves within the 'functionalized space' of the consumer society. To that extent, de Certeau arguably has a much richer sense of the *creative* potential of practice than one can find in Bourdieu's conception of practice. Now, although de Certeau's work has often given rise to a rather romanticized view of consumer resistance, de Certeau's own work is much more subtle than this. In stressing the ordinary, the quotidian, the everyday, de Certeau explicitly distances himself from studies of cultural difference that focus on direct resistance to the dominant order: subcultures (see **Hebdige**) and the 'counter-culture'. Instead, he concerns himself with the way in which everyday life invariably circumvents and subverts the dominant order from within. A useful analogy is provided by the history of Spanish colonialism in the Americas. Although they were forced into submission by the colonial powers, 'the Indians nevertheless often *made of* the rituals, representations and laws imposed on them something quite different from what their conquerors had in mind'. Even though they were largely powerless to challenge the system imposed on them directly, the colonized peoples nonetheless 'escaped it without leaving it' through the course of their everyday lives (see **Comaroff**). Similarly, in a consumer society, the everyday actions of ordinary consumers frequently 'trace out ruses of other interests and desires that are neither determined nor captured by the systems in which they develop'. In other words, the everyday bears a trace, a memory and a promise of other ways of being, of other social spaces and of insubordination to one's *habitus*. Accordingly, de Certeau regards the term 'consumer' as something of a euphemism: consumers – 'the dominated element in society' – actively *produce* their everyday lives, usually 'by *poaching* in countless ways on the property of others'.

The fact that consumers are *active* producers – that they *make* their way in the world, even *forge* their own world – is perhaps best explained by analogy with *speech-acts*. Acts of enunciation are never determined fully by the framework of grammar and vocabulary that constitute a given language. That framework is used actively and often opportunistically, as speech-acts draw upon it and move freely within it, creating meanings to suit their own purposes according to the circumstances to hand. This is, above all else, the sense in which de Certeau wishes to get us to think of consumption. Nonetheless, he is not naive enough to believe that consumption is therefore an expression of freedom after the model of consumer sovereignty. He is acutely aware of the power relations within which this kind of freedom operates, and elaborates on this context to register an important distinction between the *strategies* of the powerful and the *tactics* of the weak (compare **Simmel**).

Like Baudrillard, de Certeau believes that power infiltrates and saturates the whole of social space and daily life. We are *all* inserted as consumers into the strategic matrix of consumer society. We are *all* users of a world not of our own making (see **Barthes**). 'Marginality is today no longer limited to minority groups. Marginality is becoming universal: a silent majority.' Drawing on Michel Foucault's notion of the 'microphysics of power', de Certeau writes: 'If it is true that the grid of "discipline" is everywhere becoming clearer and more extensive, it is all the more urgent to discover how an entire society resists being reduced to it, what popular procedures (also "miniscule" and everyday) manipulate the mechanisms of discipline and conform to them only in order to evade them, deflecting their functioning by means of a multitude of "tactics" articulated in the details of everyday life.' That de Certeau should locate an exemplary kind of resistance in everyday practices such as walking, cooking, talking, reading and shopping brings this Reader full circle. The historical geography of consumption remains in the making. Its creative evolution is yet to come. For consumption is forever leading our thinking and our world astray.

31

The Gift and Potlatch[1]

Marcel Mauss

Apparently there has never existed, either in an era fairly close in time to our own, or in societies that we lump together somewhat awkwardly as primitive or inferior, anything that might resemble what is called a 'natural' economy. Through a strange but classic aberration, in order to characterize this type of economy, a choice was even made of the writings by Cook relating to exchange and barter among the Polynesians. Now, it is these same Polynesians that we intend to study here. We shall see how far removed they are from a state of nature as regards law and economics.

In the economic and legal systems that have preceded our own, one hardly ever finds a simple exchange of goods, wealth, and products in transactions concluded by individuals. First, it is not individuals but collectivities that impose obligations of exchange and contract upon each other. The contracting parties are legal entities: clans, tribes, and families who confront and oppose one another either in groups who meet face to face in one spot, or through their chiefs, or in both these ways at once. Moreover, what they exchange is not solely property and wealth, movable and immovable goods, and things economically useful. In particular, such exchanges are acts of politeness: banquets, rituals, military services, women, children, dances, festivals, and fairs, in which economic transaction is only one element, and in which the passing on of wealth is only one feature of a much more general and enduring contract. Finally, these total services and counter-services are committed to in a somewhat voluntary form by presents and gifts, although in the final analysis they are strictly compulsory, on pain of private or public warfare. We propose to call all this the *system of total services*. The purest type of such institutions seems to us to be characterized by the alliance of two phratries in Pacific or North American tribes in general, where rituals, marriages, inheritance of goods, legal ties and those of self-interest, the ranks of the military and priests – in short everything, is complementary and presumes co-operation between the two halves of the tribe. For example, their games, in particular, are regulated by both halves. The Tlingit and the Haïda, two tribes of the American Northwest, express the nature of such practices forcefully by declaring that 'the two tribal phratries show respect to each other'.

But within these two tribes of the American Northwest and throughout this region there appears what is certainly a type of these 'total services', rare but highly developed. We propose to call this form the 'potlatch', as moreover, do American authors using the Chinook term, which has become part of the everyday language of Whites and Indians from Vancouver to Alaska. The word potlatch essentially means 'to feed', 'to consume'. These tribes, which are very rich, and live on the islands, or on the coast, or in the area between the Rocky Mountains and the coast, spend the winter in a continual festival of feasts, fairs, and markets, which also constitute the solemn assembly of the tribe. The tribe is organized by hierarchical confraternities and secret societies, the latter often being confused with the former, as with the clans. Everything – clans, marriages, initiations, Shamanist seances and meetings for the worship of the great gods, the totems or the collective or individual ancestors of the clan – is woven into an inextricable network of rites, of total legal and economic services, of assignment to political ranks in the society of men, in the tribe, and in the confederations of tribes, and even internationally. Yet what is noteworthy about these tribes is the principle of rivalry and hostility that prevails in all these practices. They go as far as to fight and kill chiefs and nobles. Moreover, they even go as far as the purely sumptuary destruction of wealth that has been accumulated in order to outdo the rival chief as well as his associate (normally a grandfather, father-in-law, or son-in-law). There is total service in the sense that it is indeed the whole clan that contracts on behalf of all, for all that it possesses and for all that it does, through the person of its chief. But this act of 'service' on the part of the chief takes on an extremely marked agonistic character. It is essentially usurious and sumptuary. It is a struggle between nobles to establish a hierarchy among themselves from which their clan will benefit at a later date.

We propose to reserve the term potlatch for this kind of institution that, with less risk and more accuracy, but also at greater length, we might call: *total services of an agonistic type.*

Up to now we had scarcely found any examples of this institution except among the tribes of the American Northwest, Melanesia, and Papua. Everywhere else, in Africa, Polynesia, Malaysia, South America, and the rest of North America, the basis of exchanges between clans and families appeared to us to be the more elementary type of total services. However, more detailed research has now uncovered a quite considerable number of intermediate forms between those exchanges comprising very acute rivalry and the destruction of wealth, such as those of the American Northwest and Melanesia, and others, where emulation is more moderate but where those entering into contracts seek to outdo one another in their gifts. In the same way we vie with one another in our presents of thanks, banquets and weddings, and in simple invitations. We still feel the need to *revanchieren*, as the Germans say. We have discovered intermediate forms in the ancient Indo-European world, and especially among the Thracians.

Various themes – rules and ideas – are contained in this type of law and economy. The most important feature among these spiritual mechanisms is clearly one that obliges a person to reciprocate the present that has been received. Now, the moral and religious reason for this constraint is nowhere more apparent than in Polynesia. Let us study it in greater detail, and we will plainly see what force impels one to reciprocate the thing received, and generally to enter into real contracts.

EDITORS' NOTE

1 In line with Mauss' statement that 'the notes are only indispensable to specialists', none of the lengthy footnotes are included here.

32

Sacrifices and Wars of the Aztecs

Georges Bataille

SOCIETY OF CONSUMPTION AND SOCIETY OF ENTERPRISE

I will describe sets of social facts manifesting a general movement of the economy.

I want to state a principle from the outset: By definition, this movement, the effect of which is prodigality, is far from being equal to itself. While there is an excess of resources over needs (meaning real needs, such that a society would suffer if they were not satisfied), this excess is not always consumed to no purpose. Society can grow, in which case the excess is deliberately reserved for growth. Growth regularizes; it channels a disorderly effervescence into the regularity of productive operations. But growth, to which is tied the development of knowledge, is by nature a transitory state. It cannot continue indefinitely. Man's science obviously has to correct the perspectives that result from the historical conditions of its elaboration. Nothing is more different from man enslaved to the operations of growth than the relatively free man of stable societies. The character of human life changes the moment it ceases to be guided by fantasy and begins to meet the demands of undertakings that ensure the proliferation of given works. In the same way, the face of a man changes if he goes from the turbulence of the night to the serious business of the morning. The serious humanity of growth becomes civilized, more gentle, but it tends to confuse gentleness with the value of life, and life's tranquil duration with its poetic dynamism. Under these conditions the clear knowledge it generally has of things cannot become a full self-knowledge. It is misled by what it takes for full humanity, that is, humanity at work, living in order to work without ever fully enjoying the fruits of its labor. Of course, the man who is relatively idle or at least unconcerned about his achievements – the type discussed in both ethnography and history – is not a consummate man either. But he helps us to gauge that which we lack.

CONSUMPTION IN THE AZTEC WORLDVIEW

The Aztecs, about whom I will speak first, are poles apart from us morally. As a civilization is judged by its works, their civilization seems wretched to us. They used writing and were versed in astronomy, but all their important undertakings were useless: Their science of architecture enabled them to construct pyramids on top of which they immolated human beings.

Their worldview is singularly and diametrically opposed to the activity-oriented perspective that we have. Consumption loomed just as large in their thinking as production does in ours. They were just as concerned about *sacrificing* as we are about *working*. [. . .]

THE HUMAN SACRIFICES OF MEXICO

We have a fuller, more vivid knowledge of the human sacrifices of Mexico than we do of those of earlier times; doubtless they represent an apex of horror in the cruel chain of religious rites.

The priests killed their victims on top of the pyramids. They would stretch them over a stone altar and strike them in the chest with an obsidian knife. They would tear out the still-beating heart and raise it thus to the sun. Most of the victims were prisoners of war, which justified the idea of wars as necessary to the life of the sun: Wars meant consumption, not conquest, and the Mexicans thought that if they ceased the sun would cease to give light.

'Around Easter time' they undertook the sacrificial slaying of a young man of irreproachable beauty. He was chosen from among the captives the previous year, and from that moment he lived like a great lord. 'He went through the whole town very well dressed, with flowers in his hand and accompanied by certain personalities. He would bow graciously to all whom he met, and they all knew he was the image of Tezcatlipoca [one of the greatest gods] and prostrated themselves before him, worshipping him wherever they met him.'[1] Sometimes he could be seen in the temple on top of the pyramid of Quauchxicalco: 'Up there he would play the flute at night or in the daytime, whichever time he wished to do it. After playing the flute, he too would turn incense toward the four parts of the world, and then return home, to his room.'[2] Every care was taken to ensure the elegance and princely distinction of his life. 'If, due to the good treatment he grew stout, they would make him drink salt-water to keep slender.'[3] 'Twenty days previous to the festival they gave this youth four maidens, well prepared and educated for this purpose. During those twenty days he had carnal intercourse with these maidens. The four girls they gave him as wives and who had been reared with special care for that purpose were given names of four goddesses. . . . Five days before he was to die they gave festivities for him, banquets held in cool and gay places, and many chieftains and prominent people accompanied him. On the day of the festival when he was to die they took him to an oratory, which they called Tlacuchcalco. Before reaching it, at a place called Tlapituoaian, the women stepped aside and left him. As he got to the place where he was to be killed, he mounted the steps by himself and on each one of these he broke one of the flutes which he had played during the year.'[4] 'He was awaited at the top by the satraps or priests who were to kill him, and these now grabbed him and threw him onto the stone block, and, holding him by feet, hands and head, thrown on his back, the priest who had the stone knife buried it with a mighty thrust in the victim's breast and, after drawing it out, thrust one hand into the opening and tore out the heart, which he at once offered to the sun.'[5]

Respect was shown for the young man's body: It was carried down slowly to the temple courtyard. Ordinary victims were thrown down the steps to the bottom. The greatest violence was habitual. The dead person was flayed and the priest then clothed himself in this bloody skin. Men were thrown into a furnace and pulled out with a hook to be placed on the executioner's block still alive. More often than not the flesh consecrated by the immolation was eaten. The festivals followed one another without interruption and every year the divine service called for countless sacrifices: Twenty thousand is given as the number. One of the victims incarnating a god, he climbed to the sacrifice surrounded, like a god, by an attendance that would accompany him in death.

INTIMACY OF EXECUTIONERS AND VICTIMS

The Aztecs observed a singular conduct with those who were about to die. They treated these prisoners humanely, giving them the food and drink they asked for. Concerning a warrior who brought back a captive, then offered him in sacrifice, it was said that he had 'considered his captive as his own flesh and blood, calling him son, while the latter called him father.'[6] The victims would dance and sing with those who brought them to die. Efforts were often made to relieve their anguish. A woman incarnating the 'mother of the gods' was consoled by the healers and midwives who said to her: 'Don't be sad, fair friend; you will spend this night with the king, so you can rejoice.' It was not made clear to her that she was to be killed, because death needed to be sudden and unexpected in her case. Ordinarily the condemned prisoners were well aware of their fate and were forced to stay up the final night, singing and dancing. Sometimes they were made to drink until drunk or, to drive away the idea of impending death, they were given a concubine. [. . .]

SACRIFICE OR CONSUMPTION

This softening of the sacrificial process finally discloses a movement to which the rites of immolation were a response. This movement appears to us in its logical necessity alone and we cannot know if the sequence of acts conforms to it in detail; but in any case its coherence is evident.

Sacrifice restores to the sacred world that which servile use has degraded, rendered profane. Servile use has made a *thing* (an *object*) of that which, in a deep sense, is of the same nature as the *subject*, is in a relation of intimate participation with the subject. It is not necessary that the sacrifice actually destroy the animal or plant of which man had to make a *thing* for his use. They must at least be destroyed as things, that is, *insofar as they have become things*. Destruction is the best means of negating a utilitarian relation between man and the animal or plant. But it rarely goes to the point of holocaust. It is enough that the consumption of the offerings, or the *communion*, has a meaning that is not reducible to the shared ingestion of food. The victim of the sacrifice cannot be consumed in the same way as a motor uses fuel. What the ritual has the virtue of rediscovering is the intimate participation of the sacrificer and the victim, to which a servile use had put an end. The slave bound to labor and having become the property of another is a *thing* just as a work animal is a thing. The individual who employs the labor of his prisoner severs the tie that links him to his fellow man. He is not far from the moment when he will sell him. But the owner has not simply made a *thing*, a commodity, of this property. No one can make a *thing* of the second self that the slave is without at the same time estranging himself from his own intimate being, without giving himself the limits of a *thing*.

This should not be considered narrowly: There is no perfect operation, and neither the slave nor the master is entirely reduced to the *order of things*. The slave is a thing for the owner; he accepts this situation which he prefers to dying; he effectively loses part of his intimate value for himself, for it is not enough to be this or that: One also has to be for others. Similarly, for the slave the owner has ceased to be his fellow man; he is profoundly separated from him; even if his equals continue to see him as a man, even if he is still a man for others, he is now in a world where a man can be merely a *thing*. The same poverty then extends over human life as extends over the countryside if the weather is overcast. Overcast weather, when the sun is filtered by the clouds and the play of light goes dim, appears to 'reduce things to what they are.' The error is obvious: What is before me is never anything less than the universe; the universe is not a *thing* and I am not at all mistaken when I see its brilliance in the sun. But if the sun is hidden I more clearly see the barn, the field, the hedgerow. I no longer see the splendor of the light that played over the barn; rather I see this barn or this hedgerow like a screen between the universe and me.

In the same way, slavery brings into the world the absence of light that is the separate positing of each *thing*, reduced to the *use* that it has. Light, or brilliance, manifests the intimacy of life, that which life deeply is, which is perceived by the subject as being true to itself and as the transparency of the universe.

But the reduction of 'that which is' to the *order of things* is not limited to slavery. Slavery is abolished, but we ourselves are aware of the aspects of social life in which man is relegated to the level of *things*, and we should know that this relegation did not await slavery. From the start, the introduction of *labor* into the world replaced intimacy, the depth of desire and its free outbreaks, with rational progression, where what matters is no longer the truth of the present moment, but, rather, the subsequent results of *operations*. The first labor established the world of *things*, to which the profane world of the Ancients generally corresponds. Once the world of things was posited, man himself became one of the things of this world, at least for the time in which he labored. It is this degradation that man has always tried to escape. In his strange myths, in his cruel rites, man is *in search of a lost intimacy* from the first.

Religion is this long effort and this anguished quest: It is always a matter of detaching from the real order, from the poverty of *things*, and of restoring the *divine* order. The animal or plant that man *uses* (as if they only had value *for him* and none for themselves) is restored to the truth of the intimate world; he receives a sacred communication from it, which restores him in turn to interior freedom.

The meaning of this profound freedom is given in destruction, whose essence is to consume *profitlessly* whatever might remain in the progression of useful works. Sacrifice destroys that which it consecrates. It does not have to destroy as fire does; only the tie that connected the offering to the world of profitable activity is severed, but this separation has the sense of a definitive consumption; the consecrated offering cannot be restored to the *real* order. This principle opens the way to passionate release; it liberates violence while marking off the domain in which violence reigns absolutely.

The world of *intimacy* is as antithetical to the *real* world as immoderation is to moderation, madness to reason, drunkenness to lucidity. There is moderation only in the object, reason only in the identity of the object with itself, lucidity only in the distinct knowledge of objects. The world of the subject is the night: that changeable, infinitely suspect night which, in the sleep of reason, *produces monsters. I submit that madness itself gives a rarefied idea of the free 'subject,' unsubordinated to the 'real' order and occupied only with the present.* The *subject* leaves its own domain and subordinates itself to the *objects* of the *real* order as soon as it becomes concerned for the future. For the *subject* is consumption insofar as it is not tied down to work. If I am no longer concerned about 'what will be' but about 'what is,' what reason do I have to keep anything in reserve? I can at once, in disorder, make an instantaneous consumption of all that I possess. This useless consumption is *what suits me*, once my concern for the morrow is removed. And if I thus consume immoderately, I reveal to my fellow beings that which I am *intimately*: Consumption is the way in which *separate* beings communicate.[7] Everything shows through, everything is open and infinite between those who consume intensely. But nothing counts then; violence is released and it breaks forth without limits, as the heat increases.

What ensures the return of the *thing* to the *intimate* order is its entry into the hearth of consumption, where the violence no doubt is limited, but never without great difficulty. It is always the purpose of sacrifice to give destruction its due, to save the rest from a mortal danger of contagion. All those who have to do with sacrifice are in danger, but its limited ritual form regularly has the effect of protecting those who offer it.

Sacrifice is heat, in which the intimacy of those who make up the system of common works is rediscovered. Violence is its principle, but the works limit it in time and space; it is subordinated to the concern for uniting and preserving the commonality. The individuals break loose, but a breaking-loose that melts them and blends them indiscriminately with their fellow beings helps to connect them together in the operations of secular time. It is not yet a matter of *enterprise*, which absorbs the excess forces with a view to the unlimited development of wealth. The works in question only aim at continuance. They only predetermine the limits of the festival (whose renewal is ensured by their fecundity, which has its source in the festival itself). But the community is saved from ruination. The *victim* is given over to violence.

THE VICTIM, SACRED AND CURSED

The victim is a surplus taken from the mass of *useful* wealth. And he can only be withdrawn from it in order to be consumed profitlessly, and therefore utterly destroyed. Once chosen, he is the *accursed share*, destined for violent consumption. But the curse tears him away from the *order of things*; it gives him a recognizable figure, which now radiates intimacy, anguish, the profundity of living beings.

NOTES

1 Bernardino de Sahagún, *Historia general de las cosas de Nueva España*, Mexico City: Porrúa, 1956. Book II, Ch. 5.
2 *Ibid.*, appendix of Book II.
3 *Ibid.*, Book II, Ch. 24.
4 *Ibid.*, Book II, Ch. 5.
5 *Ibid.*, Book II, Ch. 24.
6 *Ibid.*, Book II, Ch. 21.
7 I wish to emphasize a basic fact: the separation of beings is limited to the real order. It is only if I remain attached to the order of *things* that the separation is *real*. It *is* in fact *real*, but what is real is *external*. 'Intimately, all men are one.'

Pecuniary Emulation

Thorstein Veblen

In the sequence of cultural evolution the emergence of a leisure class coincides with the beginning of ownership. This is necessarily the case, for these two institutions result from the same set of economic forces. In the inchoate phase of their development they are but different aspects of the same general facts of social structure.

It is as elements of social structure – conventional facts – that leisure and ownership are matters of interest for the purpose in hand. An habitual neglect of work does not constitute a leisure class; neither does the mechanical fact of use and consumption constitute ownership. The present inquiry, therefore, is not concerned with the beginning of indolence, nor with the beginning of the appropriation of useful articles to individual consumption. The point in question is the origin and nature of a conventional leisure class on the one hand and the beginnings of individual ownership as a conventional right or equitable claim on the other hand.

The early differentiation out of which the distinction between a leisure and a working class arises is a division maintained between men's and women's work in the lower stages of barbarism. Likewise the earliest form of ownership is an ownership of the women by the able-bodied men of the community. The facts may be expressed in more general terms, and truer to the import of the barbarian theory of life, by saying that it is an ownership of the woman by the man.

There was undoubtedly some appropriation of useful articles before the custom of appropriating women arose. The usages of existing archaic communities in which there is no ownership of women is warrant for such a view. In all communities the members, both male and female, habitually appropriate to their individual use a variety of useful things; but these useful things are not thought of as owned by the person who appropriates and consumes them. The habitual appropriation and consumption of certain slight personal effects goes on without raising the question of ownership; that is to say, the question of a conventional, equitable claim to extraneous things.

The ownership of women begins in the lower barbarian stages of culture, apparently with the seizure of female captives. The original reason for the seizure and appropriation of women seems to have been their usefulness as trophies. The practice of seizing women from the enemy as trophies gave rise to a form of ownership-marriage, resulting in a household with a male head. This was followed by an extension of slavery to other captives and inferiors, besides women, and by an extension of ownership-marriage to other women than those seized from the enemy. The outcome of emulation under the circumstances of a predatory life, therefore, has been on the one hand a form of marriage resting on coercion, and on the other hand the custom of ownership. The two institutions are not distinguishable in the initial phase of their development; both arise from the desire of the successful men to put their prowess in evidence by exhibiting some durable result of their exploits. Both also minister to that propensity for mastery which pervades all predatory communities. From the ownership of women the concept of ownership extends itself to include the products of their industry, and so there arises the ownership of things as well as of persons.

In this way a consistent system of property in goods is gradually installed. And although in the latest stages of the development, the serviceability of goods for consumption has come to be the most obtrusive element of their value, still, wealth has by no means yet lost its utility as a honorific evidence of the owner's prepotence.

Wherever the institution of private property is found, even in a slightly developed form, the economic process bears the character of a struggle between men for the possession of goods. It has been customary in economic theory, and especially among those economists who adhere with least faltering to the body of modernised classical doctrines, to construe this struggle for wealth as being substantially a struggle for subsistence. Such is, no doubt, its character in large part during the earlier and less efficient phases of industry. Such is also its character in all cases where the 'niggardliness of nature' is so strict as to afford but a scanty livelihood to the community in return for strenuous and unremitting application to the business of getting the means of subsistence. But in all progressing communities an advance is presently made beyond this early stage of technological development. Industrial efficiency is presently carried to such a pitch as to afford something appreciably more than a bare livelihood to those engaged in the industrial process. It has not been unusual for economic theory to speak of the further struggle for wealth on this new industrial basis as a competition for an increase of the comforts of life – primarily for an increase of the physical comforts which the consumption of goods affords.

The end of acquisition and accumulation is conventionally held to be the consumption of the goods accumulated – whether it is consumption directly by the owner of the goods or by the household attached to him and for this purpose identified with him in theory. This is at least felt to be the economically legitimate end of acquisition, which alone it is incumbent on the theory to take account of. Such consumption may of course be conceived to serve the consumer's physical wants – his physical comfort – or his so-called higher wants – spiritual, æsthetic, intellectual, or what not; the latter class of wants being served indirectly by an expenditure of goods, after the fashion familiar to all economic readers.

But it is only when taken in a sense far removed from its naïve meaning that consumption of goods can be said to afford the incentive from which accumulation invariably proceeds. The motive that lies at the root of ownership is emulation; and the same motive of emulation continues active in the further development of the institution to which it has given rise and in the development of all those features of the social structure which this institution of ownership touches. The possession of wealth confers honour; it is an invidious distinction. Nothing equally cogent can be said for the consumption of goods, nor for any other conceivable incentive to acquisition, and especially not for any incentive to the accumulation of wealth.

It is of course not to be overlooked that in a community where nearly all goods are private property the necessity of earning a livelihood is a powerful and ever-present incentive for the poorer members of the community. The need of subsistence and of an increase of physical comfort may for a time be the dominant motive of acquisition for those classes who are habitually employed at manual labour, whose subsistence is on a precarious footing, who possess little and ordinarily accumulate little; but it will appear in the course of the discussion that even in the case of these impecunious classes the predominance of the motive of physical want is not so decided as has sometimes been assumed. On the other hand, so far as regards those members and classes of the community who are chiefly concerned in the accumulation of wealth, the incentive of subsistence or of physical comfort never plays a considerable part. Ownership began and grew into a human institution on grounds unrelated to the subsistence minimum. The dominant incentive was from the outset the invidious distinction attaching to wealth, and, save temporarily and by exception, no other motive has usurped the primacy at any later stage of the development.

Property set out with being booty held as trophies of the successful raid. So long as the group had departed but little from the primitive communal organisation, and so long as it still stood in close contact with other hostile groups, the utility of things or persons owned lay chiefly in an invidious comparison between their possessor and the enemy from whom they were taken. The habit of distinguishing between the interests of the individual and those of the group to which he belongs is apparently a later growth. Invidious comparison between the possessor of the honorific booty and his less successful neighbours within the group was no doubt present early as an element of the utility of the things possessed, though this was not at the outset the chief element of their value. The man's prowess was still primarily the group's prowess, and the possessor of the booty felt himself to be primarily the keeper of the honour of his group. This appreciation of exploit from the communal point of view is met with also at later stages of social growth, especially as regards the laurels of war.

But so soon as the custom of individual ownership begins to gain consistency, the point of view taken in making the invidious comparison on which private property rests will begin to change. Indeed, the one change is but the reflex of the other. The initial phase of ownership, the phase of acquisition by naïve seizure and conversion, begins to pass into the subsequent stage of an incipient organisation of industry on the basis of private property (in slaves); the horde develops into a more or less self-sufficing industrial community; possessions then come to be valued not so much as evidence of successful foray, but rather as evidence of the prepotence of the possessor of these goods over other individuals within the community. The invidious comparison now becomes primarily a comparison of the owner with the other members of the group. Property is still of the nature of trophy, but, with the cultural advance, it becomes more and more a trophy of successes scored in the game of ownership carried on between the members of the group under the quasi-peaceable methods of nomadic life.

Gradually, as industrial activity further displaces predatory activity in the community's everyday life and in men's habits of thought, accumulated property more and more replaces trophies of predatory exploit as the conventional exponent of prepotence and success. With the growth of settled industry, therefore, the possession of wealth gains in relative importance and effectiveness as a customary basis of repute and esteem. Not that esteem ceases to be awarded on the basis of other, more direct evidence of prowess; not that successful predatory aggression or warlike exploit ceases to call out the approval and admiration of the crowd, or to stir the envy of the less successful competitors; but the opportunities for gaining distinction by means of this direct manifestation of superior force grow less available both in scope and frequency. At the same time opportunities for industrial aggression, and for the accumulation, of property by the quasi-peaceable methods of nomadic industry, increase in scope and availability. And it is even more to the point that property now becomes the most easily recognised evidence of a reputable degree of success as distinguished from heroic or signal achievement. It therefore becomes the conventional basis of esteem. Its possession in some amount becomes necessary in order to [achieve] any reputable standing in the community. It becomes indispensable to accumulate, to acquire property, in order to retain one's good name. When accumulated goods have in this way once become the accepted badge of efficiency, the possession of wealth presently assumes the character of an independent and definitive basis of esteem. The possession of goods, whether acquired aggressively by one's own exertion or passively by transmission through inheritance from others, becomes a conventional basis of reputability. The possession of wealth, which was at the outset valued simply as an evidence of efficiency, becomes, in popular apprehension, itself a meritorious act. Wealth is now itself intrinsically honourable and confers honour on its possessor. By a further refinement, wealth acquired passively by transmission from ancestors or other antecedents presently becomes even more honorific than wealth acquired by the possessor's own effort; but this distinction belongs at a later stage in the evolution of the pecuniary culture and will be spoken of in its place.

Prowess and exploit may still remain the basis of award of the highest popular esteem, although the possession of wealth has become the basis of commonplace reputability and of a blameless social standing. The predatory instinct and the consequent approbation of predatory efficiency are deeply ingrained in the habits of thought of those peoples who have passed under the discipline of a protracted predatory culture. According to popular award, the highest honours within human reach may, even yet, be those gained by an unfolding of extraordinary predatory efficiency in war, or by a quasi-predatory efficiency in statecraft; but for the purposes of a commonplace decent standing in the community these means of repute have been replaced by the acquisition and accumulation of goods. In order to stand well in the eyes of the community, it is necessary to come up to a certain, somewhat indefinite, conventional standard of wealth; just as in the earlier predatory stage it is necessary for the barbarian man to come up to the tribe's standard of physical endurance, cunning, and skill at arms. A certain standard of wealth in the one case, and of prowess in the other, is a necessary condition of reputability, and anything in excess of this normal amount is meritorious.

Those members of the community who fall short of this, somewhat indefinite, normal degree of prowess or of property suffer in the esteem of their fellow-men; and consequently they suffer also in their own esteem, since the usual basis of self-respect is the respect accorded by one's neighbours. Only individuals with an aberrant temperament can in the long run retain their self-esteem in the face of the disesteem of their fellows, Apparent exceptions to the rule are met with, especially among people with strong religious convictions. But these apparent exceptions are scarcely real exceptions, since such persons commonly fall back on the putative approbation of some supernatural witness of their deeds.

So soon as the possession of property becomes the basis of popular esteem, therefore, it becomes also a requisite to that complacency which we call self-respect. In any community where goods are held in severalty it is necessary, in order to his own peace of mind, that an individual should possess as large a portion of goods as others with whom he is accustomed to class himself; and it is extremely gratifying to possess something more than others. But as fast as a person makes new acquisitions, and becomes accustomed to the resulting new standard of wealth, the new standard forthwith ceases to afford appreciably greater satisfaction than the earlier standard did. The tendency in any case is constantly to make the present pecuniary standard the point of departure for a fresh increase of wealth; and this in turn gives rise to a new standard of sufficiency and a new pecuniary classification of one's self as compared with one's neighbours. So far as concerns the present question, the end sought by accumulation is to rank high in comparison with the rest of the community in point of pecuniary strength. So long as the comparison is distinctly unfavourable to himself, the normal, average individual will live in chronic dissatisfaction with his present lot; and when he has reached what may be called the normal pecuniary standard of the community, or of his class in the community, this chronic dissatisfaction will give place to a restless straining to place a wider and ever-widening pecuniary interval between himself and this average standard. The invidious comparison can never become so favourable to the individual making it that he would not gladly rate himself still higher relatively to his competitors in the struggle for pecuniary reputability.

In the nature of the case, the desire for wealth can scarcely be satiated in any individual instance, and evidently a satiation of the average or general desire for wealth is out of the question. However widely, or equally, or 'fairly', it may be distributed, no general increase of the community's wealth can make any approach to satiating this need, the ground of which is the desire of every one to excel every one else in the accumulation of goods. If, as is sometimes assumed, the incentive to accumulation were the want of subsistence or of physical comfort, then the aggregate economic wants of a community might conceivably be satisfied at some point in the advance of industrial efficiency; but since the struggle is substantially a race for reputability on the basis of an invidious comparison, no approach to a definitive attainment is possible.

What has just been said must not be taken to mean that there are no other incentives to acquisition and accumulation than this desire to excel in pecuniary standing and so gain the esteem and envy of one's fellow-men. The desire for added comfort and security from want is present as a motive at every stage of the process of accumulation in a modern industrial community; although the standard of sufficiency in these respects is in turn greatly affected by the habit of pecuniary emulation. To a great extent this emulation shapes the methods and selects the objects of expenditure for personal comfort and decent livelihood.

Besides this, the power conferred by wealth also affords a motive to accumulation. That propensity for purposeful activity and that repugnance to all futility of effort which belong to man by virtue of his character as an agent do not desert him when he emerges from the naïve communal culture where the dominant note of life is the unanalysed and undifferentiated solidarity of the individual with the group with which his life is bound up. When he enters upon the predatory stage, where self-seeking in the narrower sense becomes the dominant note, this propensity goes with him still, as the pervasive trait that shapes his scheme of life. The propensity for achievement and the repugnance to futility remain the underlying economic motive. The propensity changes only in the form of its expression and in the proximate objects to which it directs the man's activity. Under the régime of individual ownership the most available means of visibly achieving a purpose is that afforded by the acquisition and accumulation of goods; and as the self-regarding antithesis between man and man reaches fuller consciousness, the propensity for achievement – the instinct of workmanship – tends more and more to shape itself into a straining to excel others in pecuniary achievement. Relative success, tested by an invidious pecuniary comparison with other men, becomes the conventional end of action. The currently accepted legitimate end of effort becomes the achievement of a favourable comparison with other men; and therefore the repugnance to futility to a good extent coalesces with the incentive of emulation. It acts to accentuate the struggle for pecuniary reputability by visiting with a sharper disapproval all shortcoming and all evidence of shortcoming in point of pecuniary success. Purposeful effort comes to mean, primarily, effort directed to or resulting in a more creditable showing of accumulated wealth. Among the motives which lead men to accumulate wealth, the primacy, both in scope and intensity, therefore, continues to belong to this motive of pecuniary emulation.

In making use of the term 'invidious', it may perhaps be unnecessary to remark, there is no intention to extol or depreciate, or to commend or deplore any of the phenomena which the word is used to characterise. The term is used in a technical sense as describing a comparison of persons with a view to rating and grading them in respect of relative worth or value – in an æsthetic or moral sense – and so awarding and defining the relative degrees of complacency with which they may legitimately be contemplated by themselves and by others. An invidious comparison is a process of valuation of persons in respect of worth.

34

The Philosophy of Fashion

Georg Simmel

The essence of fashion consists in the fact that it should always be exercised by only a part of a given group, the great majority of whom are merely on the road to adopting it. As soon as a fashion has been universally adopted, that is, as soon as anything that was originally done only by a few has really come to be practised by all – as is the case in certain elements of clothing and in various forms of social conduct – we no longer characterize it as fashion. Every growth of a fashion drives it to its doom, because it thereby cancels out its distinctiveness. By reason of this play between the tendency towards universal acceptance and the destruction of its significance, to which this general adoption leads, fashion possesses the peculiar attraction of limitation, the attraction of a simultaneous beginning and end, the charm of newness and simultaneously of transitoriness. Fashion's question is not that of being, but rather it is simultaneously being and non-being; it always stands on the watershed of the past and the future and, as a result, conveys to us, at least while it is at its height, a stronger sense of the present than do most other phenomena.

If the momentary concentration of social consciousness upon the point which fashion signifies is also the one in which the seeds of its own death and its determined fate to be superseded already lie, so this transitoriness does not degrade it totally, but actually adds a new attraction to its existing ones. At all events, an object does not suffer degradation by being called 'fashionable', unless we reject it with disgust or wish to debase it for other material reasons; in which case, of course, fashion becomes a concept of value. In the practice of life, anything else that is similarly new and suddenly disseminated in the same manner will not be characterized as fashion, if we believe in its continuance and its *objective* justification. If, on the other hand, we are convinced that the phenomenon will vanish just as rapidly as it came into existence, then we call it fashion. Hence, among the reasons why nowadays fashion exercises such a powerful influence on our consciousness there is also the fact that the major, permanent, unquestionable convictions are more and more losing their force. Consequently, the fleeting and fluctuating elements of life gain that much more free space. The break with the past which, for more than a century, civilized human kind has been labouring unceasingly to bring about, concentrates consciousness more and more upon the present. This accentuation of the present is evidently, at the same time, an emphasis upon change and to the extent to which a particular stratum is the agent of this cultural tendency, so to that degree will it turn to fashion in all fields, and by no means merely with regard to clothing. Indeed, it is almost a sign of the *increased* power of fashion that it has overstepped the bounds of its original domain, which comprised only externals of dress, and has acquired an increasing influence over taste, theoretical convictions, and even the moral foundations of life in their changing forms.

From the fact that fashion as such can never be generally in vogue, the individual derives the satisfaction of knowing that, as adopted by him or her, it still represents something special and striking; while at the same time the individual feels inwardly supported by a broad group of persons who are striving for the same thing, and not, as is the case for other social satisfactions, by a group that is doing the same thing. Therefore the feelings which the fashionable person confronts are an apparently agreeable mixture of approval and envy. We envy the fashionable person as an individual, but approve of them as a member of a group. Yet even this envy here has a peculiar nuance. There is a nuance of envy which includes a sort of ideal participation in the envied objects. An instructive example of this is furnished by the conduct of the worker who is able to get a glimpse of the feasts of the rich. In so far as we envy an object or a person, we are no longer absolutely excluded from them, and between both there now exists some relation or other, and between both the same psychological content now exists, even though in entirely different categories and forms of sensation. This quiet personal usurpation of the envied property – which is also the pleasure of unrequited love – contains a kind of antidote, which occasionally counteracts the worst degenerations of the feeling of envy. The elements of fashion afford an especially good chance for the development of this more conciliatory shade of envy, which also gives to the envied person a better conscience because of his or her satisfaction with regard to their good fortune. This is due to the fact that, unlike many other psychological contents, these contents of fashion are not denied *absolutely* to anyone, because a change of fortune, which is never entirely out of the question, may play them into the hands of an individual who had previously been confined to the state of envy.

From all this we see that fashion is the genuine playground for individuals with dependent natures, but whose self-consciousness, however, at the same time requires a certain amount of prominence, attention, and singularity. Fashion elevates even the unimportant individual by making them the representative of a totality, the embodiment of a joint spirit. It is particularly characteristic of fashion – because, by its very essence, it can be a norm which is never satisfied by everyone – that it renders possible a social obedience that is at the same time a form of individual differentiation. In slaves to fashion (*Modenarren*) the social demands of fashion appear exaggerated to such a high degree that they completely acquire a semblance of individuality and particularity. It is characteristic of the slave to fashion that he carries the tendency of a particular fashion beyond the otherwise self-contained limits. If pointed shoes are in style, then he wears shoes that resemble spear tips; if pointed collars are all the rage, he wears collars that reach up to his ears; if it is fashionable to attend scholarly lectures, then he is never seen anywhere else, and so on. Thus he represents something totally individual, which consists in the quantitative intensification of such elements as are qualitatively common property of the given social circle. He leads the way, but all travel the same road. Representing as he does the most recently conquered heights of public taste, he seems to be marching at the head of the general process. In reality, however, what is so frequently true of the relation between individuals and groups also applies to him: that actually, the leader is the one who is led.

Democratic times obviously favour such a condition to a remarkable degree, so much so that even Bismarck and other very prominent party leaders of constitutional governments have emphasized the fact that, inasmuch as they are leaders of a group, they are bound to follow it. Such times cause persons to seek dignity and the sensation of command in this manner; they favour a confusion and ambiguity of sensations, which fail to distinguish between ruling the mass and being ruled by it. The conceit of the slave to fashion is thus the caricature of a democratically fostered constellation of the relations between the individual and the totality. Undeniably, however, the hero of fashion, through the conspicuousness gained in a purely quantitative way, but clothed in a difference of quality, represents a genuinely original state of equilibrium between the social and the individualizing impulses. This enables us to understand the outwardly so abstruse devotion to fashion of otherwise quite intelligent and broad-minded persons. It furnishes them with a combination of relations to things and human beings that, under ordinary circumstances, tend to appear separately. What is at work here is not only the mixture of individual distinctiveness and social equality, but more practically, as it were, it is the mingling of the sensation of domination and subordination. Or, formulated differently, we have here the mixing of a male and a female principle. The very fact that this mixing process only occurs in the sphere of fashion as in an ideal dilution that, as it were, only realizes the form of both elements in a content that is in itself indifferent, may lend a special attraction to fashion, especially for sensitive natures that do not care to concern themselves with robust reality. From an objective standpoint, life according to fashion consists of a mixture of destruction and construction; its content acquires its characteristics by destruction of an earlier form; it possesses a peculiar uniformity, in which the satisfaction of the destructive impulse and the drive for positive elements can no longer be separated from each other.

Because we are dealing here not with the significance of a single content or a single satisfaction, but rather with the play between both contents and their mutual distinction, it becomes evident that the same combination which extreme obedience to fashion acquires can also be won by opposition to it. Whoever consciously clothes or deports themselves in an unmodern manner does not attain the consequent sensation of individualization through any real individual qualification of his or her own, but rather through the mere negation of the social example. If modernity is the imitation of this social example, then the deliberate lack of modernity represents a similar imitation, yet under an inverted sign, but nonetheless one which offers no less a testimony of the power of the social tendency, which makes us dependent upon it in some positive or negative manner. The deliberately unmodern person accepts its forms just as much as does the slave to fashion, except that the unmodern person embodies it in another category: in that of negation, rather than in exaggeration.

Indeed, it occasionally becomes decidedly fashionable in whole circles of a large-scale society to clothe oneself in an unmodern manner. This constitutes one of the most curious social–psychological complications, in which the drive for individual conspicuousness primarily remains content, first, with a mere inversion of social imitation and, second, for its part draws its strength again from approximation to a similarly characterized narrow circle. If a club or association of club-haters were founded, then it would not be logically more impossible and psychologically more possible than the above phenomenon. Just as atheism has been made into a religion, embodying the same fanaticism, the same intolerance, the same satisfaction of emotional needs that are embraced in religion proper; and just as freedom too, by means of which a tyranny has been broken, often becomes no less tyrannical and violent; so this phenomenon of tendentious lack of modernity illustrates how ready the fundamental forms of human nature are to accept the total antithesis of contents in themselves and to show their strength and their attraction in the negation of the very thing to whose acceptance they still seemed a moment earlier to be irrevocably committed. Thus, it is often absolutely impossible to tell whether the elements of personal strength or of personal weakness have the upper hand in the complex of causes of such lack of modernity. It may result from the need not to make common cause with the crowd, a need that has as its basis, of course, not in independence from the crowd, but rather in an inner sovereign stance with respect to the latter. However, it may also be due to a weak sensibility, which causes individuals to fear that they will be unable to maintain their little piece of individuality if they adopt the forms, tastes and customs of the general public. Such opposition to the latter is by no means always a sign of personal strength. On the contrary, personal strength will be so conscious of its unique value, which is immune to any external connivance, that it will be able to submit without any unease to general forms up to and including fashion. Rather, it is precisely in this obedience that it will become conscious of the *voluntariness* of its obedience and of that which transcends obedience.

If fashion both gives expression to the impulse towards equalization and individualization, as well as to the allure of imitation and conspicuousness, this perhaps explains why it is that women, broadly speaking, adhere especially strongly to fashion. Out of the weakness of the social position to which women were condemned throughout the greatest part of history there arises their close relationship to all that is 'custom', to that which is 'right and proper', to the generally valid and approved form of existence. For those who are weak steer clear of individualization; they avoid dependence upon the self, with its responsibilities and the necessity of defending oneself unaided. Those in a weak position find protection only in the typical form of life, which prevents the strong person from exercising his exceptional powers. But resting on the firm foundation of custom, of the average, of the general level, women strive strongly for all the relative individualization and general conspicuousness that thus remains possible. Fashion offers them this very combination to the most favourable extent, for we have here, on the one hand, a sphere of general imitation, the individual floating in the broadest social current, relieved of responsibility for their tastes and their actions, and yet, on the other hand, we have a certain conspicuousness, an individual emphasis, an individual ornamentation of the personality.

It seems that there exists for each class of human beings, indeed probably for each individual, a definite quantitative relationship between the impulse towards individualization and the drive for immersion in the collectivity, so that if the satisfaction of one of these drives is denied in a certain field of life, it seeks another, in which it then fulfils the amount that it requires. Thus, it seems as though fashion were the valve, as it were, through which women's need for some measure of conspicuousness and individual prominence finds vent, when its satisfaction is more often denied in other spheres.

During the fourteenth and fifteenth centuries, Germany displayed an extraordinarily strong development of individuality. To a great extent, the collectivistic regulations of the Middle Ages were breached by the freedom of the individual person. Within this individualistic development, however, women still found no place and the freedom of personal movement and self-improvement was still denied them. They compensated for this by adopting the most extravagant and exaggerated fashions in dress. Conversely, we see that in Italy during the same epoch, women were given free play for individual development. The women of the Renaissance possessed extensive opportunities for culture, external activity, and personal differentiation such as were not offered to them again for many centuries. In the upper classes of society, especially, education and freedom of expression were almost identical for both sexes. Yet it is also reported that no particularly extravagant Italian female fashions emerged from that period. The need to exercise individuality and gain a kind of distinction in this sphere was absent, because the impulse embodied therein found sufficient satisfaction in other spheres. In general, the history of women in their outer as well as their inner life, in the individual aspect as well as in their collectivity, exhibits such a comparatively great uniformity, levelling and similarity, that they require a more lively activity at least in the sphere of fashions, which is nothing more nor less than variety, in order to add an attraction to themselves and their lives – for their own feeling as well as for others.

Just as in the case of individualism and collectivism, so there exists between the uniformity and the variety of the contents of life a definite proportion of needs, which is tossed to and fro in the different spheres and seeks to balance the refusal in the one by consent, however acquired, in the other. On the whole, we may say that the woman, compared to the man, is a more faithful creature. Now fidelity, expressing the uniformity and regularity of one's nature only according to the side of one's feelings, demands a more lively change in the outward surrounding spheres in order to establish the balance in the tendencies of life. The man, on the other hand, who in his essence is less faithful and who does not ordinarily maintain an emotional relationship that he has entered into with the same absoluteness and concentration of all his vital interests, is consequently in less need of external variation. Indeed, the lack of acceptance of changes in external fields, the indifference towards fashions in outward appearance are specifically a male quality, not because a man is more uniform, but because he is the more many-sided creature, and for that reason, can exist without external changes. Therefore the emancipated woman of the present, who seeks to approach towards the whole differentiation, personality and activity of the male sex, lays particular stress on her indifference to fashion. In a certain sense, fashion also gives women a compensation for their lack of social position in a professional group. The man who has become absorbed in a vocation has thereby entered into a relatively uniform social circle, in which he resembles many others within this stratum, and is thus often only an exemplar of the concept of this stratum or occupation. On the other hand, and as if to compensate him for this absorption, he is invested with the full importance and the objective as well as the social power of this stratum. To his individual importance is added that of his stratum, which often can cover over the defects and deficiencies of purely personal existence.

Fashion accomplishes this identical process by other means. Fashion too supplements a person's lack of importance, their inability to individualize their existence purely by their own unaided efforts, by enabling them to join a group characterized and singled out in the public consciousness by fashion alone. Here too, to be sure, the personality as such is adapted to a general formula, yet this formula itself, from a social standpoint, possesses an individual colouring, and this makes up by a socially roundabout way for precisely what is denied to the personality in a purely individual way. The fact that the *demi-monde* is so frequently the pioneer of new fashion is due to its distinctively uprooted form of life. The pariah existence to which society condemns the *demi-monde* produces an open or latent hatred against everything that has the sanction of law, against every permanent institution, a hatred that still finds its relatively most innocent expression in the striving for ever new forms of appearance. In this continual striving for new, previously unheard-of fashions, in the ruthlessness with which the one that is most opposed to the existing one is passionately adopted, there lies an aesthetic form of the destructive urge that seems to be an element peculiar to all who lead this pariah-like existence, so long as they are not inwardly completely enslaved.

And if we seek to gaze into the final and most subtle movements of the soul, which are difficult to express in words, so we find that they too exhibit this antagonistic play of the fundamental human tendencies, which seek to regain their continually lost balance by means of ever new proportions. It is in fact fundamental to fashion that it makes no distinction at all between all individualities alike, and yet it is always done in such a way that it never affects the whole human being; indeed it always remains somewhat external to the individual – even in those spheres outside mere clothing fashions. For the form of mutability in which it is presented to the individual is under all circumstances a contrast to the stability of the sense of self, and indeed the latter must become conscious of its relative duration precisely through this contrast. The changeableness of the elements of fashion can express itself as mutability and develop its attraction only through this enduring element of the sense of self. But for this very reason fashion always stands, as I have pointed out, at the very periphery of the personality, which regards itself as a *pièce de résistance* to fashion, or at least can be experienced as such in an emergency.

It is this significant aspect of fashion that is adopted by refined and special persons, in so far as they use it as a kind of mask. They consider blind obedience to the standards of the public in all externals as the conscious and desired means of reserving their personal feelings and their taste, which they are eager to keep to themselves alone; indeed, so much to themselves that they do not care to allow their feelings and tastes to be seen in a form that is visible to all. It is therefore precisely a refined feeling of modesty and reserve which, seeking not to resort to a peculiarity in externals for fear perhaps of betraying a peculiarity of their innermost soul, causes many a delicate nature to seek refuge in the masking levelling of fashion. Thereby a triumph of the soul over the given nature of existence is achieved which, at least as far as form is concerned, must be considered one of the highest and finest victories: namely, that the enemy himself is transformed into a servant, that precisely that which seemed to violate the personality is seized voluntarily, because the levelling violation is here transferred to the external strata of life in such a way that it furnishes a veil and a protection for everything that is innermost and now all the more free. The struggle between the social and the individual is resolved here in so far as the strata for both are separated. This corresponds exactly to the triviality of expression and conversation through which very sensitive and retiring people, especially women, often deceive one about the depth of the individual soul behind these expressions.

All feeling of shame rests upon the conspicuousness of the individual. It arises whenever a stress is laid upon the self, whenever the attention of a social circle is drawn to an individual, which at the same time is felt to be in some way inappropriate. For this reason retiring and weak natures are particularly inclined to feelings of shame. The moment they step into the centre of general attention, the moment they make themselves conspicuous in any way, a painful oscillation between emphasis upon and withdrawal of the sense of the self manifests itself. In so far as this individual conspicuousness, as the source of the feeling of shame, is quite independent of the particular content on the basis of which it occurs, so one is actually frequently ashamed of good and noble things. If in society, in the narrower sense of the term, banality constitutes good form, then this is due not only to a mutual regard, which causes it to be considered bad taste to stand out through some individual, singular expression that not everyone can imitate, but also to the fear of that feeling of shame which forms a self-inflicted punishment for those departing from the form and activity that is similar for, and equally accessible to, everyone.

By reason of its distinctive inner structure, fashion furnishes a conspicuousness of the individual which is always looked upon as proper, no matter how extravagant its form of appearance or manner of expression may be; as long as it is fashionable it is protected against those unpleasant reflections which the individual otherwise experiences when he or she becomes the object of attention. All mass actions are characterized by the loss of the feeling of shame. As a member of a mass, the individual will do many things which would have aroused uncontrollable repugnance in their soul had they been suggested to them alone. It is one of the most remarkable social–psychological phenomena, in which this characteristic of mass action is well exemplified, that many fashions tolerate breaches of modesty which, if suggested to the individual alone, would be angrily repudiated. But as dictates of fashion they find a ready acceptance. The feeling of shame is eradicated in matters of fashion, because it represents a mass action, in the same way that the feeling of responsibility is extinguished in participants in mass criminality, who if left to themselves as individuals would shrink from such deeds. As soon as the individual aspects of the situation begin to predominate over its social and fashionable aspects, the sense of shame immediately commences its effectiveness: many women would be embarrassed to confront a single male stranger alone in their living room with the kind of low necklines that they wear in society, according to the dictates of fashion, in front of thirty or a hundred men.

Fashion is also only one of the forms through which human beings seek to save their inner freedom all the more completely by sacrificing externals to enslavement by the general public. Freedom and dependency also belong to those pairs of opposites, whose ever-renewed struggle and endless mobility give to life much more piquancy, and permit a much greater breadth and development than could a permanent, unchangeable balance of the two. Just as, according to Schopenhauer, to each person a certain amount of joy and sorrow is given, whose measure can neither remain empty nor be filled to overflowing, but in all the diversity and vacillations of internal and external circumstances only changes its form, so – much less mystically – we may observe in each period, in each class, and in each individual either a really permanent proportion of dependency and freedom, or at least the longing for it, whereas we can only change the fields over which they are distributed. And it is the task of the higher life, to be sure, to arrange this distribution in such a way that the other values of existence thereby acquire the most favourable development. The same quantity of dependency and freedom may help, at one time, to increase moral, intellectual and aesthetic values to the highest point and, at another time, without any change in quantity but merely in distribution, it may bring about the exact opposite result. As a whole, one could say that the most favourable result for the total value of life will be obtained when all unavoidable dependency is transferred more and more to the periphery of life, to its externalities. Perhaps Goethe, in his later period, is the most eloquent example of a wholly great life, for by means of his adaptability in all externals, his strict regard for form, his willing obedience to the conventions of society, he attained a maximum of inner freedom, a complete saving of the centres of life from the unavoidable quantity of dependency. In this respect, fashion is also a social form of marvellous expediency, because, like the law, it affects only the externals of life, and hence only those sides of life which are turned towards society. It provides human beings with a formula by means of which we can unequivocally attest our dependency upon what is generally adopted, our obedience to standards established by our time, our class, and our narrower circle, and enables us to withdraw the freedom given us in life and concentrate it more and more in our innermost and fundamental elements.

Within the individual soul these relations of equalizing unification and individual demarcation are, to a certain extent, actually repeated. The antagonism of the tendencies which produce fashion is transferred, as far as form is concerned and in an entirely similar manner, even to those inner relations of some individuals who have nothing whatever to do with social obligations. The phenomenon to which I am referring exhibits the often emphasized parallelism with which the relations between individuals are repeated in the correlation between the psychological elements of individuals themselves. More or less intentionally, the individual often establishes a mode of conduct or a style for him or herself which, by the rhythm of its rise, its sway and decline, becomes characterized as fashion. Young people especially often display a sudden singularity in their behaviour; an unexpected, objectively unfounded interest arises that governs their whole sphere of consciousness, only to disappear in the same irrational manner. We might characterize this as a personal fashion, which forms a limiting case of social fashion. The former is supported, on the one hand, by the individual demand for differentiation and thereby attests to the same impulse that is active in the formation of social fashion. The needs for imitation, similarity, and for the blending of the individual into the mass, are here satisfied purely within the individuals themselves: namely through the concentration of the personal consciousness upon this one form or content, through the uniform shading his or her nature receives from that concentration, through the imitation, as it were, of their own self, which here replaces the imitation of others.

A certain intermediate stage is often realized within close social circles between individual style and personal fashion. Ordinary persons frequently adopt some expression, which they apply at every opportunity – in common with as many others as possible in the same social circle – to all manner of suitable and unsuitable objects. In one respect this is a group fashion, yet in another it is really individual, for its express purpose consists in having the *individual* make the *totality* of his or her circle of ideas subject to this formula. Brutal violence is hereby committed against the individuality of things – all nuances are blurred by the curious supremacy of this one category of expression, for example, when we designate all things that we like for any reason whatever as 'chic', or 'smart' – even though the objects in question may bear no relation whatsoever to the fields to which these expressions belong. In this manner, the inner world of the individual is made subject to fashion, and thus repeats the form of the group dominated by fashion. And this also occurs chiefly by reason of the objective absurdity of such individual manners, which illustrate the power of the formal, unifying element over the objective, rational element. In the same way, many persons and social circles only ask that they be uniformly governed, and the question as to how qualified or valuable is this domination plays a merely secondary role. It cannot be denied that, by doing violence to objects treated in this way, and by clothing them all uniformly in a category that we apply to them, the individual exercises an authority over them, and gains an individual feeling of power, an emphasis of the self over against these objects.

The phenomenon that appears here in the form of a caricature is noticeable to lesser degrees everywhere in the relationship of people to objects. It is only the most noble human beings who find the greatest depth and power of their ego precisely in the fact that they respect the individuality inherent in things. The hostility which the soul bears to the supremacy, independence and indifference of the universe continuously gives rise, as it were – beside the loftiest and most valuable strivings of humanity – to attempts to oppress things externally. The self prevails against them not by absorbing and moulding their powers, not by recognizing their individuality in order to make them serviceable, but by forcing them outwardly to subjugate themselves to some subjective formula. To be sure, in reality the self has not gained control of the things themselves, but only of its own falsified fantasy image of them. However, the feeling of power, which originates from this, reveals its lack of foundation and its illusory nature by the rapidity with which such expressions of fashions pass away. It is just as illusory as the feeling of the uniformity of being, that springs for the moment out of this schematization of all expressions.

We have seen that in fashion, as it were, the different dimensions of life acquire a peculiar convergence, that fashion is a complex structure in which all the leading antithetical tendencies of the soul in one way or another are represented. This makes it abundantly clear that the total rhythm in which individuals and groups move will also exert an important influence upon their relationships to fashion; that the various strata of a group, quite aside from their different contents of life and external possibilities, will exhibit different relationships to fashion simply because their contents of life are evolved either in a conservative or in a rapidly varying form. On the one hand, the lower masses are difficult to set in motion and are slow to develop. On the other hand, it is the highest strata, as everyone knows, who are the most conservative, and who are frequently enough quite archaic. They frequently dread every movement and transformation, not because they have an antipathy to the contents or because the latter are injurious to them, but simply because it is transformation as such and because they regard every modification of the whole, which after all provides them with the highest position, as suspicious and dangerous. No change can bring to them any additional power; at most they have something to fear from each change, but nothing more to hope for from any transformation.

The real variability of historical life is therefore vested in the middle classes and, for this reason, the history of social and cultural movements has taken on an entirely different pace since the *tiers état* assumed power. This is why fashion, itself the changing and contrasting form of life, has since then become much broader and more animated. This is also because of the transformations in immediate political life, for people require an ephemeral tyrant the moment they have rid themselves of an absolute and permanent one. The frequent changes in fashion constitute a tremendous subjugation of the individual and in that respect form one of the necessary complements to increased social and political freedom. A class which is inherently so much more variable, so much more restless in its rhythms than the lowest classes with their silently unconscious conservatism, and the highest classes with their consciously desired conservatism, is the totally appropriate location for a form of life in which the moment of an element's triumph marks the beginning of its decline. Classes and individuals who demand constant change – because precisely the rapidity of their development secures them an advantage over others – find in fashion something that keeps pace with their own inner impulses. And social advancement must be directly favourable to the rapid advance of fashion, because it equips the lower strata so much more quickly to imitate the upper strata and thus the process characterized above – according to which every higher stratum throws aside a fashion the moment a lower one adopts it – acquires a breadth and vitality never dreamed of before.

This fact has an important bearing upon the content of fashion. Above all, it creates a situation in which fashions can no longer be so expensive and, therefore, obviously can no longer be so extravagant as they were in earlier times, where there was a compensation in the form of the longer duration of their domination for the costliness of the first acquisition or the difficulties in transforming conduct and taste. The more an article becomes subject to rapid changes of fashion, the greater the demand for *cheap* products of its kind. This is not only because the larger and therefore poorer classes nevertheless have enough purchasing power to regulate industry and demand objects which at least bear the outward and precarious semblance of modernity, but rather also because even the higher strata of society could not afford to adopt the rapid changes in fashion forced upon them by the pressure of the lower classes if its objects were not relatively cheap. The speed of development is of such importance in genuine articles of fashions that it even withdraws them from certain economic advances that have been won gradually in other fields. It has been noticed, especially in the older branches of production in modern industry, that the speculative element gradually ceases to play an influential role. The movements of the market can be better observed, requirements can be better foreseen and production can be more accurately regulated than before, so that the rationalization of production makes greater and greater inroads upon the fortuitousness of market opportunities and upon the unplanned fluctuations of supply and demand. Only pure articles of fashion seem to be excluded from this. The polar fluctuations, which in many cases the modern economy knows how to avoid and from which it is visibly striving towards entirely new economic orders and formations, still predominate in the fields immediately subject to fashion. The form of feverish change is so essential here that fashion stands, as it were, in a logical contradiction to the developmental tendencies of modern societies.

35

Classes and Classifications

Pierre Bourdieu

> If I have to choose the lesser of two evils,
> I choose neither
>
> (Karl Kraus)

Taste is an acquired disposition to 'differentiate' and 'appreciate',[1] as Kant says – in other words, to establish and mark differences by a process of distinction which is not (or not necessarily) a distinct knowledge, in Leibniz's sense, since it ensures recognition (in the ordinary sense) of the object without implying knowledge of the distinctive features which define it.[2] The schemes of the habitus, the primary forms of classification, owe their specific efficacy to the fact that they function below the level of consciousness and language, beyond the reach of introspective scrutiny or control by the will. Orienting practices practically, they embed what some would mistakenly call *values* in the most automatic gestures or the apparently most insignificant techniques of the body – ways of walking or blowing one's nose, ways of eating or talking – and engage the most fundamental principles of construction and evaluation of the social world, those which most directly express the division of labour (between the classes, the age groups and the sexes) or the division of the work of domination, in divisions between bodies and between relations to the body which borrow more features than one, as if to give them the appearances of naturalness, from the sexual division of labour and the division of sexual labour. Taste is a practical mastery of distributions which makes it possible to sense or intuit what is likely (or unlikely) to befall – and therefore to befit – an individual occupying a given position in social space. It functions as a sort of social orientation, a 'sense of one's place', guiding the occupants of a given place in social space towards the social positions adjusted to their properties, and towards the practices or goods which befit the occupants of that position. It implies a practical anticipation of what the social meaning and value of the chosen practice or thing will probably be, given their distribution in social space and the practical knowledge the other agents have of the correspondence between goods and groups.

Thus, the social agents whom the sociologist classifies are producers not only of classifiable acts but also of acts of classification which are themselves classified. Knowledge of the social world has to take into account a practical knowledge of this world which pre-exists it and which it must not fail to include in its object, although, as a first stage, this knowledge has to be constituted *against* the partial and interested representations provided by practical knowledge. To speak of habitus is to include in the object the knowledge which the agents, who are part of the object, have of the object, and the contribution this knowledge makes to the reality of the object. But it is not only a matter of putting back into the real world that one is endeavouring to know, a knowledge of the real world that contributes to its reality (and also to the force it exerts). It means conferring on this knowledge a genuinely constitutive power, the very power it is denied when, in the name of an objectivist conception of objectivity, one makes common knowledge or theoretical knowledge a mere reflection of the real world.

Those who suppose they are producing a materialist theory of knowledge when they make knowledge a passive recording and abandon the 'active aspect' of knowledge to idealism, as Marx complains in the *Theses on Feuerbach*, forget that all knowledge, and in particular all knowledge of the social world, is an act of construction implementing schemes of thought and expression and that between conditions of existence and practices or representations there intervenes the structuring activity of the agents, who, far from reacting mechanically to mechanical stimulations, respond to the invitations or threats of a world whose meaning they have helped to produce. However, the principle of this structuring activity is not, as an intellectualist and anti-genetic idealism would have it, a system of universal forms and categories but a system of internalized, embodied schemes which, having been constituted in the course of collective history, are acquired in the course of individual history and function in their *practical* state, *for practice* (and not for the sake of pure knowledge).

EMBODIED SOCIAL STRUCTURES

This means, in the first place, that social science, in constructing the social world, takes note of the fact that agents are, in their ordinary practice, the subjects of acts of construction of the social world; but also that it aims, among other things, to describe the social genesis of the principles of construction and seeks the basis of these principles in the social world.[3] Breaking with the anti-genetic prejudice which often accompanies recognition of the active aspect of knowledge, it seeks in the objective distributions of properties especially material ones (brought to light by censuses and surveys which all presuppose selection and classification), the basis of the systems of classification which agents apply to every sort of thing, not least to the distributions themselves. In contrast to what is sometimes called the 'cognitive' approach which, both in its ethnological form (structural anthropology, ethnoscience, ethnosemantics, ethnobotany, etc.) and in its sociological form (interactionism, ethnomethodology, etc.), ignores the question of the genesis of mental structures and classifications, social science enquires into the relationship between the principles of division and the social divisions (between the generations, the sexes, etc.) on which they are based, and into the variations of the use made of these principles according to the position occupied in the distributions (questions which all require the use of statistics).

The cognitive structures which social agents implement in their practical knowledge of the social world are internalized, 'embodied' social structures. The practical knowledge of the social world that is presupposed by 'reasonable' behaviour within it implements classificatory schemes (or 'forms of classification', 'mental structures' or 'symbolic forms' – apart from their connotations, these expressions are virtually interchangeable), historical schemes of perception and appreciation which are the product of the objective division into classes (age groups, genders, social classes) and which function below the level of consciousness and discourse. Being the product of the incorporation of the fundamental structures of a society, these principles of division are common to all the agents of the society and make possible the production of a common, meaningful world, a common-sense world.

All the agents in a given social formation share a set of basic perceptual schemes, which receive the beginnings of objectification in the pairs of antagonistic adjectives commonly used to classify and qualify persons or objects in the most varied areas of practice. The network of oppositions between high (sublime, elevated, pure) and low (vulgar, low, modest), spiritual and material, fine (refined, elegant) and coarse (heavy, fat, crude, brutal), light (subtle, lively, sharp, adroit) and heavy (slow, thick, blunt, laborious, clumsy), free and forced, broad and narrow, or, in another dimension, between unique (rare, different, distinguished, exclusive, exceptional, singular, novel) and common (ordinary, banal, commonplace, trivial, routine), brilliant (intelligent) and dull (obscure, grey, mediocre), is the matrix of all the commonplaces which find such ready acceptance because behind them lies the whole social order. The network has its ultimate source in the opposition between the 'élite' of the dominant and the 'mass' of the dominated, a contingent, disorganized multiplicity, interchangeable and innumerable, existing only statistically. These mythic roots only have to be allowed to take their course in order to generate, at will, one or another of the tirelessly repeated themes of the eternal sociodicy, such as apocalyptic denunciations of all forms of 'levelling', 'trivialization' or 'massification', which identify the decline of societies with the decadence of bourgeois houses, i.e., a fall into the homogeneous, the undifferentiated, and betray an obsessive fear of number, of undifferentiated hordes indifferent to difference and constantly threatening to submerge the private spaces of bourgeois exclusiveness.[4]

The seemingly most formal oppositions within this social mythology always derive their ideological strength from the fact that they refer back, more or less discreetly, to the most fundamental oppositions within the social order: the opposition between the dominant and the dominated, which is inscribed in the division of labour, and the opposition, rooted in the division of the labour of domination, between two principles of domination, two powers, dominant and dominated, temporal and spiritual, material and intellectual, etc. It follows that the map of social space previously put forward can also be read as a strict table of the historically constituted and acquired categories which organize the idea of the social world in the minds of all the subjects belonging to that world and shaped by it. The same classificatory schemes (and the oppositions in which they are expressed) can function, by being specified, in fields organized around polar positions, whether in the field of the dominant class, organized around an opposition homologous to the opposition constituting the field of the social classes, or in the field of cultural production, which is itself organized around oppositions which reproduce the structure of the dominant class and are homologous to it (e.g., the opposition between bourgeois and avant-garde theatre). So the fundamental opposition constantly supports second, third or nth rank oppositions (those which underlie the 'purest' ethical or aesthetic judgements, with their high or low sentiments, their facile or difficult notions of beauty, their light or heavy styles, etc.), while euphemizing itself to the point of misrecognizability.

Thus, the opposition between the heavy and the light, which, in a number of its uses, especially scholastic ones, serves to distinguish popular or petit-bourgeois tastes from bourgeois tastes, can be used by theatre criticism aimed at the dominant fraction of the dominant class to express the relationship between 'intellectual' theatre, which is condemned for its 'laborious' pretensions and 'oppressive' didacticism, and 'bourgeois' theatre, which is praised for its tact and its art of skimming over surfaces. By contrast, 'intellectual' criticism, by a simple inversion of values, expresses the relationship in a scarcely modified form of the same opposition, with lightness, identified with frivolity, being opposed to profundity. Similarly, it can be shown that the opposition between right and left, which, in its basic form, concerns the relationship between the dominant and the dominated, can also, by means of a first transformation, designate the relations between dominated fractions and dominant fractions within the dominant class; the words right and left then take on a meaning close to the meaning they have in expressions like 'right-bank' theatre or 'left-bank' theatre. With a further degree of 'de-realization', it can even serve to distinguish two rival tendencies within an avant-garde artistic or literary group, and so on.

It follows that, when considered in each of their uses, the pairs of qualifiers, the system of which constitutes the conceptual equipment of the judgement of taste, are extremely poor, almost indefinite, but, precisely for this reason, capable of eliciting or expressing the sense of the indefinable. Each particular use of one of these pairs only takes on its full meaning in relation to a universe of discourse that is different each time and usually implicit – since it is a question of the system of self-evidences and presuppositions that are taken for granted in the field in relation to which the speakers' strategies are defined. But each of the couples specified by usage has for undertones all the other uses it might have – because of the homologies between the fields which allow transfers from one field to another – and also all the other couples which are interchangeable with it, within a nuance or two (e.g. fine/crude for light/heavy), that is, in slightly different contexts.

The fact that the semi-codified oppositions contained in ordinary language reappear, with very similar values, as the basis of the dominant vision of the social world, in all class-divided social formations (consider the tendency to see the 'people' as the site of totally uncontrolled appetites and sexuality), can be understood once one knows that, reduced to their formal structure, the same fundamental relationships, precisely those which express the major relations of order (high/low, strong/weak, etc.) reappear in all class-divided societies. And the recurrence of the triadic structure studied by Georges Dumézil, which Georges Duby shows in the case of feudal society to be rooted in the social structures it legitimates, may well be, like the invariant oppositions in which the relationship of domination is expressed, simply a necessary outcome of the intersection of the two principles of division which are at work in all class-divided societies – the division between the dominant and the dominated, and the division between the different fractions competing for dominance in the name of different principles, *bellatores* (warriors) and *oratores* (scholars) in feudal society, businessmen and intellectuals now.[5]

[. . .]

THE REALITY OF REPRESENTATION AND THE REPRESENTATION OF REALITY

The classifying subjects who classify the properties and practices of others, or their own, are also classifiable objects which classify themselves (in the eyes of others) by appropriating practices and properties that are already classified (as vulgar or distinguished, high or low, heavy or light, etc. – in other words, in the last analysis, as popular or bourgeois) according to their probable distribution between groups that are themselves classified. The most classifying and best classified of these properties are, of course, those which are overtly designated to function as signs of distinction or marks of infamy, stigmata, especially the names and titles expressing class membership whose intersection defines social identity at any given time – the name of a nation, a region, an ethnic group, a family name, the name of an occupation, an educational qualification, honorific titles, and so on. Those who classify themselves or others, by appropriating or classifying practices or properties that are classified and classifying, cannot be unaware that, through distinctive objects or practices in which their 'powers' are expressed and which, being appropriated by and appropriate to classes, classify those who appropriate them, they classify themselves in the eyes of other classifying (but also classifiable) subjects, endowed with classificatory schemes analogous to those which enable them more or less adequately to anticipate their own classification.

Social subjects comprehend the social world which comprehends them. This means that they cannot be characterized simply in terms of material properties, starting with the body, which can be counted and measured like any other object in the physical world. In fact, each of these properties, be it the height or volume of the body or the extent of landed property, when perceived and appreciated in relation to other properties of the same class by agents equipped with socially constituted schemes of perception and appreciation, functions as a symbolic property. It is therefore necessary to move beyond the opposition between a 'social physics' – which uses statistics in objectivist fashion to establish distributions (in both the statistical and economic senses), quantified expressions of the differential appropriation of a finite quantity of social energy by a large number of competing individuals, identified through 'objective indicators' – and a 'social semiology' which seeks to decipher meanings and bring to light the cognitive operations whereby agents produce and decipher them. We have to refuse the dichotomy between, on the one hand, the aim of arriving at an objective 'reality', 'independent of individual consciousnesses and wills', by breaking with common representations of the social world (Durkheim's 'pre-notions'), and of uncovering 'laws' – that is, significant (in the sense of non-random) relationships between distributions – and, on the other hand, the aim of grasping, not 'reality', but agents' representations of it, which are the whole 'reality' of a social world conceived 'as will and representation'.

In short, social science does not have to choose between that form of social physics, represented by Durkheim – who agrees with social semiology in acknowledging that one can only know 'reality' by applying logical instruments of classification[6] – the idealist semiology which, undertaking to construct 'an account of accounts', as Harold Garfinkel puts it, can do no more than record the recordings of a social world which is ultimately no more than the product of mental, i.e. linguistic, structures. What we have to do is to bring into the science of scarcity, and of competition for scarce goods, the practical knowledge which the agents obtain for themselves by producing – on the basis of their experience of the distributions, itself dependent on their position in the distributions – divisions and classifications which are no less objective than those of the balance-sheets of social physics. In other words, we have to move beyond the opposition between objectivist theories which identify the social classes (but also the sex or age classes) with discrete groups, simple countable populations separated by boundaries objectively drawn in reality, and subjectivist (or marginalist) theories which reduce the 'social order' to a sort of collective classification obtained by aggregating the individual classifications or, more precisely, the individual strategies, classified and classifying, through which agents class themselves and others.[7]

One only has to bear in mind that goods are converted into distinctive signs, which may be signs of distinction but also of vulgarity, as soon as they are perceived relationally, to see that the representation which individuals and groups inevitably project through their practices and properties is an integral part of social reality. A class is defined as much by its *being-perceived* as by its *being*, by its consumption – which need not be conspicuous in order to be symbolic – as much as by its position in the relations of production (even if it is true that the latter governs the former). The Berkeleian – i.e. petit-bourgeois – vision which reduces social being to perceived being, to seeming, and which, forgetting that there is no need to give theatrical performances (*représentations*) in order to be the object of mental representations, reduces the social world to the sum of the (mental) representations which the various groups have of the theatrical performances put on by the other groups, has the virtue of insisting on the relative autonomy of the logic of symbolic representations with respect to the material determinants of socio-economic condition. The individual or collective classification struggles aimed at transforming the categories of perception and appreciation of the social world and, through this, the social world itself, are indeed a forgotten dimension of the class struggle. But one only has to realize that the classificatory schemes which underlie agents' practical relationship to their condition and the representation they have of it are themselves the product of that condition, in order to see the limits of this autonomy. Position in the classification struggle depends on position in the class structure; and social subjects – including intellectuals, who are not those best placed to grasp that which defines the limits of their thought of the social world, that is, the illusion of the absence of limits – are perhaps never less likely to transcend 'the limits of their minds' than in the representation they have and give of their position, which defines those limits.

NOTES

1 I. Kant, *Anthropology from a Pragmatic Point of View* (Carbondale and Edwardsville, Southern Illinois University Press, 1978), p. 141.

2 G. W. Leibniz, 'Meditationes de cognitione, veritate et ideis' in *Opuscula Philosophical Selecta* (Paris, Boivin, 1939), pp. 1–2 (see also *Discours de Métaphysique*, par. 24). It is remarkable that to illustrate the idea of 'clear but confused' knowledge, Leibniz evokes, in addition to the example of colours, tastes and smells which we can distinguish 'by the simple evidence of the senses and not by statable marks', the example of painters and artists who can recognize a good or bad work but cannot justify their judgement except by invoking the presence or absence of a 'je ne sais quoi'.

3 It would be the task of a genetic sociology to establish how this sense of possibilities and impossibilities, proximities and distances is constituted.

4 Just as the opposition between the unique and the multiple lies at the heart of the dominant philosophy of history, so the opposition, which is a transfigured form of it, between the brilliant, the visible, the distinct, the distinguished, the 'outstanding', and the obscure, the dull, the greyness of the undifferentiated, indistinct, inglorious mass is one of the fundamental categories of the dominant perception of the social world.

5 See G. Duby, *Les trois ordres ou l'imaginaire du féodalisme* (Paris, Gallimard, 1978).

6 One scarcely needs to point out the affinity between social physics and the positivist inclination to see classifications either as arbitrary, 'operational' divisions (such as age groups or income brackets) or as 'objective' cleavages (discontinuities in distributions or bends in curves) which only need to be recorded.

7 Here is a particularly revealing expression (even in its metaphor) of this social marginalism: 'Each individual is responsible for the demeanour image of himself and the deference image of others, so that for a complete man to be expressed, individuals must hold hands in a chain of ceremony, each giving deferentially with proper demeanour to the one on the right what will be received deferentially from the one on the left.' E. Goffman, 'The Nature of Deference and Demeanour', *American Anthropologist*, 58 (June 1956), 473–502; 'routinely the question is that of whose opinion is voiced most frequently and most forcibly, who makes the minor ongoing decisions apparently required for the coordination of any joint activity, and whose passing concerns have been given the most weight. And however trivial some of these *little gains and losses* may appear to be, *by summing them all up* across all the social situations in which they occur, we can see that their total effect is enormous. The expression of subordination and domination through this swarm of situational means is more than a mere tracing or symbol or ritualistic affirmation of the social hierarchy. These expressions *considerably constitute* the hierarchy.' E. Goffman, 'Gender Display'. (Paper presented at the Third International Symposium, 'Female Hierarchies', Harry Frank Guggenheim Foundation, April 3–5, 1974); italics mine.

36

Production, Consumption, Distribution, Exchange (Circulation)

Karl Marx

THE GENERAL RELATION OF PRODUCTION TO DISTRIBUTION, EXCHANGE, CONSUMPTION

Before going further in the analysis of production, it is necessary to focus on the various categories which the economists line up next to it.

The obvious, trite notion: in production the members of society appropriate (create, shape) the products of nature in accord with human needs; distribution determines the proportion in which the individual shares in the product; exchange delivers the particular products into which the individual desires to convert the portion which distribution has assigned to him; and finally, in consumption, the products become objects of gratification, of individual appropriation. Production creates the objects which correspond to the given needs; distribution divides them up according to social laws; exchange further parcels out the already divided shares in accord with individual needs; and finally, in consumption, the product steps outside this social movement and becomes a direct object and servant of individual need, and satisfies it in being consumed. Thus production appears as the point of departure, consumption as the conclusion, distribution and exchange as the middle, which is however itself twofold, since distribution is determined by society and exchange by individuals. The person objectifies himself in production, the thing subjectifies itself in the person;[1] in distribution, society mediates between production and consumption in the form of general, dominant determinants; in exchange the two are mediated by the chance characteristics of the individual.

Distribution determines the relation in which products fall to individuals (the amount); exchange determines the production[2] in which the individual demands the portion allotted to him by distribution.

Thus production, distribution, exchange and consumption form a regular syllogism; production is the generality, distribution and exchange the particularity, and consumption the singularity in which the whole is joined together. This is admittedly a coherence, but a shallow one. Production is determined by general natural laws, distribution by social accident, and the latter may therefore promote production to a greater or lesser extent; exchange stands between the two as formal social movement; and the concluding act, consumption, which is conceived not only as a terminal point but also as an end-in-itself, actually belongs outside economics except in so far as it reacts in turn upon the point of departure and initiates the whole process anew.

The opponents of the political economists – whether inside or outside its realm – who accuse them of barbarically tearing apart things which belong together, stand either on the same ground as they, or beneath them. Nothing is more common than the reproach that the political economists view production too much as an end in itself, that distribution is just as important. This accusation is based precisely on the economic notion that the spheres of distribution and of production are independent, autonomous neighbours. Or that these moments were not grasped in their unity. As if this rupture had made its way not from reality into the textbooks, but rather from the textbooks into reality, and as if the task were the dialectic balancing of concepts, and not the grasping of real relations!

[Consumption and Production]

Production is also immediately consumption. Twofold consumption, subjective and objective: the individual not only develops his abilities in production, but also expends them, uses them up in the act of production, just as natural procreation is a consumption of life forces. Secondly: consumption of the means of production, which become worn out through use, and are partly (e.g. in combustion) dissolved into their elements again. Likewise, consumption of the raw material, which loses its natural form and composition by being used up. The act of production is therefore in all its moments also an act of consumption. But the economists admit this. Production as directly identical with consumption, and consumption as directly coincident with production, is termed by them *productive consumption*. This identity of production and consumption amounts to Spinoza's thesis: *determinatio est negatio.*[3]

But this definition of productive consumption is advanced only for the purpose of separating consumption as identical with production from consumption proper, which is conceived rather as the destructive antithesis to production. Let us therefore examine consumption proper.

Consumption is also immediately production, just as in nature the consumption of the elements and chemical substances is the production of the plant. It is clear that in taking in food, for example, which is a form of consumption, the human being produces his own body. But this is also true of every kind of consumption which in one way or another produces human beings in some particular aspect. Consumptive production. But, says economics, this production which is identical with consumption is secondary, it is derived from the destruction of the prior product. In the former, the producer objectified himself, in the latter, the object he created personifies itself. Hence this consumptive production – even though it is an immediate unity of production and consumption – is essentially different from production proper. The immediate unity in which production coincides with consumption and consumption with production leaves their immediate duality intact.

Production, then, is also immediately consumption, consumption is also immediately production. Each is immediately its opposite. But at the same time a mediating movement takes place between the two. Production mediates consumption; it creates the latter's material; without it, consumption would lack an object. But consumption also mediates production, in that it alone creates for the products the subject for whom they are products. The product only obtains its 'last finish'[4] in consumption. A railway on which no trains run, hence which is not used up, not consumed, is a railway only δυνάμει,[5] and not in reality. Without production, no consumption; but also, without consumption, no production; since production would then be purposeless. Consumption produces production in a double way, (1) because a product becomes a real product only by being consumed. For example, a garment becomes a real garment only in the act of being worn; a house where no one lives is in fact not a real house; thus the product, unlike a mere natural object, proves itself to be, *becomes*, a product only through consumption. Only by decomposing the product does consumption give the product the finishing touch; for the product is production not as[6] objectified activity, but rather only as object for the active subject; (2) because consumption creates the need for *new* production, that is it creates the ideal, internally impelling cause for production, which is its presupposition. Consumption creates the motive for production; it also creates the object which is active in production as its determinant aim. If it is clear that production offers consumption its external object, it is therefore equally clear that consumption *ideally posits* the object of production as an internal image, as a need, as drive and as purpose. It creates the objects of production in a still subjective form. No production without a need. But consumption reproduces the need.

Production, for its part, correspondingly (1) furnishes the material and the object for consumption.[7] Consumption without an object is not consumption; therefore, in this respect, production creates, produces consumption. (2) But the object is not the only thing which production creates for consumption. Production also gives consumption its specificity, its character, its finish. Just as consumption gave the product its finish as product, so does production give finish to consumption. *Firstly*, the object is not an object in general, but a specific object which must be consumed in a specific manner, to be mediated in its turn by production itself. Hunger is hunger, but the hunger gratified by cooked meat eaten with a knife and fork is a different hunger from that which bolts down raw meat with the aid of hand, nail and tooth. Production thus produces not only the object but also the manner of consumption, not only objectively but also subjectively. Production thus creates the consumer. (3) Production not only supplies a material for the need, but it also supplies a need for the material. As soon as consumption emerges from its initial state of natural crudity and immediacy – and, if it remained at that stage, this would be because production itself had been arrested there – it becomes itself mediated as a drive by the object. The need which consumption feels for the object is created by the perception of it. The object of art – like every other product – creates a public which is sensitive to art and enjoys beauty. Production thus not only creates an object for the subject, but also a subject for the object. Thus production produces consumption (1) by creating the material for it; (2) by determining the manner of consumption; and (3) by creating the products, initially posited by it as objects, in the form of a need felt by the consumer. It thus produces the object of consumption, the manner of consumption and the motive of consumption. Consumption likewise produces the producer's *inclination* by beckoning to him as an aim-determining need.

The identities between consumption and production thus appear threefold:

1 *Immediate identity:* Production is consumption, consumption is production. Consumptive production. Productive consumption. The political economists call both productive consumption. But then make a further distinction. The first figures as reproduction, the second as productive consumption. All investigations into the first concern productive or unproductive labour; investigations into the second concern productive or non-productive consumption.
2 [In the sense] that one appears as a means for the other, is mediated by the other: this is expressed as their mutual dependence; a movement which relates them to one another, makes them appear indispensable to one another, but still leaves them external to each other. Production creates the material, as external object, for consumption; consumption creates the need, as internal object, as aim, for production. Without production no consumption; without consumption no production. [This identity] figures in economics in many different forms.
3 Not only is production immediately consumption and consumption immediately production, not only is production a means for consumption and consumption the aim of production, i.e. each supplies the other with its object (production supplying the external object of consumption, consumption the conceived object of production); but also, each of them, apart from being immediately the other, and apart from mediating the other, in addition to this creates the other in completing itself, and creates itself as the other. Consumption accomplishes the act of production only in completing the product as product by dissolving it, by consuming its independently material form, by raising the inclination developed in the first act of production, through the need for repetition, to its finished form; it is thus not only the concluding act in which the product becomes product, but also that in which the producer becomes producer. On the other side, production produces consumption by creating the specific manner of consumption; and, further, by creating the stimulus of consumption, the ability to consume, as a need. This last identity, as determined under (3), [is] frequently cited in economics in the relation of demand and supply, of objects and needs, of socially created and natural needs.

Thereupon, nothing simpler for a Hegelian than to posit production and consumption as identical. And this has been done not only by socialist belletrists but by prosaic economists themselves, e.g. Say;[8] in the form that when one looks at an entire people, its production is its consumption. Or, indeed, at humanity in the abstract. Storch[9] demonstrated Say's error, namely that e.g. a people does not consume its entire product, but also creates means of production, etc., fixed capital, etc. To regard society as one single subject is, in addition, to look at it wrongly; speculatively. With a single subject, production and consumption appear as moments of a single act. The important thing to emphasize here is only that, whether production and consumption are viewed as the activity of one or of many individuals, they appear in any case as moments of one process, in which production is the real point of departure and hence also the predominant moment. Consumption as urgency, as need, is itself an intrinsic moment of productive activity. But the latter is the point of departure for realization and hence also its predominant moment; it is the act through which the whole process again runs its course. The individual produces an object and, by consuming it, returns to himself, but returns as a productive and self-reproducing individual. Consumption thus appears as a moment of production.

In society, however, the producer's relation to the product, once the latter is finished, is an external one, and its return to the subject depends on his relations to other individuals. He does not come into possession of it directly. Nor is its immediate appropriation his purpose when he produces in society. *Distribution* steps between the producers and the products, hence between production and consumption, to determine in accordance with social laws what the producer's share will be in the world of products.

Now, does distribution stand at the side of and outside production as an autonomous sphere?

NOTES

1 *Marx-Engles Werke* (hereafter *MEW*) XIII substitutes 'in consumption'.
2 *MEW* XIII substitutes 'products'.
3 'Determination is negation', i.e. given the undifferentiated self-identity of the universal world substance, to attempt to introduce particular determinations is to negate this self-identity. (Spinoza, *Letters*, No. 50, to J. Jelles, 2 June 1674.)
4 In English in the original.
5 'Potentially'. Cf. Aristotle, *Metaphysics* Bk VIII, Ch. 6, 2.
6 The manuscript has: 'for the product is production not only as . . .'. *MEW* XIII substitutes: 'for the product is a product not as . . .'.
7 The manuscript has 'for production'.
8 Jean-Baptiste Say (1767–1832), 'the inane Say', who 'superficially condensed political economy into a textbook' (Marx), a businessman who popularized and vulgarized the doctrines of Adam Smith in his *Traité d'économie politique*, Paris, 1803.
9 Heinrich Friedrich Storch (1766–1835), Professor of Political Economy in the Russian Academy of Sciences at St Petersburg. Say issued Storch's work *Cours d'économie politique* with critical notes in 1823; he attacked Say's interpretation of his views in *Considérations sur la nature du revenu national*, Paris, 1824, pp. 144–159.

The Ideological Genesis of Needs[1]

Jean Baudrillard

The rapturous satisfactions of consumption surround us, clinging to objects as if to the sensory residues of the previous day in the delirious excursion of a dream. As to the logic that regulates this strange discourse – surely it compares to what Freud uncovered in *The Interpretation of Dreams*? But we have scarcely advanced beyond the explanatory level of naive psychology and the medieval dreambook. We believe in 'Consumption': we believe in a real subject, motivated by needs and confronted by real objects as sources of satisfaction. It is a thoroughly vulgar metaphysic. And contemporary psychology, sociology and economic science are all complicit in the fiasco. So the time has come to deconstruct all the assumptive notions involved – object, need, aspiration, *consumption* itself – for it would make as little sense to theorize the quotidian from surface evidence as to interpret the manifest discourse of a dream: it is rather the dream-work and the dream-processes that must be analyzed in order to recover the unconscious logic of a more profound discourse. And it is the workings and processes of an unconscious social logic that must be retrieved beneath the consecrated ideology of consumption.

CONSUMPTION AS A LOGIC OF SIGNIFICATIONS

The empirical 'object,' given in its contingency of form, color, material, function and discourse (or, if it is a cultural object, in its aesthetic finality), is a myth. How often it has been wished away! But the object is *nothing*. It is nothing but the different types of relations and significations that converge, contradict themselves, and twist around it, as such – the hidden logic that not only arranges this bundle of relations, but directs the manifest discourse that overlays and occludes it.

The logical status of objects

Insofar as I make use of a refrigerator as a machine, it is not an object. It is a refrigerator. Talking about refrigerators or automobiles in terms of 'objects' is something else. That is, it has nothing to do with them in their 'objective' relation to keeping things cold or transportation. It is to speak of the object as functionally decontextualized:

1 Either as an object of psychic investment[2] and fascination, of passion and projection – qualified by its exclusive relation with the subject, who then cathects it as if it were his own body (a borderline case). Useless and sublime, the object then loses its common name, so to speak, and assumes the title of Object as generic proper name. For this reason, the collector never refers to a statuette or a vase as a beautiful statuette, vase, etc., but as 'a beautiful Object.' This status is opposed to the generic dictionary meaning of the word, that of the 'object' plain and simple: 'Refrigerator: an object that refrigerates. . . .'
2 Or (between the Object, as proper name and projective equivalent of the subject, and the object, with the status of a common name and implement) as an object specified by its trademark, charged with differential connotations of status, prestige and fashion. *This* is the 'object of consumption.' It can just as easily be a vase as a refrigerator, or, for that

matter, a whoopee cushion. Properly speaking, it has no more existence than a phoneme has an absolute meaning in linguistics. This object does not assume meaning either in a symbolic relation with the subject (the Object) or in an operational relation to the world (object-as-implement): it finds meaning with other objects, in difference, according to a hierarchical code of significations. This alone, at the risk of the worst confusion, defines the object of consumption.

Of symbolic exchange 'value'

In symbolic exchange, of which the gift is our most proximate illustration, the object is not an object: it is inseparable from the concrete relation in which it is exchanged, the transferential pact that it seals between two persons: it is thus not independent as such. It has, properly speaking, neither use value nor (economic) exchange value. The object given has symbolic exchange value. This is the paradox of the gift: it is on the one hand (relatively) arbitrary: it matters little what object is involved. Provided it is given, it can fully signify the relation. On the other hand, once it has been given – and *because* of this – it is *this* object and not another. The gift is unique, specified by the people exchanging and the unique moment of the exchange. It is arbitrary, and yet absolutely singular.

As distinct from language, whose material can be dissassociated from the subjects speaking it, the material of symbolic exchange, the objects given, are not autonomous, hence not codifiable as signs. Since they do not depend on economic exchange, they are not amenable to systematization as commodities and exchange value.

What constitutes the object as value in symbolic exchange is that one separates himself from it in order to give it, to throw it at the feet of the other, under the gaze of the other (*ob-jicere*); one divests himself as if of a part of himself – an act which is significant in itself as the basis, simultaneously, of both the mutual presence of the terms of the relationship, and their mutual absence (their distance). The ambivalence of all symbolic exchange material (looks, objects, dreams, excrement) derives from this: the gift is a medium of relation *and* distance; it is always love and aggression.[3]

From symbolic exchange to sign value

It is from the (theoretically isolatable) moment when the exchange is no longer purely transitive, when the object (the material of exchange) is immediately presented as such, that it is reified into a sign. Instead of abolishing itself in the relation that it establishes, and thus assuming symbolic value (as in the example of the gift), the object becomes autonomous, intransitive, opaque, and so begins to signify the abolition of the relationship. Having become a sign object, it is no longer the mobile signifier of a lack between two beings, it is 'of' and 'from' the reified relation (as is the commodity at another level, in relation to reified labor power). Whereas the symbol refers to lack (to absence) as a virtual relation of desire, the sign object only refers to the absence of relation itself, and to isolated individual subjects.

The sign object is neither given nor exchanged: it is appropriated, withheld and manipulated by individual subjects as a sign, that is, as coded difference. Here lies the object of consumption. And it is always of and from a reified, abolished social relationship that is 'signified' in a code.

What we perceive in the symbolic object (the gift, and also the traditional, ritual and artisanal object) is not only the concrete manifestation of a total relationship (ambivalent, and total because it is ambivalent) of desire; but also, through the singularity of an object, the transparency of social relations in a dual or integrated group relationship. In the commodity, on the other hand, we perceive the opacity of social relations of production and the reality of the division of labor. What is revealed in the contemporary profusion of sign objects, objects of consumption, is precisely this opacity, the *total constraint of the code* that governs social value: it is the specific weight of *signs* that regulates the social logic of exchange.

The object-become-sign no longer gathers its meaning in the concrete relationship between two people. It assumes its meaning in its differential relation to other signs. Somewhat like Lévi-Strauss' myths, sign-objects exchange among themselves. Thus, only when objects are autonomized as differential signs and thereby rendered systematizable can one speak of consumption and of objects of consumption.

A logic of signification

So it is necessary to distinguish the logic of consumption, which is a logic of the sign and of difference, from several other logics that habitually get entangled with it in the welter of evidential considerations. (This confusion is echoed by all the naive and authorized literature on the question.) Four logics would be concerned here:

1 A functional logic of use value;
2 An economic logic of exchange value;
3 A logic of symbolic exchange;
4 A logic of sign value.

The first is a logic of practical operations, the second one of equivalence, the third, ambivalence, and the fourth, difference.

Or again: a logic of utility, a logic of the market, a logic of the gift, and a logic of status. Organized in accordance with one of the above groupings, the object assumes respectively the status of an *instrument*, a *commodity*, a *symbol*, or a *sign*.

Only the last of these defines the specific field of consumption. Let us compare two examples:

The wedding ring: This is a unique object, symbol of the relationship of the couple. One would neither think of changing it (barring mishap) nor of wearing several. The symbolic object is made to last and to witness in its duration the permanence of the relationship. Fashion plays as negligible a role at the strictly symbolic level as at the level of pure instrumentality.

The ordinary ring is quite different: it does not symbolize a relationship. It is a non-singular object, a personal gratification, a sign in the eyes of others. I can wear several of them. I can substitute them. The ordinary ring takes part in the play of my accessories and the constellation of fashion. It is an object of consumption.

Living accommodations: The house, your lodgings, your apartment: these terms involve semantic nuances that are no doubt linked to the advent of industrial production or to social standing. But, whatever one's social level in France today, one's domicile is not necessarily perceived as a 'consumption' good. The question of residence is still very closely associated with patrimonial goods in general, and its symbolic scheme remains largely that of the body. Now, for the logic of consumption to penetrate here, the exteriority of the sign is required. The residence must cease to be hereditary, or interiorized as an organic family space. One must avoid the appearance of filiation and identification if one's debut in the world of fashion is to be successful.

In other words, domestic practice is still largely a function of determinations, namely: symbolic (profound emotional investment, etc.), and economic (scarcity).

Moreover, the two are linked: only a certain 'discretionary income' permits one to play with objects as status signs – a stage of fashion and the 'game' where the symbolic and the utilitarian are both exhausted. Now, as to the question of residence – in France at least – the margin of free play for the mobile combinatory of prestige or for the game of substitution is limited. In the United States, by contrast, one sees living arrangements indexed to social mobility, to trajectories of careers and status. Inserted into the global constellation of status, and subjugated to the same accelerated obsolescence of any other object of luxury, the house truly becomes an object of consumption.

This example has a further interest: it demonstrates the futility of any attempt to define the object empirically. Pencils, books, fabrics, food, the car, curios – are these objects? Is a house an object? Some would contest this. The decisive point is to establish whether the symbolism of the house (sustained by the shortage of housing) is irreducible, or if even this can succumb to the differential and reified connotations of fashion logic: for if this is so, then the home becomes an object of consumption – as any other object will, if it only answers to the same definition: being, cultural trait, ideal, gestural pattern, language, etc. – anything can be made to fit the bill. The definition of an object of consumption is entirely independent of objects themselves and *exclusively a function of the logic of significations*.

An object is not an object of consumption unless it is released from its psychic determinations as *symbol*; from its functional determinations as *instrument*; from its commercial determinations as *product*; and is thus *liberated as a sign* to be recaptured by the formal logic of fashion, i.e. by the logic of differentiation.

The order of signs and social order

There is no object of consumption before the moment of its substitution, and without this substitution having been determined by the social law, which demands not only the renewal of distinctive material, but the obligatory registration of individuals on the scale of status, through the mediation of their group and as a function of their relations with other groups. *This scale is properly the social order,* since the acceptance of this hierarchy of differential signs and the interiorization by the individual of signs in general (i.e. of the norms, values, and social imperatives that signs are) constitutes the fundamental, decisive form of social control – more so even than acquiescence to ideological norms.

It is now clear that there is no autonomous problematic of objects, but rather the much more urgent need for a theory of social logic, and of the codes that it puts into play (sign systems and distinctive material).

The common name, the proper name, and the brand name

Let us recapitulate the various types of status of the object according to the specific and (theoretically) exhaustive logics that may penetrate it:

1 The refrigerator is specified by its function and irreplaceable in this respect. There is a necessary relation between the object and its function. The arbitrary nature of the sign is not involved. But all refrigerators are interchangeable in regard to this function (their objective 'meaning').
2 By contrast, if the refrigerator is taken as an element of comfort or of luxury (standing), then in principle any other such element can be substituted for it. The object tends to the status of sign, and each social status will be signified by an entire constellation of exchangeable signs. No necessary relation to the subject or the world is involved. There is only a systematic relation obligated to all other signs. And in this combinatory abstraction lie the elements of a code.
3 In their symbolic relationship to the subject (or in reciprocal exchange), all objects are potentially interchangeable. Any object can serve as a doll for the little girl. But once cathected, it is *this* one and not another. The symbolic material is relatively arbitrary, but the subject–object relation is fused. Symbolic discourse is an idiom.

NOTES

1 This piece first appeared in *Cahiers Internationaux de Sociologie*, 1969.
2 *Investissement*: this is the standard, and literal, French equivalent of Freud's *Besetzung*, which also means investment in ordinary German. The English, however, have insisted on rendering this concept by coining a word that sounds more technical: cathexis, to cathect, etc. The term has been used here mainly to draw attention to the psychoanalytic sense, which varies in intensity and precision, of Baudrillard's *investissement, investir*. Loosely, Freud's concept involves the quantitative transfer of psychic energy to parts of the psyche, images, objects, etc. – Translator's footnote.
3 Thus the structure of exchange (cf. Lévi-Strauss) is never that of simple reciprocity. It is not two simple terms, but two *ambivalent* terms that exchange, and the exchange establishes their relationship as ambivalent.

38

The Practice of Everyday Life

Michel de Certeau

This essay is part of a continuing investigation of the ways in which users – commonly assumed to be passive and guided by established rules – operate. The point is not so much to discuss this elusive yet fundamental subject as to make such a discussion possible; that is, by means of inquiries and hypotheses, to indicate pathways for further research. This goal will be achieved if everyday practices, 'ways of operating' or doing things, no longer appear as merely the obscure background of social activity, and if a body of theoretical questions, methods, categories, and perspectives, by penetrating this obscurity, make it possible to articulate them.

The examination of such practices does not imply a return to individuality. The social atomism which over the past three centuries has served as the historical axiom of social analysis posits an elementary unit – the individual – on the basis of which groups are supposed to be formed and to which they are supposed to be always reducible. This axiom, which has been challenged by more than a century of sociological, economic, anthropological, and psychoanalytic research (although in history that is perhaps no argument), plays no part in this study. Analysis shows that a relation (always social) determines its terms, and not the reverse, and that each individual is a locus in which an incoherent (and often contradictory) plurality of such relational determinations interact. Moreover, the question at hand concerns modes of operation or schemata of action, and not directly the subjects (or persons) who are their authors or vehicles. It concerns an operational logic whose models may go as far back as the age-old ruses of fishes and insects that disguise or transform themselves in order to survive, and which has in any case been concealed by the form of rationality currently dominant in Western culture. The purpose of this work is to make explicit the systems of operational combination (*les combinatoires d'opérations*) which also compose a 'culture,' and to bring to light the models of action characteristic of users whose status as the dominated element in society (a status that does not mean that they are either passive or docile) is concealed by the euphemistic term 'consumers.' Everyday life invents itself by *poaching* in countless ways on the property of others.

1 CONSUMER PRODUCTION

Since this work grew out of studies of 'popular culture' or marginal groups,[1] the investigation of everyday practices was first delimited negatively by the necessity of not locating cultural *difference* in groups associated with the 'counter-culture' – groups that were already singled out, often privileged, and already partly absorbed into folklore – and that were no more than symptoms or indexes. Three further, positive determinations were particularly important in articulating our research.

Usage, or consumption

Many, often remarkable, works have sought to study the representations of a society, on the one hand, and its modes of behavior, on the other. Building on our knowledge of these social phenomena, it seems both possible and necessary to determine the *use* to which they are put by groups or individuals. For example, the analysis of the images broadcast by television (representation) and of the time spent watching television (behavior) should be complemented by a study of what the cultural consumer 'makes' or 'does' during this time and with these images. The same goes for the use of urban space, the products purchased in the supermarket, the stories and legends distributed by the newspapers, and so on.

The 'making' in question is a production, a *poiēsis*[2] – but a hidden one, because it is scattered over areas defined and occupied by systems of 'production' (television, urban development, commerce, etc.), and because the steadily increasing expansion of these systems no longer leaves 'consumers' any *place* in which they can indicate what they *make* or *do* with the products of these systems. To a rationalized, expansionist and at the same time centralized, clamorous, and spectacular production corresponds *another* production, called 'consumption.' The latter is devious, it is dispersed, but it insinuates itself everywhere, silently and almost invisibly, because it does not manifest itself through its own products, but rather through its *ways of using* the products imposed by a dominant economic order.

For instance, the ambiguity that subverted from within the Spanish colonizers' 'success' in imposing their own culture on the indigenous Indians is well known. Submissive, and even consenting to their subjection, the Indians nevertheless often *made of* the rituals, representations, and laws imposed on them something quite different from what their conquerors had in mind; they subverted them not by rejecting or altering them, but by using them with respect to ends and references foreign to the system they had no choice but to accept. They were *other* within the very colonization that outwardly assimilated them; their use of the dominant social order deflected its power, which they lacked the means to challenge; they escaped it without leaving it. The strength of their difference lay in procedures of 'consumption.' To a lesser degree, a similar ambiguity creeps into our societies through the use made by the 'common people' of the culture disseminated and imposed by the 'elites' producing the language.

The presence and circulation of a representation (taught by preachers, educators, and popularizers as the key to socioeconomic advancement) tells us nothing about what it is for its users. We must first analyze its manipulation by users who are not its makers. Only then can we gauge the difference or similarity between the production of the image and the secondary production hidden in the process of its utilization.

Our investigation is concerned with this difference. It can use as its theoretical model the *construction* of individual sentences with an *established* vocabulary and syntax. In linguistics, 'performance' and 'competence' are different: the act of speaking (with all the enunciative strategies that implies) is not reducible to a knowledge of the language. By adopting the point of view of enunciation – which is the subject of our study – we privilege the act of speaking; according to that point of view, speaking operates within the field of a linguistic system; it effects an appropriation, or reappropriation, of language by its speakers; it establishes a *present* relative to a time and place; and it posits a *contract with the other* (the interlocutor) in a network of places and relations. These four characteristics of the speech act[3] can be found in many other practices (walking, cooking, etc.). An objective is at least adumbrated by this parallel, which is, as we shall see, only partly valid. Such an objective assumes that (like the Indians mentioned above) users make (*bricolent*) innumerable and infinitesimal transformations of and within the dominant cultural economy in order to adapt it to their own interests and their own rules. We must determine the procedures, bases, effects, and possibilities of this collective activity.

The procedures of everyday creativity

A second orientation of our investigation can be explained by reference to Michel Foucault's *Discipline and Punish*. In this work, instead of analyzing the apparatus exercising power (i.e. the localizable, expansionist, repressive, and legal institutions), Foucault analyzes the mechanisms (*dispositifs*) that have sapped the strength of these institutions and surreptitiously reorganized the functioning of power: 'miniscule' technical procedures acting on and with details, redistributing a discursive space in order to make it the means of a generalized 'discipline' (*surveillance*).[4] This approach raises a new and different set of problems to be investigated. Once again, however, this 'microphysics of power' privileges the productive apparatus (which produces the 'discipline'), even though it discerns in 'education' a system of 'repression' and shows how, from the wings as it were, silent technologies determine or short-circuit institutional stage directions. If it is true that the grid of 'discipline' is everywhere becoming clearer and more extensive, it is all the more urgent to discover how an entire society resists being reduced to it, what popular procedures (also 'miniscule' and quotidian) manipulate the mechanisms of discipline and conform to them only in order to evade them, and finally, what 'ways of operating' form the counterpart, on the consumer's (or 'dominee's'?) side, of the mute processes that organize the establishment of socioeconomic order.

These 'ways of operating' constitute the innumerable practices by means of which users reappropriate the space organized by techniques of sociocultural production. They pose questions at once analogous and contrary to those dealt with in Foucault's book: analogous, in that the goal is to perceive and analyze the microbe-like operations proliferating within technocratic structures and deflecting their functioning by means of a multitude of 'tactics' articulated in the details of everyday life; contrary, in that the goal is not to make clearer how the violence of order is transmuted into a disciplinary technology, but rather to bring to light the clandestine forms taken by the dispersed, tactical, and makeshift creativity of groups or individuals already caught in the nets of 'discipline.' Pushed to their ideal limits, these procedures and ruses of consumers compose the network of an antidiscipline[5] which is the subject of this book.

The formal structure of practice

It may be supposed that these operations – multiform and fragmentary, relative to situations and details, insinuated into and concealed within devices whose mode of usage they constitute, and thus lacking their own ideologies or institutions – conform to certain rules. In other words, there must be a logic of these practices. We are thus confronted once again by the ancient problem: What is an *art* or 'way of making'? From the Greeks to Durkheim, a long tradition has sought to describe with precision the complex (and not at all simple or 'impoverished') rules that could account for these operations. From this point of view, 'popular culture,' as well as a whole literature called 'popular,'[6] take on a different aspect: they present themselves essentially as 'arts of making' this or that, i.e. as combinatory or utilizing modes of consumption. These practices bring into play a 'popular' *ratio*, a way of thinking invested in a way of acting, an art of combination which cannot be dissociated from an art of using.

In order to grasp the formal structure of these practices I have carried out two sorts of investigations. The first, more descriptive in nature, has concerned certain ways of making that were selected according to their value for the strategy of the analysis, and with a view to obtaining fairly differentiated variants: readers' practices, practices related to urban spaces, utilizations of everyday rituals, re-uses and functions of the memory through the 'authorities' that make possible (or permit) everyday practices, etc. In addition, two related investigations have tried to trace the intricate forms of the operations proper to the recompositon of a space (the Croix-Rousse quarter in Lyons) by familial practices, on the one hand, and on the other, to the tactics of the art of cooking, which simultaneously organizes a network of relations, poetic ways of 'making do' (*bricolage*), and a re-use of marketing structures.[7]

The second series of investigations has concerned the scientific literature that might furnish hypotheses allowing the logic of unselfconscious thought to be taken seriously. Three areas are of special interest. First, sociologists, anthropologists, and indeed historians (from E. Goffman to P. Bourdieu, from Mauss to M. Détienne, from J. Boissevain to E. O. Laumann) have elaborated a theory of such practices, mixtures of rituals and makeshifts (*bricolages*), manipulations of spaces, operators of networks.[8] Second, in the wake of J. Fishman's work, the ethnomethodological and sociolinguistic investigations of H. Garfinkel, W. Labov, H. Sachs, E. A. Schegloff, and others, have described the procedures of everyday interactions relative to structures of expectation, negotiation, and improvisation proper to ordinary language. [9]

Finally, in addition to the semiotics and philosophies of 'convention' (from O. Ducrot to D. Lewis),[10] we must look into the ponderous formal logics and their extension, in the field of analytical philosophy, into the domains of action (G. H. von Wright, A. C. Danto, R. J. Bernstein),[11] time (A.N. Prior, N. Rescher and J. Urquhart),[12] and modalization (G. E. Hughes and M. J. Cresswell, A. R. White).[13] These extensions yield a weighty apparatus seeking to grasp the delicate layering and plasticity of ordinary language, with its almost orchestral combinations of logical elements (temporalization, modalization, injunctions, predicates of action, etc.) whose dominants are determined in turn by circumstances and conjunctural demands. An investigation analogous to Chomsky's study of the oral uses of language must seek to restore to everyday practices their logical and cultural legitimacy, at least in the sectors – still very limited – in which we have at our disposal the instruments necessary to account for them.[14] This kind of research is complicated by the fact that these practices themselves alternately exacerbate and disrupt our logics. Its regrets are like those of the poet, and like him, it struggles against oblivion: 'And I forgot the element of chance introduced by circumstances, calm or haste, sun or cold, dawn or dusk, the taste of strawberries or abandonment, the half-understood message, the front page of newspapers, the voice on the telephone, the most anodyne conversation, the most anonymous man or woman, everything that speaks, makes noise, passes by, touches us lightly, meets us head on.'[15]

The marginality of a majority

These three determinations make possible an exploration of the cultural field, an exploration defined by an investigative problematics and punctuated by more detailed inquiries located by reference to hypotheses that remain to be verified. Such an exploration will seek to situate the types of *operations* characterizing consumption in the framework of an economy, and to discern in these practices of appropriation indexes of the creativity that flourishes at the very point where practice ceases to have its own language.

Marginality is today no longer limited to minority groups, but is rather massive and pervasive; this cultural activity of the non-producers of culture, an activity that is unsigned, unreadable, and unsymbolized, remains the only one possible for all those who nevertheless buy and pay for the showy products through which a productivist economy articulates itself. Marginality is becoming universal. A marginal group has now become a silent majority.

That does not mean the group is homogeneous. The procedures allowing the re-use of products are linked together in a kind of obligatory language, and their functioning is related to social situations and power relationships. Confronted by images on television, the immigrant worker does not have the same critical or creative elbow-room as the average citizen. On the same terrain, his inferior access to information, financial means, and compensations of all kinds elicits an increased deviousness, fantasy, or laughter. Similar strategic deployments, when acting on different relationships of force, do not produce identical effects. Hence the necessity of differentiating both the 'actions' or 'engagements' (in the military sense) that the system of products effects within the consumer grid, *and* the various kinds of room to maneuver left for consumers by the situations in which they exercise their 'art.'

The relation of procedures to the fields of force in which they act must therefore lead to a *polemological* analysis of culture. Like law (one of its models), culture articulates conflicts and alternately legitimizes, displaces or controls the superior force. It develops in an atmosphere of tensions, and often of violence, for which it provides symbolic balances, contracts of compatibility and compromises, all more or less temporary. The tactics of consumption, the ingenious ways in which the weak make use of the strong, thus lend a political dimension to everyday practices.

2 THE TACTICS OF PRACTICE

In the course of our research, the scheme, rather too neatly dichotomized, of the relations between consumers and the mechanisms of production has been diversified in relation to three kinds of concerns: the search for a problematics that could articulate the material collected; the description of a limited number of practices (reading, talking, walking, dwelling, cooking, etc.) considered to be particularly significant; and the extension of the analysis of these everyday operations to scientific fields apparently governed by another kind of logic. Through the presentation of our investigation along these three lines, the overly schematic character of the general statement can be somewhat nuanced.

Trajectories, tactics, and rhetorics

As unrecognized producers, poets of their own acts, silent discoverers of their own paths in the jungle of functionalist rationality, consumers produce through their signifying practices something that might be considered similar to the 'wandering lines' ('*lignes d'erre*') drawn by the autistic children studied by F. Deligny:[16] 'indirect' or 'errant' trajectories obeying their own logic. In the technocratically constructed, written, and functionalized space in which the consumers move about, their trajectories form unforeseeable sentences, partly unreadable paths across a space. Although they are composed with the vocabularies of established languages (those of television, newspapers, supermarkets, or museum sequences) and although they remain subordinated to the prescribed syntactical forms (temporal modes of schedules, paradigmatic orders of spaces, etc.), the trajectories trace out the ruses of other interests and desires that are neither determined nor captured by the systems in which they develop.[17]

Even statistical investigation remains virtually ignorant of these trajectories, since it is satisfied with classifying, calculating, and putting into tables the 'lexical' units which compose them but to which they cannot be reduced, and with doing this in reference to its own categories and taxonomies. Statistical investigation grasps the material of these practices, but not their *form*; it determines the elements used, but not the 'phrasing' produced by the *bricolage* (the artisan-like inventiveness) and the discursiveness that combine these elements, which are all in general circulation and rather drab. Statistical inquiry, in breaking down these 'efficacious meanderings' into units that it defines itself, in reorganizing the results of its analyses according to its own codes, 'finds' only the homogenous. The power of its calculations lies in its ability to divide, but it is precisely through this analytic fragmentation that it loses sight of what it claims to seek and to represent.[18]

'Trajectory' suggests a movement, but it also involves a plane projection, a flattening out. It is a transcription. A graph (which the eye can master) is substituted for an operation; a line which can be reversed (i.e. read in both directions) does duty for an irreversible temporal series, a tracing for acts. To avoid this reduction, I resort to a distinction between *tactics* and *strategies*.

I call a 'strategy' the calculus of force-relationships which becomes possible when a subject of will and power (a proprietor, an enterprise, a city, a scientific institution) can be isolated from an 'environment.' A strategy assumes a place that can be circumscribed as *proper* (*propre*) and thus serve as the basis for generating relations with an exterior distinct from it (competitors, adversaries, 'clientèles,' 'targets,' or 'objects' of research). Political, economic, and scientific rationality has been constructed on this strategic model.

I call a 'tactic,' on the other hand, a calculus which cannot count on a 'proper' (a spatial or institutional localization), nor thus on a borderline distinguishing the other as a visible totality. The place of a tactic belongs to the other.[19] A tactic insinuates itself into the other's place, fragmentarily, without taking it over in its entirety, without being able to keep it at a distance. It has at its disposal no base where it can capitalize on its advantages, prepare its expansions and secure independence with respect to circumstances. The 'proper' is a victory of space over time. On the contrary, because it does not have a place, a tactic depends on time – it is always on the watch for opportunities that must be seized 'on the wing.' Whatever it wins, it does not keep. It must constantly manipulate events in order to turn them into 'opportunities.' The weak must continually turn to their own ends forces alien to them. This is achieved in the propitious moments when they are able to combine heterogeneous elements (thus, in the supermarket, the housewife confronts heterogeneous and mobile data – what she has in the refrigerator, the tastes, appetites, and moods of her guests, the best buys and their possible combinations with what she already has on hand at home, etc.); the intellectual synthesis of these given elements takes the form, however, not of a discourse, but of the decision itself, the act and manner in which the opportunity is 'seized.'

Many everyday practices (talking, reading, moving about, shopping, cooking, etc.) are tactical in character. And so are, more generally, many 'ways of operating': victories of the 'weak' over the 'strong' (whether the strength be that of powerful people or the violence of things or of an imposed order, etc.), clever tricks, knowing how to get away with things, 'hunter's cunning,' maneuvers, polymorphic simulations, joyful discoveries, poetic as well as warlike. The Greeks called these 'ways of operating' *mētis*.[20] But they go much further back, to the immemorial intelligence displayed in the tricks and imitations of plants and fishes. From the depths of the ocean to the streets of modern megalopolises, there is a continuity and permanence in these tactics.

In our societies, as local stabilities break down, it is as if, no longer fixed by a circumscribed community, tactics wander out of orbit, making consumers into immigrants in a system too vast to be their own, too tightly woven for them to escape from it. But these tactics introduce a Brownian movement into the system. They also show the extent to which intelligence is inseparable from the everyday struggles and pleasures that it articulates. Strategies, in contrast, conceal beneath objective calculations their connection with the power that sustains them from within the stronghold of its own 'proper' place or institution.

The discipline of rhetoric offers models for differentiating among the types of tactics. This is not surprising, since, on the one hand, it describes the 'turns' or tropes of which language can be both the site and the object, and, on the other hand, these manipulations are related to the ways of changing (seducing, persuading, making use of) the will of another (the audience).[21] For these two reasons, rhetoric, the science of the 'ways of speaking,' offers an array of figure-types for the analysis of everyday ways of acting even though such analysis is in theory excluded from scientific discourse. Two logics of action (the one tactical, the other strategic) arise from these two facets of practicing language. In the space of a language (as in that of games), a society makes more explicit the formal rules of action and the operations that differentiate them.

In the enormous rhetorical corpus devoted to the art of speaking or operating, the Sophists have a privileged place, from the point of view of tactics. Their principle was, according to the Greek rhetorician Corax, to make the weaker position seem the stronger, and they claimed to have the power of turning the tables on the powerful by the way in which they made use of the opportunities offered by the particular situation.[22] Moreover, their theories inscribe tactics in a long tradition of reflection on the relationships between reason and particular actions and situations. Passing by way of *The Art of War* by the Chinese author Sun Tzu[23] or the Arabic anthology, *The Book of Tricks*,[24] this tradition of a logic articulated on situations and the will of others continues into contemporary sociolinguistics.

Reading, talking, dwelling, cooking, etc.

To describe these everyday practices that produce without capitalizing, that is, without taking control over time, one starting point seemed inevitable because it is the 'exorbitant' focus of contemporary culture and its consumption: *reading*. From TV to newspapers, from advertising to all sorts of mercantile epiphanies, our society is characterized by a cancerous growth of vision, measuring everything by its ability to show or be shown and transmuting communication into a visual journey. It is a sort of *epic* of the eye and of the impulse to read. The economy itself, transformed into a 'semeiocracy,'[25] encourages a hypertrophic development of reading. Thus, for the binary set production–consumption, one would substitute its more general equivalent: writing–reading. Reading (an image or a text), moreover, seems to constitute the maximal development of the passivity assumed to characterize the consumer, who is conceived of as a voyeur (whether troglodytic or itinerant) in a 'showbiz society'.[26]

In reality, the activity of reading has on the contrary all the characteristics of a silent production: the drift across the page, the metamorphosis of the text effected by the wandering eyes of the reader, the improvisation and expectation of meanings inferred from a few words, leaps over written spaces in an ephemeral dance. But since he is incapable of stockpiling (unless he writes or records), the reader cannot protect himself against the erosion of time (while reading, he forgets himself and he forgets what he has read) unless he buys the object (book, image) which is no more than a substitute (the spoor or promise) of moments 'lost' in reading. He insinuates into another person's text the ruses of pleasure and appropriation: he poaches on it, is transported into it, pluralizes himself in it like the internal rumblings of one's body. Ruse, metaphor, arrangement, this production is also an 'invention' of the memory. Words become the outlet or product of silent histories. The readable transforms itself into the memorable: Barthes reads Proust in Stendhal's text;[27] the viewer reads the landscape of his childhood in the evening news. The thin film of writing becomes a movement of strata, a play of spaces. A different world (the reader's) slips into the author's place.

This mutation makes the text habitable, like a rented apartment. It transforms another person's property into a space borrowed for a moment by a transient. Renters make comparable changes in an apartment they furnish with their acts and memories; as do speakers, in the language into which they insert both the messages of their native tongue and, through their accent, through their own 'turns of phrase,' etc., their own history; as do pedestrians, in the streets they fill with the forests of their desires and goals. In the same way the users of social codes turn them into metaphors and ellipses of their own quests. The ruling order serves as a support for innumerable productive activities, while at the same time blinding its proprietors to this creativity (like those 'bosses' who simply *can't* see what is being created within their own enterprises).[28] Carried to its limit, this order would be the equivalent of the rules of meter and rhyme for poets of earlier times: a body of constraints stimulating new discoveries, a set of rules with which improvisation plays.

Reading thus introduces an 'art' which is anything but passive. It resembles rather that art whose theory was developed by medieval poets and romancers: an innovation infiltrated into the text and even into the terms of a tradition. Imbricated within the strategies of modernity (which identify creation with the invention of a personal language, whether cultural or scientific), the procedures of contemporary consumption appear to constitute a subtle art of 'renters' who know how to insinuate their countless differences into the dominant text. In the Middle Ages, the text was framed by the four, or seven, interpretations of which it was held to be susceptible. And it was a book. Today, this text no longer comes from a tradition. It is imposed by the generation of a productivist technocracy. It is no longer a referential book, but a whole society made into a book, into the writing of the anonymous law of production.

It is useful to compare other arts with this art of readers. For example, the art of conversationalists: the rhetoric of ordinary conversation consists of practices which transform 'speech situations,' verbal productions in which the interlacing of speaking positions weaves an oral fabric without individual owners, creations of a communication that belongs to no one. Conversation is a provisional and collective effect of competence in the art of manipulating 'commonplaces' and the inevitability of events in such a way as to make them 'habitable.'[29]

NOTES

1 See M. de Certeau, *La Prise de parole* (Paris: DDB, 1968); *La Possession de Loudun* (Paris: Julliard-Gallimard, 1970); *L'Absent de l'histoire* (Paris: Mame,1973); *La Culture au pluriel* (Paris: UGE 10/18, 1974); *Une Politique de la langue* (with D. Julia and J. Revel) (Paris: Gallimard, 1975); etc.

2 From the Greek *poiein* 'to create, invent, generate.'

3 See Emile Benveniste, *Problèmes de linguistique générale* (Paris: Gallimard, 1966), I, 251–266.

4 Michel Foucault, *Surveiller et punir* (Paris: Gallimard, 1975); *Discipline and Punish*, trans. A. Sheridan (New York: Pantheon, 1977).

5 From this point of view as well, the works of Henri Lefebvre on everyday life constitute a fundamental source.

6 For this literature, see the booklets mentioned in *Le Livre dans la vie quotidienne* (Paris: Bibliothèque Nationale, 1975) and in Geneviève Bollème, *La Bible bleue, Anthologie d'une littérature 'populaire'* (Paris: Flammarion, 1975), 141–379.

7 The first of these two monographs was written by Pierre Mayol, the second by Luce Giard (on the basis of interviews made by Marie Ferrier). See *L'Invention du quotidien*, II, Luce Giard and Pierre Mayol, *Habiter, cuisiner* (Paris: UGE 10/18, 1980).

8 By Erving Goffman, see especially *Interaction Rituals* (Garden City, N.Y.: Anchor Books, 1976); *The Presentation of Self in Everyday Life* (Woodstock, N.Y.: The Overlook Press, 1973); *Frame Analysis* (New York: Harper & Row, 1974). By Pierre Bourdieu, see *Esquisse d'une théorie de la pratique. Précédée de trois études d'ethnologie kabyle* (Genève: Droz, 1972); 'Les Stratégies matrimoniales,' *Annales: économies, sociétés, civilisations* 27 (1972), 1105–1127: 'Le Langage autorisé,' *Actes de la recherche en sciences sociales* 5–6 (November 1975), 184–190; 'Le Sens pratique,' *Actes de la recherche en sciences sociales* 1 (February 1976), 43–86. By Marcel Mauss, see especially 'Techniques du corps,' in *Sociologie et anthropologie* (Paris: PUF, 1950). By Marcel Détienne and Jean-Pierre Vernant, *Les Ruses de l'intelligence. La métis des Grecs* (Paris:Flammarion, 1974). By Jeremy Boissevain, *Friends of Friends. Networks, Manipulators and Coalitions* (Oxford: Blackwell, 1974). By Edward O. Laumann, *Bonds of Pluralism. The Form and Substance of Urban Social Networks* (New York: John Wiley, 1973).

9 Joshua A. Fishman, *The Sociology of Language* (Rowley, Mass.: Newbury, 1972). See also the essays in *Studies in Social Interaction*, ed. David Sudnow (New York: The Free Press, 1972); William Labov, *Sociolinguistic Patterns* (Philadelphia: University of Pennsylvania Press, 1973); etc.

10 Oswald Ducrot, *Dire et ne pas dire* (Paris: Hermann, 1972); and David K. Lewis, *Convention: a Philosophical Study* (Cambridge, Mass.: Harvard University Press, 1974), and *Counterfactuals* (Cambridge, Mass.: Harvard University Press, 1973).

11 Georg H. von Wright, *Norm and Action* (London: Routledge & Kegan Paul, 1963); *Essay in Deontic Logic and the General Theory of Action* (Amsterdam: North Holland, 1968); *Explanation and Understanding* (Ithaca, N.Y.: Cornell University Press, 1971). And A. C. Danto, *Analytical Philosophy of Action* (Cambridge: Cambridge University Press, 1973); Richard J. Bernstein, *Praxis and Action* (London: Duckworth, 1972); and *La Sémantique de l'action*, ed. Paul Ricoeur and Doriane Tiffeneau (Paris: CNRS, 1977).

12 A. N. Prior, *Past, Present and Future: a Study of 'Tense Logic'* (Oxford: Oxford University Press, 1967) and *Papers on Tense and Time* (Oxford: Oxford University Press, 1968); N. Rescher and J. Urquhart, *Temporal Logic* (Oxford: Oxford University Press, 1975).

13 Alan R. White, *Modal Thinking* (Ithaca, N.Y.: Cornell University Press, 1975); G. E. Hughes and M. J. Cresswell, *An Introduction to Modal Logic* (Oxford: Oxford University Press, 1973); I. R. Zeeman, *Modal Logic* (Oxford: Oxford University Press, 1975); S. Haacker, *Deviant Logic* (Cambridge: Cambridge University Press, 1976); *Discussing Language with Chomsky, Halliday, etc.*, ed. H. Parret (The Hague: Mouton, 1975).

14 As it is more technical, the study concerning the logics of action and time, as well as modalization, will be published elsewhere.

15 Jacques Sojcher, *La Démarche poétique* (Paris: UGE 10/18, 1976), 145.

16 See Fernand Deligny, *Les Vagabonds efficaces* (Paris: Maspero, 1970); *Nous et l'innocent* (Paris: Maspero, 1977); etc.

17 See M. de Certeau, *La Culture au pluriel*, 283–308; and 'Actions culturelles et stratégies politiques,' *La Revue nouvelle*, April 1974, 351–360.

18 The analysis of the principles of isolation allows us to make this criticism both more nuanced and more precise. See *Pour une histoire de la statistique* (Paris: INSEE, 1978), I, in particular Alain Desrosières, 'Eléments pour l'histoire des nomenclatures socio-professionnelles,' 155–231.

19 The works of P. Bourdieu and those of M. Détienne and J-P. Vernant make possible the notion of' 'tactic' more precise, but the sociolinguistic investigations of H. Garfinkel, H. Sacks *et al.* also contribute to this clarification, See notes 8 and 9.

20 M. Détienne and J-P. Vernant, *Les Ruses de l'intelligence.*

21 See S. Toulmin, *The Uses of Argument* (Cambridge: Cambridge University Press, 1958); Ch. Perelman and L. Ollbrechts-Tyteca, *Traité de l'argumentation* (Bruxelles: Université libre, 1970); J. Dubois *et al.*, *Rhétorique générale* (Paris: Larousse, 1970); etc.

22 The works of Corax, said to be the author of the earliest Greek text on rhetoric, are lost; on this point, see Aristotle, *Rhetoric*, II, 24, 1402a. See W. K. C. Guthrie, *The Sophists* (Cambridge: Cambridge University Press, 1971), 178–179.

23 Sun Tzu, *The Art of War*, trans. S. B. Griffith (Oxford: The Clarendon Press, 1963). Sun Tzu (Sun Zi) should not be confused with the later military theorist Hsün Tzu (Xun Zi).

24 *Le Livre des ruses. La Stratégie politique des Arabes*, ed. R. K. Khawam (Paris: Phébus, 1976).

25 See Jean Baudrillard, *Le Système des objets* (Paris: Gallimard, 1968); *La Société de consommation* (Paris: Denoël, 1970); *Pour une critique de l'économie politique du signe* (Paris: Gallimard, 1972).

26 Guy Debord, *La Société du spectacle* (Paris: Buchet-Chastel, 1967).

27 Roland Barthes, *Le Plaisir du texte* (Paris: Seuil, 1973), 58; *The Pleasure of the Text*, trans. R. Miller (New York: Hill & Wang, 1975).

28 See Gérard Mordillat and Nicolas Philibert, *Ces Patrons éclairés qui craignent la lumière* (Paris: Albatros, 1979).

29 See the essays of H. Sacks, E. A. Schegloff, etc., quoted above. This analysis, entitled *Arts de dire*, will be published separately.

ACKNOWLEDGEMENTS AND COPYRIGHT INFORMATION

The editors would like to thank Ann Michael, Melanie Attridge and Andrew Mould for their help and advice in compiling this Reader.

Permission given by the following authors and/or copyright holders is gratefully acknowledged. While considerable effort has been made to trace and contact copyright holders prior to publication, the editors and publishers apologise for any oversights or omissions and if notified will endeavour to remedy these at the earliest opportunity.

1 HISTORY

Joyce Appleby, 'Consumption in early modern social thought' from J. Brewer and R. Porter (eds), *Consumption and the World of Goods*, pp. 162–173. Copyright © 1993 by Routledge.

Neil McKendrick, 'The consumer revolution of eighteenth-century England' from N. McKendrick, J. Brewer and J. H. Plumb, *The Birth of a Consumer Society: The Commercialization of Eighteenth-century England*, pp. 9–11 (London, Europa Publications, 1982). Copyright © 1982 by Europa Publications.

Ben Fine and Ellen Leopold, 'Consumerism and the Industrial Revolution', *Social History* 15(2), 1990, pp. 165–173. Copyright © 1990 by Taylor & Francis Ltd. *http://www.tandf.co.uk*

Colin Campbell, 'Traditional and modern hedonism' from C. Campbell, *The Romantic Ethic and the Spirit of Modern Consumerism*, pp. 69–76, 242–243. Copyright © 1987 by Colin Campbell.

Zygmunt Bauman, 'Industrialism, consumerism and power', *Theory, Culture and Society* 1(3), 1983, pp. 34–41, 43. Copyright © 1983 by Sage Publications.

Alan Hunt, 'The governance of consumption: sumptuary laws and shifting forms of regulation', *Economy and Society* 25(3), 1996, pp. 410–427. Copyright © 1996 by Taylor & Francis Ltd. *http://www.tandf.co.uk*

Martin Purvis, 'Societies of consumers and consumer societies: co-operation, consumption and politics in Britain and continental Europe *c.* 1850–1920', *Journal of Historical Geography* 24(2), 1998, pp. 154–169. Reprinted by permission of the publisher, Academic Press. Copyright © 1998 by Academic Press.

2 GEOGRAPHY

Wolfgang Schivelbusch, 'Night life' from W. Schivelbusch, *Disenchanted Night: The Industrialization of Light in the Nineteenth Century*, pp. 137–154. Copyright © 1988 by Wolfgang Schivelbusch.

Peter J. Taylor, 'What's modern about the modern world-system? Introducing ordinary modernity through world hegemony', *Review of International Political Economy* 3(2), 1996, pp. 260–286. Copyright © 1996 by Taylor & Francis Ltd. *http://www.tandf.co.uk*

Jean Comaroff, 'The empire's old clothes: fashioning the colonial subject' from D. Howes (ed.), *Cross-cultural Consumption: Global Markets, Local Realities*, pp. 24–32, 38, 195–208. Copyright © 1996 by Routledge.

Michael Wildt, 'Plurality of taste: food and consumption in West Germany during the 1950s', *History Workshop Journal* 39, 1995, pp. 23–41. Reprinted by permission of Oxford University Press. Copyright © 1995 by Oxford University Press.

Ian Cook and **Philip Crang**, 'The world on a plate: culinary culture, displacement and geographical knowledges', *Journal of Material Culture* 1(2), 1996, extracts from pp. 137–140, 149–153. Copyright © 1996 by Sage Publications.

John Urry, 'The "consumption" of tourism', *Sociology* 24(1), 1990, pp. 23–35. Copyright © 1990 by Cambridge University Press.

Paul A. Redfern, 'A new look at gentrification: 1. Gentrification and domestic technologies', *Environment and Planning A* 29, 1997, pp. 1288–1293, 1294–1296. Copyright © 1997 by Pion Ltd.

Sharon Zukin, 'Urban lifestyles: diversity and standardisation in spaces of consumption', *Urban Studies* 35(5–6), 1998, pp. 825–839. Copyright © 1998 by Taylor & Francis Ltd. *http://www.tandf.co.uk*

3 SUBJECTS AND IDENTITY

Simon Mohun, 'Consumer sovereignty' from F. Green and P. Nore (eds), *Economics: An Anti-text* pp. 57–75 (London, Macmillan Press Ltd, 1977). Reproduced with permission of Palgrave. Copyright © 1977 by Palgrave.

Mary Douglas, 'The consumer's revolt' from M. Douglas, *Thought Styles: Critical Essays on Good Taste*, pp. 106–125. Copyright © 1996 by Sage Publications.

Dick Hebdige, 'Subculture and style' from *Subculture: The Meaning of Style*, pp. 1–4, 172, 174. Copyright © 1979 by Routledge.

Andrew Bennett, 'Subcultures or neo-tribes? Rethinking the relationship between youth, style and musical taste', *Sociology* 33(3), 1999, pp. 599–617. Copyright © 1999 by Cambridge University Press.

Rosemary Crompton, 'Consumption and class analysis' from S. Edgell, K. Hetherington and A. Warde (eds), *Consumption Matters: The Production and Experience of Consumption* (Special Issue of *The Sociological Review*), pp. 113–132. Copyright © 1996 by The Editorial Board of *The Sociological Review*.

Mike Featherstone, 'The body in consumer culture', *Theory, Culture and Society* 1(2), 1982, extracts from pp. 27–30, 31–33. Copyright © 1982 by Sage Publications.

Rachel Bowlby, 'Commerce and femininity' from *Just Looking: Consumer Culture in Dreiser, Gissing & Zola*, pp. 18–24, 29–34, 157–158. Copyright © 1985 by Routledge.

Hillel Schwartz, 'The three-body problem and the end of the world' from M. Feher (ed.), *Fragments for a History of the Human Body: Part Two*, pp. 411–420, 452–454 (New York, Zone Books, 1989). Copyright © 1989 by Urzone, Inc.

4 OBJECTS AND TECHNOLOGY

Pasi Falk, 'The genealogy of advertising' from P. Sulkunen, J. Holmwood, H. Radner and G. Schulze (eds), *Constructing the New Consumer Society*, pp. 81–90, 105–107 (New York, St Martin's Press). Copyright © 1997 by Pekka Sulkunen, John Holmwood, Hilary Radner and Gerhard Schulze. Reprinted with permission of Macmillan Ltd.

Bruno Latour, 'Where are the missing masses? The sociology of a few mundane artifacts' from W. E. Bijker and J. Law (eds), *Shaping Technology/Building Society: Studies in Sociotechnical Change* (Boston, MIT Press, 1992), pp. 225–227, 255, 309–325. Copyright © 1992 by Massachusetts Institute of Technology.

Christina Kiaer, 'Rodchenko in Paris', *October* 75 (winter 1996), excerpt pp. 3–9. Copyright © 1996 by October magazine Ltd and Massachusetts Institute of Technology.

Roland Barthes, 'Toys' from *Mythologies* by Roland Barthes, translated by Annette Lavers, used by permission of the Estate of Roland Barthes, the translator, Jonathan Cape as publishers, The Random House Group Limited. *Mythologies* was first published in French by du Seuil in 1957. Printed in the USA in *Mythologies* by Roland Barthes, translated by Annette Lavers. English translation copyright © 1973 by Annette Lavers. Reprinted by permission of Hill & Wang, a division of Farrar, Straus and Giroux, LLC.

Roger Miller, 'The Hoover® in the garden: middle-class women and suburbanization, 1850–1920', *Environment and Planning D: Society and Space* 1, 1983, pp. 82–87. Copyright © 1983 by Pion Ltd.

Elaine S. Abelson, 'The two-way mirror' from *When Ladies Go A-thieving: Middle-class Shoplifters in the Victorian Department Store*. Copyright © 1992 by Elaine S. Abelson. Used by permission of Oxford University Press, Inc.

Jon Goss, '"We know who you are and we know where you live:" the instrumental rationality of geodemographic systems', *Economic Geography* 71(2), April 1995, pp. 187–191, 194–198. Copyright © 1995 by *Economic Geography*.

5 THEORY

Marcel Mauss, 'Introduction' from *The Gift: The Form and Reason for Exchange in Archaic Societies*, pp. 5–7, 85–87. Copyright © 1990 by Routledge.

Georges Bataille, 'Sacrifices and wars of the Aztecs' from *The Accursed Share: An Essay on General Economy, Vol. 1*, translated by Robert Hurley, pp. 45–46, 49–51, 55–59, 192 (New York, Zone Books, 1988). Copyright © 1988 by Urzone, Inc.

Thorstein Veblen, 'Pecuniary emulation' from *The Theory of the Leisure Class: An Economic Study in the Evolution of Institutions*. Public domain.

Georg Simmel, 'The philosophy of fashion', translated by M. Ritter and D. Frisby, from D. Frisby and M. Featherstone (eds), *Simmel on Culture*, pp. 192–203. Copyright © 1997 by Sage Publications.

Pierre Bourdieu, 'Classes and classifications' from *Distinction: A Social Critique of the Judgement of Taste*, pp. 466–470, 482–484, 596–597. Copyright © 1986 by Routledge.

Karl Marx, 'Production, consumption, distribution, exchange (circulation)' from *Grundrisse: Foundations of the Critique of Political Economy*, pp. 88–94, translated by Martin Nicolaus (London, Pelican Books). Copyright © 1973 by *New Left Review*.

Jean Baudrillard, 'The ideological genesis of needs' from J. Baudrillard, *For a Critique of the Political Economy of the Sign*, trans, C. Levin pp. 63–68. Copyright © 1981 by Telos Press Ltd.

Michel de Certeau, 'General introduction' from *The Practice of Everyday Life*, edited/translated by Steven Rendall, pp xi–xxii and 205–207. Copyright © 1984 by The Regents of the University of California.

KEY REFERENCES

[**Emboldened** entries included in the body of the Reader.]

Economics and marketing

Ando, A. and Modigliani, F. (1963) 'The "life cycle" hypothesis of saving: aggregate implications and tests'. *American Economic Review* 3(1), 55–84.

Becker, G. S. and Murphy, K. M. (1988) 'A theory of rational addiction'. *Journal of Political Economy* 96(4), 675–700.

Belk, R. W. (1995) 'Studies in the new consumer behaviour' in Miller, D. (ed.) *Acknowledging Consumption: A Review of New Studies*. Routledge, London, 58–95.

Drakopoulos, S. A. (1992) 'Keynes' economic thought and the theory of consumer behaviour'. *Scottish Journal of Political Economy* 39(3), 318–336.

Duesenberry, J. S. (1949) *Income, Saving and the Theory of Consumer Behaviour*. Harvard University Press, Cambridge.

Dyer, A. W. (1997) 'Prelude to a theory of homo absurdus: variations on themes from Thorstein Veblen and Jean Baudrillard'. *Cambridge Journal of Economics* 21, 45–53.

Endres, A. M. (1991) 'Marshall's analysis of economizing behaviour with particular reference to the consumer'. *European Economic Review* 35, 333–341.

Fine, B. (1995) 'From political economy to consumption' in Miller, D. (ed.) *Acknowledging Consumption: A Review of New Studies*. Routledge, London, 127–163.

Fine, B. and Leopold, E. (1993) *The World of Consumption*. Routledge, London.

Firat, A. F. and Venkatesh, A. (1995) 'Liberatory postmodernism and the reenchantment of consumption'. *Journal of Consumer Research* 22, 239–267.

Foxall, G. R. (1988) *Consumer Behaviour: A Practical Guide*. Routledge, London.

Frank, R. H. (1999) *Luxury Fever: Why Money Fails to Satisfy in an Era of Excess*. Free Press, New York.

Friedman, M. (1957) *A Theory of the Consumption Function*. Princeton University Press, Princeton, NJ.

Galbraith, J. K. (1958) *The Affluent Society*. Riverside Press, Cambridge.

Galbraith, J. K. (1967) *The New Industrial State*. Penguin, Harmondsworth.

Hamilton, D. B. (1987) 'Institutional economics and consumption'. *Journal of Economic Issues* 21, 1531–1554.

Herden, G., Knoche, K., Seidl, C. and Trockel, W. (eds) (1999) *Mathematical Utility Theory: Utility Functions, Models and Applications in the Social Sciences*. Springer-Verlag, New York.

Hicks, J. and Allen, R. J. (1934) 'A reconsideration of the theory of value: Part II – a mathematical theory of individual demand functions'. *Economica* 1(2), 196–219.

Hirsch, F. (1976) *The Social Limits to Growth*. Harvard University Press, Cambridge.

Holbrook, M. B. (1995) *Consumer Research: Introspective Essays on the Study of Consumption*. Sage, London.

Howard, J. A. and Sheth, J. N. (1969) *The Theory of Buyer Behaviour*. Wiley, New York.

Keynes, J. M. (1936) *The General Theory of Employment, Interest and Money*. Macmillan, London.

Kuznets, S. (1946) *National Product since 1869*. National Bureau of Economic Research, New York.

Lancaster, K. (1966a) 'Change and innovation in the technology of consumption'. *American Economic Review* 56, 14–23.

Lancaster, K. (1966b) 'A new approach to consumer theory'. *Journal of Political Economy* 74, 132–157.
Leibenstein, H. (1950) 'Bandwagon, snob and Veblen effects in the theory of consumers' demand'. *Quarterly Journal of Economics* 44(2), 183–207.
Lunt, P. K. and Livingstone, S. M. (1992) *Mass Consumption and Personal Identity: Everyday Economic Experience.* Open University Press, Milton Keynes.
McCloskey, D. N. (1985) *The Rhetoric of Economics.* Wheatsheaf, Brighton.
McCormick, K. (1983) 'Duesenberry and Veblen: the demonstration effect revisited'. *Journal of Economic Issues* 17, 1125–1129.
Marx, K. (1967) *Capital. A Critique of Political Economy.* International Publishers, New York.
Mátyás, A. (1985) *History of Modern Non-Marxian Economics: From Marginalist Revolution Through the Keynesian Revolution to Contemporary Monetarist Counter-revolution.* Macmillan, London.
Mishan, E. J. (1961) 'Theories of consumption: a cynical view'. *Economica* 28(109), 1–11.
Modigliani, F. and Brumberg, R. E. (1955) 'Utility analysis and the consumption function: an interpretation of cross-sectional data' in Kurihara, K. K. (ed.) *Post-Keynesian Economics.* Allen & Unwin, London, 388–436.
Mohun, S. (1977) 'Consumer sovereignty' in Green, F. and Nore, P. (eds) *Economics: An Anti-text.* Macmillan, London, 57–75.
Pollack, R. (1970) 'Habit formation and dynamic demand functions'. *Journal of Political Economy* 78, 745–763.
Pollack, R. (1976) 'Interdependent preferences'. *American Economic Review* 66, 309–320.
Preteceille, E. and Terrail, J–P. (1985) *Capitalism, Consumption and Needs.* Blackwell, Oxford.
Samuelson, P. (1948) 'Consumer theory in terms of revealed preferences'. *Economica* 15(60), 243–253.
Scitovsky, T. (1976) *The Joyless Economy: An Inquiry into Human Satisfaction and Consumer Dissatisfaction.* Oxford University Press, Oxford.
Smith, A. (1937) *An Inquiry into the Nature and Causes of the Wealth of Nations.* Modern Library, New York.
Veblen, T. (1994) *The Theory of the Leisure Class: An Economic Study in the Evolution of Institutions.* Dover Press, New York.
von Neumann, J. and Morgenstern, O. (1947) *Theory of Games and Economic Behaviour* (2nd edn). Princeton University Press, Princeton, NJ.

Sociology, anthropology and cultural studies

Abelson, E. S. (1992) *When Ladies Go A-thieving: Middle-class Shoplifters in the Victorian Department Store.* Oxford University Press, Oxford.
Appadurai, A. (ed.) (1986) *The Social Life of Things: Commodities in Cultural Perspective.* Cambridge University Press, Cambridge, 64–91.
Baker, A. (ed.) (2000) *Serious Shopping: Essays in Psychotherapy and Consumerism.* Free Association Press, London.
Barthes, R. (1973) *Mythologies.* Paladin, London.
Barthes, R. (1990) *The Fashion System.* University of California Press, Berkeley.
Bataille, G. (1988) *The Accursed Share: An Essay on General Economy. Volume I. Consumption.* Zone Books, New York.
Baudrillard, J. (1975) *The Mirror of Production.* Telos, St Louis.
Baudrillard, J. (1981) *For a Critique of the Political Economy of the Sign.* Telos, St Louis.
Baudrillard, J. (1996) *The System of Objects.* Verso, London.
Baudrillard, J. (1998a) *The Consumer Society: Myths and Structures.* Sage, London.
Baudrillard, J. (1998b) *Paroxysm.* Verso, London.
Bauman, Z. (1983) 'Industrialism, consumerism and power'. *Theory, Culture and Society* 1(3), 32–43.
Bauman, Z. (1987) *Legislators and Interpreters: On Modernity, Post-modernity and Intellectuals.* Polity Press, Cambridge.
Bauman, Z. (1992) 'Survival as a social construct'. *Theory, Culture and Society* 9: 1–36.
Bauman, Z. (1996) 'From pilgrim to tourist – or a short history of identity' in Hall, S. and du Gay, P. (eds) *Questions of Cultural Identity.* Sage, London, 18–36.
Bauman, Z. (1998) *Work, Consumerism and the New Poor.* Open University Press, Milton Keynes.
Bauman, Z. (2001) 'Consuming life'. *Journal of Consumer Culture* 1: 9–29.
Benjamin, W. (1999) *The Arcades Project.* Belknap Press, Cambridge, MA.
Bennett, A. (1999) 'Subcultures or neo-tribes? Rethinking the relationship between youth, style and musical taste'. *Sociology* 33(3), 599–617.
Berking, H. (1999) *Sociology of Giving.* Sage, London.

Bois, Y-A. and Krauss, R. (1997) *Formless: A User's Guide*. Zone, New York.

Bordo, S. (1992) 'Anorexia nervosa: psychopathology as the crystallization of culture' in Curtin, D. W. and Heldke, A. M. (eds) *Cooking, Eating, Thinking: Transformative Philosophies of Food*. Indiana University Press, Bloomington, 28–55.

Bourdieu, P. (1977) *Outline of a Theory of Practice*. Cambridge University Press, Cambridge.

Bourdieu, P. (1986) *Distinction: A Social Critique of the Judgement of Taste*. Routledge, London.

Bowlby, R. (1985) *Just Looking: Consumer Culture in Dreisser, Gissing and Zola*. Methuen, London.

Bowlby, R. (1993) *Shopping with Freud*. Routledge, London.

Bowlby, R. (2001) *Carried Away: The Invention of Modern Shopping*. Faber and Faber, London.

Buck-Morss, S. (1989) *The Dialectics of Seeing: Walter Benjamin and the Arcades Project*. MIT Press, Cambridge, MA.

Campbell, C. (1987) *The Romantic Ethic and the Spirit of Modern Consumerism*. Blackwell, Oxford.

Campbell, C. (1995) 'The sociology of consumption' in Miller, D. (ed.) *Acknowledging Consumption: A Review of New Studies*. Routledge, London, 96–126.

Campbell, C. (1997) 'When the meaning is not a message: a critique of the consumption as communication thesis' in Nava, M., Blake, A., McRury, I. and Richards, B. (eds) *Buy this Book: Studies in Advertising and Consumption*. Routledge, London, 340–351.

Carrier, J. (1991) 'Gifts, commodities and social relations: a Maussian view of exchange'. *Sociological Forum* 6(1), 119–136.

Carrier, J. (1994) *Gifts and Commodities: Exchange and Western Capitalism since 1700*. Routledge, London.

Carter, E. (1984) 'Alice in the consumer wonderland' in McRobbie, A. and Nava, M. (eds) *Gender and Generation*. Macmillan, London, 185–214.

Chaney, D. (1996) *Lifestyles*. Routledge, London.

Corrigan, P. (1997) *The Sociology of Consumption: An Introduction*. Sage, London.

Crompton, R. (1996) 'Consumption and class analysis' in Edgell, S., Hetherington, K. and Warde, A. (eds) *Consumption Matters: The Production and Experience of Consumption*. Blackwell, Oxford, 113–132.

Csikszentmihályi, M. and Rochberg-Halton, E. (1987) *The Meaning of Things: Domestic Symbols and the Self*. Cambridge University Press, Cambridge.

de Certeau, M. (1984) *The Practice of Everyday Life*. University of California Press, Berkeley.

Deleuze, G. and Guattari, F. (1984) *Anti-Oedipus: Capitalism and Schizophrenia*. Athlone, London.

Deleuze, G. and Guattari, F. (1988) *A Thousand Plateaus: Capitalism and Schizophrenia*. Athlone, London.

Derrida, J. (1992) *Given Time: 1. Counterfeit Money*. University of Chicago Press, Chicago, IL.

Derrida, J. (1994) *Specters of Marx: The State of the Debt, the Work of Mourning, and the New International*. Routledge, London.

de Saussure, F. (1959) *Course in General Linguistics*. The Philosophical Library, New York.

Dittmar, H. (1992) *The Social Psychology of Material Possessions: To Have is to Be*. Harvester Wheatsheaf, Hemel Hempstead.

Douglas, M. (1996) *Thought Styles: Critical Essays on Good Taste*. Sage, London.

Douglas, M. and Isherwood, B. (1996) *The World of Goods: Towards an Anthropology of Consumption* (new edn). Routledge, London.

DuCille, A. (1994) 'Dyes and dolls: multicultural Barbie and the merchandising of difference'. *Differences* 6(1), 46–68.

du Gay, P. (1996) *Consumption and Identity at Work*. Sage, London.

Dumont, L. (1977) *From Mandeville to Marx: The Genesis and Triumph of Economic Ideology*. University of Chicago Press, Chicago, IL.

Dumont, L. (1986) *Essays on Individualism: Modern Ideology in Anthropological Perspective*. University of Chicago Press, Chicago, IL.

Durkheim, E. (1947) *The Division of Labour in Society*. Free Press, Glencoe, IL.

Edwards, T. (1997) *Men in the Mirror: Men's Fashion, Masculinity and Consumer Society*. Cassell, London.

Falk, P. (1994) *The Consuming Body*. Sage, London.

Falk, P. (1997) 'The genealogy of advertising' in Sulkunen, P., Holmwood, J., Radner, H. and Schulze, G. (eds) *Constructing the New Consumer Society*. St Martin's Press, New York, 81–107.

Featherstone, M. (1982) 'The body in consumer culture'. *Theory, Culture and Society* 1(2), 18–33.

Featherstone, M. (1991) *Consumer Culture and Postmodernism*. Sage, London.

Frisby, D. and Featherstone, M. (eds) (1997) *Simmel on Culture*. Sage, London.

George, A. and Murcott, A. (1992) 'Monthly strategies for discretion: shopping for sanitary towels and tampons'. *The Sociological Review* 40(1), 146–162.

Giddens, A. (1994) 'Living in a post-traditional society' in Beck, U., Giddens, A. and Lash, S. (eds) *Reflexive Modernization: Politics, Tradition and Aesthetics in the Modern Social Order*. Polity Press, Cambridge, 56–109.

Goss, J. (1995) '"We know who you are and we know where you live": the instrumental rationality of geodemographic systems'. *Economic Geography* 71(2), 171–198.

Haug, W. F. (1986) *Critique of Commodity Aesthetics: Appearance, Sexuality and Advertising in Capitalist Society*. Polity Press, Cambridge.

Hearn, J. and Roseneil, S. (eds) (1999) *Consuming Cultures: Power and Resistance*. Macmillan, London.

Hebdige, D. (1979) *Subculture: The Meaning of Style*. Methuen, London.

Hebdige, D. (1996) *Hiding in the Light: On Images and Things*. Routledge, London.

Jackson, P. (1994) 'Black male: advertising and the cultural politics of masculinity'. *Gender, Place and Culture* **1**, 49–59.

Jackson, S. and Moores, S. (eds) (1995) *The Politics of Domestic Consumption: Critical Readings*. Prentice Hall, London.

Jameson, F. (1983) 'Postmodernism and consumer society' in Foster, H. (ed.) *The Anti-aesthetic: Essays on Postmodern Culture*. Bay Press, Port Townsend, 111–125.

Kiaer, C. (1996) 'Rodchenko in Paris'. *October* 75, 3–35.

Lamont, M. and Molnár, V. (2001) 'How blacks use consumption to shape their collective identity: evidence from marketing specialists'. *Journal of Consumer Culture* **1**(1), 31–45.

Latour, B. (1992) 'Where are the missing masses? The sociology of a few mundane artifacts' in Bijker, W. E. and Law, J. (eds) *Shaping Technology/Building Society: Studies in Sociotechnical Change*. MIT Press, Cambridge, 225–258.

Lee, M. J. (1992) *Consumer Culture Reborn: The Cultural Politics of Consumption*. Routledge, London.

Lury, C. (1996) *Consumer Culture*. Polity Press, Cambridge.

Lyon, D. (1994) *The Electronic Eye: The Rise of Surveillance Society*. Polity Press, Cambridge.

Lyotard, J-F. (1993) *Libidinal Economy*. Indiana University Press, Bloomington.

Lyotard, J-F. (1998) *The Assassination of Experience by Painting – Monory*. Black Dog, London.

MacClancey, J. (1992) *Consuming Culture*. London, Paul Chapman.

McClintock, A. (1994) 'Soft-soaping empire: commodity racism and imperial advertising' in Robinson, G., Mash, M., Tickner, L., Bird, J. and Curtis, B. A. (eds) *Travellers' Tales: Narratives of Home and Displacement*. Routledge, London, 131–155.

McCracken, G. (1990) *Culture and Consumption: New Approaches to the Symbolic Character of Consumer Goods and Activities*. Indiana University Press, Bloomington.

McRobbie, A. (1997) 'Bridging the gap: feminism, fashion and consumption'. *Feminist Review* 55, 73–89.

Maffesoli, M. (1996) *The Time of the Tribes: The Decline of Individualism in Mass Society*. Sage, London.

Marcuse, H. (1964) *One-dimensional Man: Studies in the Ideology of Advanced Industrial Society*. Sphere, London.

Marx, K. (1973) *Grundrisse: Foundations of the Critique of Political Economy*. Penguin, Harmondsworth.

Mauss, M. (1990) *The Gift: The Form and Reason for Exchange in Archaic Societies*. Routledge, London.

Miles, S. (1998) *Consumerism: As a Way of Life*. Sage, London.

Miller, D. (1987) *Material Culture and Mass Consumption*. Blackwell, Oxford.

Miller, D. (ed.) (1995) *Acknowledging Consumption: A Review of New Studies*. Routledge, London.

Miller, R. (1983) 'The Hoover® in the garden: middle-class women and suburbanization, 1850–1920'. *Environment and Planning D: Society and Space* 1, 73–87.

Mort, F. (1996) *Cultures of Consumption: Masculinities and Social Space in Late Twentieth-century Britain*. Routledge, London.

Nava, M. (1992) *Changing Cultures: Feminism, Youth and Consumerism*. Sage, London.

Nixon, S. (1996) *Hard Looks: Masculinities, Spectatorship and Contemporary Consumption*. UCL Press, London.

Poster, M. (1990) *The Mode of Information: Poststructuralism and Social Context*. Polity Press, Cambridge.

Radner, H. (1995) *Shopping Around: Feminine Culture and the Pursuit of Pleasure*. Routledge, London.

Ritzer, G. (1992) *The McDonaldization of Society: An Investigation into the Changing Character of Contemporary Social Life.* Sage, London.

Ritzer, G. (1997) *The McDonaldization Thesis: Explorations and Extensions.* Sage, London.

Ritzer, G. (ed.) (2002) *McDonaldization: The Reader.* Sage, London.

Schivelbusch, W. (1993) *Tastes of Paradise: A Social History of Spices, Stimulants, and Intoxicants.* Vintage, London.

Schwartz, H. (1989) 'The three-body problem and the end of the world' in Feher, M. (ed.) *Fragments for a History of the Human Body: Part Two.* Zone Books, New York, 406–465.

Shields, R. (ed.) (1992) *Lifestyle Shopping: The Subject of Consumption.* Routledge, London.

Slater, D. (1997) *Consumer Culture and Modernity.* Polity Press, Cambridge.

Smart, B. (ed.) (1999) *Resisting McDonaldization.* Sage, London.

Storey, J. (1999) *Consumer Culture and Everyday Life.* Arnold, London.

Tester, K. (ed.) (1994) *The Flâneur.* Routledge, London.

Warde, A. (1996) 'The future of the sociology of consumption' in Edgell, S., Hetherington, K. and Warde, A. (eds) *Consumption Matters: The Production and Experience of Consumption.* Blackwell, Oxford, 302–312.

Warde, A. (1997) *Consumption, Food and Taste: Culinary Antinomies and Commodity Culture.* Sage, London.

Weber, M. (1976) *The Protestant Ethic and the Spirit of Capitalism.* Allen & Unwin, London.

Weinbaum, B. and Bridges, A. (1979) 'The other side of the paycheck: monopoly capital and the structure of consumption' in Eisenstein, Z. R. (ed.) *Capitalist Patriarchy and the Case for Socialist Feminism.* Monthly Review Press, New York, 190–205.

Willis, S. (1990) 'I want the black one: is there a place for Afro-American culture in commodity culture?', *New Formations* 10, 77–97.

Wilson, E. (1992) 'The invisible *flâneur*'. *New Left Review* 191, 90–110.

Wolff, J. (1985) 'The invisible *flâneuse*: women and the literature of modernity'. *Theory, Culture and Society* 2(3), 37–45.

Wolff, K. (ed.) (1950) *The Sociology of Georg Simmel.* Free Press, New York.

History and geography

Adburgham, A. (1981) *Shops and Shopping, 1800–1914: Where and in What Manner the Well-dressed Englishwoman Bought Her Clothes.* Allen & Unwin, London.

Alexander, D. (1970) *Retailing in England during the Industrial Revolution.* Athlone, London.

Appleby, J. (1978) *Economic Thought in Seventeenth-century England.* Princeton University Press, Princeton, NJ.

Appleby, J. (1993) 'Consumption in early modern social thought' in Brewer, J. and Porter, R. (eds) *Consumption and the World of Goods.* Routledge, London, 162–173.

Bauman, Z. (1982) *Memories of Class: The Pre-history and After-life of Class.* Routledge & Kegan Paul, London.

Bauman, Z. (1983) 'Industrialism, consumerism and power'. *Theory, Culture and Society* 1(3), 32–43.

Bell, D. and Valentine, G. (1997) *Consuming Geographies: We Are Where We Eat.* Routledge, London.

Benson, J. (1994) *The Rise of Consumer Society in Britain, 1880–1980.* Longman, London.

Benson, S. P. (1986) *Counter Cultures: Saleswomen, Managers, and Customers in American Department Stores, 1890–1940.* University of Illinois Press, Chicago, IL.

Bermingham, A. and Brewer, J. (eds) (1995) *The Consumption of Culture: Image, Object, Text.* Routledge, London.

Berry, B. J. L. (1967) *Geography of Market Centers and Retail Distribution.* Prentice-Hall, Englewood Cliffs, NJ.

Berry, B. J. L. and Parr, J. (1988) *Market Centres and Retail Location: Theory and Applications.* Prentice Hall, Englewood Cliffs, NJ.

Berry, C. J. (1994) *The Idea of Luxury: A Conceptual and Historical Investigation.* Cambridge University Press, Cambridge.

Breen, T. H. (1988) '"Baubles of Britain": the American Consumer Revolution of the eighteenth century'. *Past and Present* 119, 73–104.

Brewer, J. and Porter, R. (eds) (1993) *Consumption and the World of Goods.* Routledge, London.

Bronner, S. J. (ed.) (1983) *Consuming Visions: Accumulation and Display of Goods in America, 1880–1920.* Norton, New York.

Butler, T. (1997) *Gentrification and the Middleclasses.* Aldershot, Ashgate.

Campbell, C. (1987) *The Romantic Ethic and the Spirit of Modern Consumerism*. Blackwell, Oxford.

Charney, L. and Schwartz, V. (eds) (1995) *Cinema and the Invention of Modern Life*. University of California Press, Berkeley.

Clammer, J. (1997) *Contemporary Urban Japan: A Sociology of Consumption*. Blackwell, Oxford.

Clarke, D. B. (2003) *The Consumer Society and the Postmodern City*. Routledge, London.

Comaroff, J. (1996) 'The empire's old clothes: fashioning the colonial subject' in Howes, D. (ed.) *Cross-cultural Consumption: Global Markets, Local Realities*. Routledge, London, 19–38.

Cook, I. and Crang, P. (1996) 'The world on a plate: culinary culture, displacement and geographical knowledges'. *Journal of Material Culture* 1(2), 131–153.

Crawford, M. (1992) 'The world in a shopping mall' in Sorkin, M. (ed.) *Variations on a Theme Park: The New American City and the End of Public Space*. Noonday Press, New York, 3–30.

Crewe, L. and Gregson, N. (1997) 'The bargain, the knowledge, and the spectacle: making sense of consumption in the space of the car boot sale'. *Environment and Planning D: Society and Space* 15, 87–112.

Crewe, L. and Gregson, N. (1998) 'Tales of the unexpected: exploring car boot sales as marginal spaces of contemporary consumption'. *Transactions of the Institute of British Geographers* 23, 39–54.

Cross, G. (1993) *Time and Money: The Making of a Consumer Culture*. Routledge, London.

Crossick, G. and Jaumain, S. (eds) (1999) *Cathedrals of Consumption*. Ashgate, Aldershot.

Davis, D. (1966) *Fairs, Shops and Supermarkets: A History of English Shopping*. University of Toronto Press, Toronto.

Dawson, J. A. (ed.) (1980) *Retail Geography*. Croom Helm, Beckenham.

Dyer, C. (1989) 'The consumer and the market in the later middle ages'. *Economic History Review* 42, 305–327.

Falk, P. and Campbell, C. (eds) (1997) *The Shopping Experience*. Sage, London.

Fine, B. and Leopold, E. (1990) 'Consumerism and the Industrial Revolution'. *Social History* 15(2), 151–179.

Foucault, M. (1980) *Power/Knowledge: Selected Interviews and Other Writings by Michel Foucault, 1972–1977* (ed. C. Gordon). Harvester Press, Brighton.

Fox, R. W. and Lears, T. J. J. (eds) (1983) *The Culture of Consumption: Critical Essays in American History, 1880–1980*. Pantheon Press, New York.

Fraser, W. (1981) *The Coming of the Mass Market, 1850–1914*. Hamish Hamilton, London.

Furlough, E. (1991) *Consumer Cooperation in Modern France: The Politics of Consumption*. Cornell University Press, Ithaca, NY.

Furlough, E. and Strikwerda, C. (eds) (1999) *Consumers Against Capitalism? Consumer Co-operation in Europe, North America and Japan, 1840–1990*. Rowman and Littlefield, Lanham, MD.

Glennie, P. D. (1995) 'Consumption within historical studies' in Miller, D. (ed.) *Acknowledging Consumption: A Review of New Studies*. Routledge, London, 164–203.

Glennie, P. D. and Thrift, N. J. (1992) 'Modernity, urbanism and modern consumption'. *Environment and Planning D: Society and Space* 10, 423–443.

Glickman, L. B. (ed.) (1999) *Consumer Society in American History: A Reader*. Cornell University Press, Ithaca, NY.

Goss, J. (1993) 'The "magic of the mall": an analysis of form, function and meaning in the contemporary retail built environment'. *Annals of the Association of American Geographers* 83, 18–47.

Goss, J. (1995) '"We know who you are and we know where you live": the instrumental rationality of geodemographic systems'. *Economic Geography* 71(2), 170–198.

Gurney, P. (1996) *Co-operative Culture and the Politics of Consumption in England, 1870–1930*. Manchester University Press, Manchester.

Hartwick, E. (1998) 'Geographies of consumption: a commodity-chain approach'. *Environment and Planning D: Society and Space* 16, 423–437.

Harvey, D. (1987) 'Flexible accumulation through urbanization: reflections on postmodernism in the American city'. *Antipode* 19, 260–286.

Hewison, R. (1987) *The Heritage Industry*. Butterworth, London.

Hill, C. (1955) *The English Revolution, 1640*. Lawrence & Wishart, London.

Hopkins, J. S. P. (1991) 'West Edmonton Mall: landscapes of myth and elsewhereness'. *Canadian Geographer* 34, 2–17.

Hunt, A. (1996a) *Governance of the Consuming Passions: A History of Sumptuary Law*. Macmillan, London.

Hunt, A. (1996b) 'The governance of consumption:

sumptuary laws and shifting forms of regulation'. *Economy and Society* 25(3), 410–427.

Jackson, P. and Thrift, N. (1995) 'Geographies of consumption' in Miller, D. (ed.) *Acknowledging Consumption: A Review of New Studies*. Routledge, London, 204–237.

Jefferys, J. B. (1954) *Retail Trading in Great Britain, 1850–1950*. Cambridge University Press, Cambridge.

Kowinski, W. S. (1985) *The Malling of America: An Inside Look at the Great Consumer Paradise*. William Morrow, New York.

Laermans, R. (1993) 'Learning to consume: early department stores and the shaping of modern consumer culture, 1860–1914'. *Theory, Culture and Society* 10, 79–102.

Lancaster, B. (1995) *The Department Store: A Social History*. Leicester University Press, Leicester.

Lawrence, J. C. (1992) 'Geographical space, social space, and the realm of the department store'. *Urban History* 19, 64–83.

Leach, W. R. (1984) 'Transformations in a culture of consumption: women and department stores, 1890–1925'. *Journal of American History* 71, 319–342.

Lehtonen, T-K. and Mäenpää, P. (1997) 'Shopping in the East Centre Mall' in Falk, P. and Campbell, C. (eds) *The Shopping Experience*. Sage, London, 136–165.

Ley, D. (1996) *The New Middle Class and the Remaking of the Central City*. Oxford University Press, Oxford.

Lovejoy, A. O. (1961) *The Great Chain of Being: A Study of the History of an Idea*. Harvard University Press, Cambridge, MA.

Lowe, M. and Wrigley, N. (eds) (2002) *Reading Retail: A Geographical Perspective on Retailing and Consumption Spaces*. Arnold, London.

Lowenthal, D. (1996) *The Heritage Crusade and the Spoils of History*. Viking, London.

MacCannell, D. (1992) *Empty Meeting Grounds. The Tourist Papers, Volume 1*. Routledge, London.

McKendrick, N., Brewer, J. and Plumb, J. H. (1982) *The Birth of a Consumer Society: The Commercialization of Eighteenth-century England*. Europa, London.

Miller, D., Jackson, P., Thrift, N. J., Holbrook, B. and Rowlands, M. (1998) *Shopping, Place and Identity*. Routledge, London.

Miller, M. (1981) *The Bon Marché: Bourgeois Culture and the Department Store, 1869–1920*. Princeton University Press, London.

Mokyr, J. (1977) 'Supply *vs.* demand in the industrial revolution'. *Journal of Economic History* 37, 981–1008.

Moore, J. S. (1980) 'Probate inventories: problems and prospects' in Riden, P. (ed.) *Probate Records and the Local Community*. Alan Sutton, Gloucestershire, 11–28.

Morris, M. (1988) 'Things to do with shopping centres' in Sheridan, S. (ed.) *Grafts: Feminist Cultural Criticism*. Verso, London, 193–225.

Muckerji, C. (1983) *Graven Images: Patterns of Modern Materialism*. Columbia University Press, New York.

Mui, H-C. and Mui, L. H. (1989) *Shops and Shopkeeping in Eighteenth-century England*. Routledge, London.

Purvis, M. (1998) 'Societies of consumers and consumer societies: co-operation, consumption and politics in Britain and continental Europe *c.* 1850–1920'. *Journal of Historical Geography* 24(2), 147–169.

Redfern, P. A. (1997) 'A new look at gentrification: 1. Gentrification and domestic technologies'. *Environment and Planning A* 29, 1275–1296.

Reekie, G. (1993) *Temptations: Sex, Selling and the Department Store*. Allen & Unwin, London.

Richards, T. S. (1991) *The Commodity Culture of Victorian England: Advertising and Spectacle, 1851–1914*. Verso, London.

Sack, R. D. (1993) *Place, Modernity and the Consumer's World: A Relational Framework for Geographical Analysis*. Johns Hopkins University Press, Baltimore, MD.

Schama, S. (1988) *The Embarrassment of Riches: An Interpretation of Dutch Culture in the Golden Age*. University of California Press, Berkeley.

Schivelbusch, W. (1988) *Disenchanted Night: The Industrialization of Light in the Nineteenth Century*. University of California Press, Berkeley.

Schivelbusch, W. (1992) *Tastes of Paradise: A Social History of Spices, Stimulants, and Intoxicants*. Vintage, New York.

Sekora, J. (1977) *Luxury: The Concept in Western Thought, Eden to Smollett*. Johns Hopkins University Press, Baltimore, MD.

Shammas, C. (1990) *The Pre-industrial Consumer in England and America*. Oxford University Press, Oxford.

Shepherd, I. D. H. and Thomas, C. J. (1980) 'Urban consumer behaviour' in Dawson, J. A. (ed.) *Retail Geography*. Croom Helm, Beckenham, 18–94.

Sherratt, A. G. (1998) 'The human geography of Europe: a prehistoric perspective' in Butlin, R. A. and Dodgshon, R. A. (eds) *An Historical Geography of Europe*. Clarendon Press, Oxford, 1–25.

Shields, R. (1989) 'Social spatialization and the built environment: the West Edmonton Mall'. *Environment and Planning D: Society and Space* 7, 147–164.

Shields, R. (1994) 'The logic of the mall' in Riggins, S. H. (ed.) *The Socialness of Things: Essays on the Socio-semiotics of Objects*. Mouton de Gruyter, New York, 203–229.

Smith, N. (1996) *The New Urban Frontier: Gentrification and the Revanchist City*. Routledge, London.

Taylor, P. J. (1996) 'What's modern about the modern world-system? Introducing ordinary modernity through world hegemony'. *Review of International Political Economy* 3(2), 260–286.

Thirsk, J. (1978) *Economic Policy and Projects: The Development of a Consumer Society in Early-modern England*. Clarendon Press, Oxford.

Thompson, E. P. (1963) *The Making of the English Working Class*. Methuen, London.

Thrift, N. J. and Glennie, P. D. (1993) 'Historical geographies of urban life and modern consumption' in Kearns, G. and Philo, C. (eds) *Selling Places: The City as Cultural Capital, Past and Present*. Pergamon Press, London, 33–48.

Tiersten, L. (1993) 'Redefining consumer culture: recent literature on consumption and the bourgeoisie in western Europe'. *Radical History Review* 57, 116–159.

Urry, J. (1990) 'The "consumption" of tourism'. *Sociology* 24(1), 23–35.

Urry, J. (1995) *Consuming Places*. Routledge, London.

van der Woude, A. and Schuurman, A. (eds) (1980) *Probate Inventories: A New Source for the Historical Study of Wealth, Material Culture, and Agricultural Development A. A. G. Bijdragen* 23, Wageningen, The Netherlands.

Weatherhill, L. (1988) *Consumer Behaviour and Material Culture in Britain, 1660–1760*. Routledge, London.

Wildt, M. (1995) 'Plurality of taste: food and consumption in West Germany during the 1950s'. *History Workshop Journal* 39, 23–41.

Williams, R. H. (1982) *Dream Worlds: Mass Consumption in Late Nineteenth-century France*. University of California Press, Berkeley.

Williamson, J. (1992) 'I-less and gaga in the West Edmonton Mall: towards a pedestrian feminist reading' in Currie, D. H. and Raoul, V. (eds) *The Anatomy of Gender: Women's Struggles for the Body*. Carelton University Press, Ottawa, 79–115.

Winstanley, M. (1983) *The Shopkeeper's World, 1830–1914*. Manchester University Press, Manchester.

Zukin, S. (1988) *Loft Living: Culture and Capital in Urban Change*. Hutchinson/Radius, London.

Zukin, S. (1990) 'Socio-spatial prototypes of a new organization of consumption: the role of real cultural capital'. *Sociology* 24(1), 37–56.

Zukin, S. (1995) *The Cultures of Cities*. Blackwell, Oxford.

Zukin, S. (1998) 'Urban lifestyles: diversity and standardisation in spaces of consumption'. *Urban Studies* 35(5–6), 825–839.

Index